Ayşe Asiltürk
Ayşegül Sarıoğlu Kemer
Editors

Disaster Management in the Digital Age and Emerging Technologies

nova
science publishers
www.novapublishers.com

DOI: https://doi.org/10.52305/HKIF2441

NOTICE TO THE READER

The Publisher has taken reasonable care in the preparation of this book but makes no expressed or implied warranty of any kind and assumes no responsibility for any errors or omissions. No liability is assumed for incidental or consequential damages in connection with or arising out of information contained in this book. The Publisher shall not be liable for any special, consequential, or exemplary damages resulting, in whole or in part, from the readers' use of, or reliance upon, this material. Any parts of this book based on government reports are so indicated and copyright is claimed for those parts to the extent applicable to compilations of such works.

Independent verification should be sought for any data, advice or recommendations contained in this book. In addition, no responsibility is assumed by the Publisher for any injury and/or damage to persons or property arising from any methods, products, instructions, ideas or otherwise contained in this publication.

This publication is designed to provide accurate and authoritative information with regards to the subject matter covered herein. It is sold with the clear understanding that the Publisher is not engaged in rendering legal or any other professional services. If legal or any other expert assistance is required, the services of a competent person should be sought. FROM A DECLARATION OF PARTICIPANTS JOINTLY ADOPTED BY A COMMITTEE OF THE AMERICAN BAR ASSOCIATION AND A COMMITTEE OF PUBLISHERS.

Library of Congress Cataloging-in-Publication Data

ISBN: 979-8-89530-213-2 (Hardcover)
ISBN: 979-8-89530-331-3 (eBook)

Published by Nova Science Publishers, Inc. † New York

Contents

Preface

Natural disasters are events that countries have inevitably had to face in world history. In addition to slowly developing types such as severe cold, drought and famine, sudden natural disasters such as earthquakes, floods, landslides, rockfalls, avalanches, storms, tornadoes, volcanic eruptions and fires leave societies in a difficult situation. When disasters are mentioned, we come across not only natural disasters, but also man-made disasters such as war, terrorist attacks, mining accidents and migration situations.

In recent years, there has been an increase in the number of natural disasters and man-made disasters around the world. According to the United Nations report, 7348 major disasters occurred in the world between 2000 and 2019. There has been an increase in the number of earthquakes and tsunamis; 1.23 million people lost their lives. Some of the natural disasters experienced in world history are the South Asian Tsunami, Hurricane Katrina (2004), Kashmir earthquake (2005), California forest fires (2007), China Wenchuan earthquake (2008), Italy earthquake (2009), Haiti and Indonesia earthquakes (2010), Covid-19 pandemic (2020), Kahramanmaraş Pazarcık and Elbistan earthquakes (2023). Approximately 80% of the earthquakes in the world occur in Indonesia, China, Japan, Pakistan, Chile, Russia, Türkiye, Iran, Peru and Italy.

Since disasters often cause multiple losses such as loss of life and property, social losses, and psychological losses, rapid and effective interventions are required in the disaster area. Disasters that cause great damage can be turned into an opportunity for managers and society to question their crisis management skills, identify deficiencies in practices and improve them.

Management of disasters includes topics such as preventing disasters, reducing the damage caused by disasters, intervening in disaster areas when disasters occur, rescuing disaster victims and meeting their needs. It is seen that traditional perspectives and methods are inadequate in reducing and eliminating the damage caused by disasters. In the digital age, humanity's most powerful weapon in combating disasters is technological developments. With the development of science and technology around the world, scientists and researchers have begun to make high-tech inventions that will support disaster risk reduction. It can be argued that the use of technology will play an important role in combating disasters with its contributions to disaster management stages such as "risk reduction, intervention, mitigation, rehabilitation and reconstruction," solving disaster management problems and increasing disaster management effectiveness.

The increase in number and diversity of disasters encountered in the world motivates researchers to develop technologies that can be integrated into disaster management. In reducing the number and impacts of future disasters, there is a need for coordinated research activities involving all stakeholders, especially communities at risk, and a better understanding and use of new technologies. This book's chapters consist of up-to-date information that covers the basic issues in disaster management from a broad perspective and is thought to make a significant contribution to the literature. This book is a study that brings together interdisciplinary researchers and experts who want to develop solutions in the field of emergency aid and disaster management and aims to develop the potential uses of new technologies in terms of disaster management. At the same time, based on data obtained from recent disasters, researchers open a new window to governments, decision makers and disaster experts in the digital age. This book will be of interest to researchers, policy makers, students and practitioners interested in disaster management, disaster response, emergency management, digitalization, new technologies.

Editors
Ayşe Asiltürk
Ayşegül Sarıoğlu Kemer

About the Editors

Ayşe ASİLTÜRK, PhD – Associate Professor

Dr. Asiltürk earned her bachelor's degree at Selçuk University, Faculty of Economics and Administrative Sciences, Department of Business Administration (1999-2003), her master's degree at Selçuk University Social Sciences Institute, Department of Business Administration, Thesis Master's Program (2003-2006), and her doctorate degree at Selçuk University Social Sciences Institute, Department of Business Administration (2003-2006). She completed her studies in the Department of Business Administration (2006-2011). She has been working as a faculty member at Trabzon University, School of Applied Sciences, Department of Management Information Systems since 2021. She is responsible for preparing content and curricula for courses in business administration, management and organization, human resources management, organizational behavior and digital transformation in the Department of Management Information Systems. She conducts relevant courses as an associate professor. She acts as the secretariat of the Management Information Systems Department Science Advisory Board by communicating with leading companies, especially companies in the IT sector. She has supervised 16 theses on various sectors and managerial issues. She has studies in interdisciplinary areas such as Industry 4.0 and its elements, and IT activities.

Ayşegül SARIOĞLU KEMER, PhD – Assistant Professor

Dr. Kemer graduated from Atatürk University, School of Nursing in 2009. In 2014, she completed the Nursing Management Master Program at Atatürk University, The Institute of Health Sciences. In 2017, she completed a doctorate program at Karadeniz Technical University, The Institute of Health Sciences, Department of Nursing. Between 2009 and 2011, she worked as a registered nurse at Atatürk University Health Research and Application Hospital, the Anesthesia and Reanimation Intensive Care Unit. Between 2011 and 2014, she worked as a research assistant at Atatürk University's Faculty of Nursing, and between 2014 and 2016 she worked in a similar role at Karadeniz Technical University's Faculty of Health Sciences. From 2017-2021, she worked as a faculty member at Atatürk University's Faculty of Nursing. In 2018, she became an assistant professor doctor. She has been working as an assistant professor at Trabzon University, Department of Emergency Aid and Disaster Management. She continues her studies in the subjects of disaster, disaster management, first aid, leadership, innovativeness, organizational citizenship behaviour, professional commitment and legislative regulations.

Chapter 1

Natural Disasters

Fatma Gündüz*, PhD

Boyabat Vocational School Higher Education, Sinop University, Boyabat, Sinop, Türkiye

Abstract

Natural disasters affect millions of people on a global scale every year. Although there are many classifications in the literature on disaster types, it is possible to evaluate disasters as sudden and continuous disasters. Disasters are categorized as nature-induced disasters and human-induced disasters according to the cause of their occurrence. Examples of natural disasters are avalanches, floods, earthquakes, droughts, cold waves, heat waves, hail, hurricanes, landslides, lightning, river floods, volcanic activities, forest fires, tornadoes, rock falls, and tornadoes. The potential of natural hazards to turn into disaster risks varies depending on the structure of the population that may be affected, sociocultural and economic values, infrastructure and building stock factors, and the geographical characteristics of the regions. Although the subject of natural disasters affecting every geography and society is broad and inclusive, it reveals the need for all disciplines to work effectively and for technological developments to be evaluated within the disaster management cycle. In this chapter, the potential of natural hazards to become natural disasters and the factors affecting this will be presented within the framework of lessons learned, and technological developments in disaster management will be evaluated.

Keywords: disaster, disaster management, natural disasters, technology

* Corresponding Author's Email: fgunduz@sinop.edu.tr.

In: Disaster Management in the Digital Age and Emerging Technologies
Editors: Ayşe Asiltürk and Ayşegül Sarıoğlu Kemer
ISBN: 979-8-89530-213-2

Introduction

Türkiye is a country that faces many types of natural disasters. Disaster is considered to be the result of events that create changes in the normal functioning of society, including physical, social, economic, cultural, and environmental impacts that exceed the coping capacity of the affected society or communities (Aslan and Şahinöz, 2023; Usta et al., 2023). According to the Emergency and Disaster Management Presidency, a natural disaster is defined as the general name given to the consequences of natural events such as earthquakes, floods, landslides, avalanches, droughts, storms, hailstorms, tornadoes, droughts, meteorite strikes, etc., the occurrence of which cannot be prevented, resulting from geological, meteorological, hydrological, climatological, biological, and extraterrestrial hazards (AFAD, 2014a). Türkiye is a country of 780 thousand square kilometers, with a large part of its territory located on the Asian continent and a bridge between Europe and Asia. Due to its topography and geological structure, Türkiye is frequently exposed to natural disasters. Global damage from disasters and their frequency is intensifying due to the accelerated impact of human activities and climate change (Usta, 2021; Usta et al., 2023). Disasters destroy people's habitats, businesses, infrastructure, and agricultural areas, killing people and animals. At the same time, disasters force people to be internally displaced, to leave their homes or their usual places of residence, and to seek refuge elsewhere for short or long periods of time (Kaya et al., 2023). Furthermore, many people around the world are displaced by natural disasters or the impacts of the climate crisis (Karaman, 2016:3; Özer, 2017).

In the World Disaster Report (2020), 83% of all disasters triggered by natural hazards in the last decade were weather and climate-related, such as floods, storms, and heat waves. More than 410,000 people have lost their lives due to climate and weather-related disasters. The highest number of casualties occurred in heat waves and storms (IFRC 2020:2).

According to the CRED 2022 September report (issue no. 68), the countries most affected by the types of disasters that resulted in the highest number of deaths were floods (India, Pakistan, South Africa, Brazil, Bangladesh, and Niger), storms (Megi) in the Philippines, and storms (Batsirai) in Madagascar and earthquakes in Afghanistan (CRED, 2023). According to the EMDAT database, for an event to be considered a disaster, at least 10 people must die, at least 100 people must be affected, a state of emergency must be declared, and at least one of the following criteria must be met by an international call for assistance (EM-DAT, t.y.a).

According to the report published in 2022, 176 floods, 108 storms, 31 earthquakes, 22 droughts, 17 landslides, 15 wildfires, 12 extreme temperatures, 5 volcanic activity, and 0 mass movement (dry) were reported (CRED, 2023). According to the type of disaster, extreme temperatures had the highest number of deaths (CRED, 2023).

A hazard is a physical event or phenomenon caused by nature, technology, or human beings that occurs in a certain time or geography and threatens life, and has the potential to harm the socioeconomic order and activities of society, the natural environment, historical, and cultural resources. In other words, hazard refers to all events that are caused by nature, technology, or human beings and may lead to physical, economic, and social losses (AFAD, 2014b). Exposure is defined as "the presence of people, structures, and systems in disaster zones and risky production areas under the threat of potential loss" (AFAD, 2014b). The concept of vulnerability is defined as "The measure of physical, social, economic, or environmental damages and losses that people and their living environments may suffer in the face of hazards of different types and magnitudes" (AFAD, 2014b). The potential of natural hazards to become disaster risks and disasters interact with exposure and vulnerability components. In the transformation of a hazard into a disaster risk, physical, social, economic, cultural, and environmental vulnerability factors and the elements and values that can be affected are important.

While the potential and impacts of natural hazards that turn into disasters are significant, technological developments in the global world are expected to guide the reduction of natural disaster risks. The impact of natural disasters can vary depending on factors such as the potential for hazards to turn into disaster risks, the ability to be prepared for disasters, individual, institutional, and social capacity, and sociocultural realities (Dinçer, 2023:177-178; Atalay and Kaya, 2023:190).

Disruption of ecological balance, impact on flora and fauna, climate change effects that are gradually increasing their impact and severity, drought, famine, and related climate migrations may cause disasters to re-occur and increase their severity. Two important factors can be mentioned in the transformation of natural events into disasters. One of them is the magnitude, severity, and scope of the event, and the other is the structure and capacity of the affected community (Demiröz Yıldırım and Yılmaz, 2019). The integration of a resilient society and technological developments is important to prevent natural hazards from becoming disasters. While an event of the same magnitude and severity may not become a disaster in some countries, it may become a disaster in others. In this section, the potential of natural hazards to become disasters, the factors affecting them, and technological developments will be evaluated. Nature-induced disasters are disaster events resulting from the interaction of natural hazards with other components in addition to human activities.

In the International Disaster Database of the Research Center for Epidemiology of Disasters, nature-induced disasters are classified into six main categories:

- *Geophysical:* Disasters originate on the Earth and are used interchangeably with the term geological hazard. These are earthquakes, mass movements (dry), and volcanic activity.
- *Biological:* Disasters caused by the exposure of living organisms to toxic substances or microbes. These are animal accidents, epidemics, and insect infestations.
- *Hydrological:* Disasters caused by the formation, movement, and distribution of surface and subsurface freshwater and saline water. These are; flood, landslide, and wave action.
- *Meteorological:* Disasters caused by atmospheric processes and extreme weather events that can last minutes or days. These are storms, extreme temperatures, and fog.
- *Climatological:* Disasters caused by long-lived, medium- to macro-scale atmospheric processes. These are drought, glacial lake outbursts, and wildfire.
- *Extraterrestrial:* Disasters caused by asteroids, meteors, and comets passing near or striking the Earth, or by any change in interplanetary conditions. These are impacts and space weather (EM-DAT, t.y.b).

Geophysical Disasters

Disasters originate from the earth and are used interchangeably with the term geological hazard. These are earthquakes, mass movement (dry), and volcanic activity.

Earthquakes

The largest earthquake belt in the world, known as the 1st earthquake belt, extends from Chile northward along the coasts of South America, Central America, Mexico, the west coast of the USA and the Aleutian Islands south of Alaska, Japan, the Philippines, New Guinea, the South Pacific Islands, and New Zealand. Belt 2 (Alpine) starts in Indonesia (Java-Sumatra) and reaches the Atlantic Ocean via the Himalayas and the Mediterranean Sea. The 3rd belt (Atlantic) extends along the plate boundary (Atlantic Ocean Ridge) in the middle of the Atlantic Ocean (Kandilli Rasathanesi ve Deprem Araştırma Enstitüsü, 2024). The countries with the highest earthquake hazard in Europe are Italy, Greece, Türkiye, and Romania (Gülsoy and Uyan, 2023:7).

Türkiye is located on one of the most active earthquake zones (Alpine-Himalayan belt) in the world. It is known that 92% of our country is located in earthquake zones, 95% of our population lives under earthquake hazards, and 98% of large industrial centers and 93% of our dams are located in earthquake zones (Kandilli Rasathanesi ve Deprem Araştırma Enstitüsü, 2024). An earthquake is defined as an event in which the vibrations that occur suddenly due to fractures in the Earth's crust spread in waves and shake the Earth and the environments they pass through (Kandilli Rasathanesi ve Deprem Araştırma Enstitüsü, 2024). Looking at the history of earthquakes, Türkiye is a country that has faced earthquakes in many periods. As a result of the earthquakes experienced historically, many lives and property have been lost and social, economic, and physical losses have been incurred.

When we look at the most devastating earthquakes in the history of the world, China (Shaanxi) is the first place, where an estimated 830,000 people lost their lives in a magnitude 8 earthquake in 1556, and more than 97 counties in China were reported to have been affected. Secondly, Haiti (Port-au-Prince) was hit by a magnitude 7 earthquake in 2010, killing an estimated 316,000 people, and the death toll is still disputed. Thirdly, Türkiye (Antakya; the ancient city of Antioch in present-day Antakya) was hit by a 7.5 magnitude earthquake in 115 BC that killed an estimated 260,000 people and devastated the surrounding area, even triggering a localized tsunami off the coast of Lebanon. Fourthly, Türkiye (Antakya) reportedly lost an estimated 250,000 people in a magnitude 7 earthquake in 525 BC. On the other hand, fires and strong winds in the aftermath of the earthquake have been reported to have compounded the destruction. Fifth, a 7.5 magnitude earthquake in China (Tangshan) in 1976 killed an estimated 242,769 people and destroyed 85% of the buildings. Sixth, in Azerbaijan (Gyzndja), in 1139, a 6.5 magnitude earthquake killed 230,000 people. Seventh, Indonesia (Sumatra) was hit by a 9.1 magnitude earthquake in 2004, killing 227,899 people. After this earthquake, tsunamis reached heights of 15-30 meters were experienced. It is also known that there are no tsunami warning systems for tsunamis in the region. Eighth, estimated 200,000 people were reportedly killed in a 7.9 magnitude earthquake in Iran (Damgan) in 856 BC. It was also reported that 200,000 people were killed in an 8.3 magnitude earthquake in China (Gansu) in 1920, with the death toll increasing due to cold weather conditions. Ninth, the earthquake in Armenia (Dvin) in 893 BC is estimated to have killed 150,000 people. There is no clear information on the magnitude of the earthquake. Finally, Japan (Tokyo) was hit by a 7.9 magnitude earthquake in 1923, killing 142,807 people. There were also fires, hurricanes, and tsunamis after the earthquake (Our World in Data,2024).

When the historical data on earthquakes are examined, many countries in the world geography have suffered devastation due to earthquakes. Factors affecting the potential for earthquake hazards to become disasters should be identified, and risks should be reduced or eliminated using technology. The UN Platform for Space-borne Information for Disaster Management and Emergency Response (UNSPIDER), led by the United Nations Office for Space Affairs, is a program that aims to make critical satellite data available to disaster management agencies and authorities around the world. UN-SPIDER can also provide information on space technologies that can be used to image different types of natural hazards, products developed by the space community for disaster management applications, links to satellite imagery sites, and software packages for processing such imagery (IFRC, 2020:182).

Volcanic Activity

Volcanic movements have caused great destruction due to materials such as lava, ash, and gas vapor from the past to the present. The most important action to be taken to prevent volcanic activities, one of the natural hazards, from becoming a disaster risk is not to build human settlements and populations in areas with active volcanism. Volcanic activities are defined as natural hazards because they occur independently of human actions. For this reason, it is necessary to take initiatives to prevent and mitigate volcanic hazards from becoming disaster risks. There are approximately 1400 active volcanoes around the world. Volcanoes in Turkey that are known to have been active in the recent past and are believed to have the potential to operate; Acıgöl-

Nevşehir, Ağrı (Ararat), Erciyes Mountain, Göllü Mountain, Hasan Mountain, Karaca Mountain, Karapınar volcanic area, Kula, Nemrut Mountain, Tendürek Mountain. Volcanoes that have been active in the last 10,000 years but have not been confirmed; Girekol Hill, Gölcük, Kars Plateau, and Süphan Mountain. The volcano that has not erupted for thousands of years is Cappadocia (Maden Tetkik ve Arama Genel Müdürlüğü, 2024).

The volcanic eruptions that took place in the last 200 years and caused the most destruction are as follows: The volcanic eruptions that took place in the last 200 years and caused the most destruction are as follows: 92,000 deaths in Indonesia (Tambora) in 1985, 5,500 deaths in Indonesia (Galunggung) in 1822, 3,000 deaths in Indonesia (Awu) in 1826, 36,500 deaths in Indonesia (Krakatau) in 1883, Martinique in 1902 29,000 deaths in (Mount Pele), 6,000 deaths in Guatemala (Santa Maria) in 1902, 2,950 deaths in Papua New Guinea (Lamington) in 1951, 1,900 deaths in Indonesia (Agung) in 1963, 1,700 deaths and deaths occurred in Mexico (El Chichon) in 1982, and it was recorded that 25 thousand people deaths as a result of the explosion that occurred on November 13, 1985, at the Nevado del Ruiz Volcano in Colombia (Ekici, 2023).

The fact that society or communities are in danger due to volcanic movements should be evaluated separately for each geography. Dangers caused by nature can occur in all geography. The main point is to reduce risks and be prepared before dangers turn into disasters. Volcanic activity can be monitored, and society can be made resilient with early warning systems. At this point, the integration of developing technology and the integration of society with approaches in the digitalization age are important.

Mass Movement

AFAD defines mass movement as "the slow or rapid mass displacement of soil or rock masses down the slopes under the influence of gravity" (AFAD,2014b). One of the natural phenomena that change the shape of the Earth is mass movements. Factors affecting mass movements; excavations on the slopes or erosion by water; excessive loading of the slope by filling or debris; underground and surface waters; geological effects; erosion of rocks; change in vegetation; and climatological effects (Taşdemiroğlu, 1970). According to the EMDAT 2022 data, no mass movement notification was made in 2022 (CRED, 2023).

Biological Disasters

Disasters are caused by the exposure of living organisms to toxic substances or microbes. These are epidemics and insect infestations.

Epidemics

Natural hazards that expose living organisms or their toxins to vector-borne diseases are known as biological hazards. Biological hazards arise from nature or are caused by human activities.

Throughout human history, it is known that there have been epidemics affecting countries socially, economically, politically, and culturally. From plague to typhoid, cholera, influenza, smallpox, and HIV/AIDS, humanity is familiar with many epidemics. Epidemics have caused loss of life throughout history and left deep pain in the minds. Justinian's Plague, which started in the 6th century A.D. and continued until the 8th century, the Plague epidemic called the Black Death starting in the 14th century and continued until the 19th century, Influenza epidemics, Sars and Mers epidemics are always in the minds due to the losses they caused (Turan and Çelikyay, 2020). The Black Plague is known as the epidemic that started in China and Central Asia and spread to Europe in the ongoing process, causing millions of people to die. It is known to have started in China, followed the Silk Road, and spread to Crimea and then to Europe in 1347. It is known that the plague outbreak was caused by the fleas that lived in the rats on the ships that were carried to the port cities and that the disease was transmitted from person to person by fleas that bit people or from person to person and turned into an epidemic (Sussman, 2011).

The cholera epidemic of 1847-1848, which was caused by the land between the Ganges and Brahmaputra rivers in India and became a global threat, was first seen in the Ottoman lands in 1822 and caused epidemics in Japan, Moscow, Berlin, Paris, and London (Afanasiev, 2021). In the 21st century, SARS and MERS outbreaks were caused by coronaviruses. Human coronaviruses often cause mild to moderate upper respiratory tract diseases. There are four subgroups of coronaviruses known as alpha, beta, gamma, and delta. Coronaviruses (CoVs) can cause everything from the common cold to more serious illnesses such as Middle East Respiratory Syndrome (MERS-CoV) and Severe Acute Respiratory Syndrome (SARS-CoV) (Machhi et al., 2020; Aleebrahim-Dehkordi et al., 2021). Introduced in 2003, SARS-CoV has killed hundreds of people in the current century. Countries such as Canada, China, the USA, Hong Kong, Singapore, Taiwan, Thailand, Vietnam, Hong Kong, Singapore, Singapore, Taiwan, Thailand, and Vietnam have been among the countries most affected by the virus (Peiris et al., 2004; Hui and Zumla, 2019). As the Covid 19 outbreak, which started in China, spread to the world in a short time, it was declared a pandemic, and many countries were affected by this pandemic socially, economically, culturally, and environmentally (WHO, 2024).

With the development of technology, communities can quickly access global information and verify information. Notifying biological hazards and recognizing early warnings by societies will be an important approach to risk management in disasters.

Insect Infestations

According to the Emergency and Disaster Management Presidency, "An infestation caused by an excessive increase of harmful insects in one area or by the transportation of these insects to another area. The most common and harmful of these is the disaster caused by locusts, which increase in number under certain atmospheric conditions and affect other regions in swarms" (AFAD, 2014b). Changes in temperatures, atmospheric conditions, and climate change have led to insect infestations in some geographical areas. Deterioration of climatic conditions and ecosystems, along with the lack of forest hygiene, leads to an increase in insect populations that can cause serious damage. Most often, the spread of insects over long distances and countries is due to the transportation of infested goods (UNDRR, 2024).

Hydrological Disasters

Disasters are caused by the formation, movement, and distribution of surface and subsurface freshwater and saline water. These are flood, landslide, and wave action.

Floods

One of the disasters that cause the most destruction globally is floods. According to the Emergency and Disaster Management Presidency, "The phenomenon of water rising in the area or coming from another place and covering the usually dry surfaces is called a flood" (AFAD, 2014b). Floods are classified as slow-developing, fast-developing, and sudden floods according to their formation speed, and floods are also classified as coastal floods, city floods, dry stream floods, dam/pond floods, and stream (stream and river) floods according to their formation location. Societies face flood disasters due to the impact of climate change. The Türkiye Mediterranean, Aegean, Black Sea, and Eastern Anatolia regions are more prone to flood risk. According to the EM DAT 2022 report; Pakistan, India, Bangladesh, China, Nigeria, South Africa, Brazil, and Eastern Australia were the countries most affected by the flood. In the age of digitalization, meteorological data is shared worldwide, and early warning systems are available (CRED, 2023). There are many actions that can prevent a natural hazard from becoming a flood risk or reduce the risks. Disaster policies need to be developed considering the structure of society and the characteristics of geography.

Landslides

According to the Emergency and Disaster Management Presidency, "A situation in which rock, soil or pieces of landslide or move down a noticeable slope due to the effect of gravity or external factors such as earthquakes, excessive rainfall, landslides" (AFAD, 2014b). According to the EMDAT report, 17 landslides were reported (CRED, 2023). In addition to deaths and economic losses, landslides damage urbanization and have negative impacts on agriculture and the quality of water resources (Gökçeoğlu and Ercanoğlu, 2001). In Türkiye, studies are being carried out to reduce the losses caused by landslides, to provide a better understanding of the spatial distribution of landslides and the environmental factors that control them, and to establish a national digital landslide database at a certain standard. Furthermore, landslide hazard maps and susceptibility analyses were prepared (Çan et al., 2013; Alptekin and Yakar, 2020).

Meteorological Disasters

Disasters are caused by atmospheric processes and extreme weather events that can last minutes or days. These include storms and extreme temperatures.

Storms

According to the Emergency and Disaster Management Presidency, a storm "is defined as a wind with a speed varying between 23 and 26 m/s that harms nature and people. As wind speed and intensity increase, the damage it causes to nature and people also increases. On the other hand, the increase in speed and intensity further increases the risk of storm disasters. Rain, snow, hail, sand, etc. along with strong winds" (AFAD, 2014b). When these elements occur, they are called snowstorms, sandstorms, and dust storms. Severe winds are called storms, tornadoes, cyclones, and tornadoes, depending on their wind speed. Tropical storms are small-scale intense cyclones that develop in regions where surface water is very warm (tropical ocean). The 1922 China cyclone (100,000 deaths), the 1970 Bangladesh cyclone (304,000 deaths), the 1991 Bangladesh cyclone (146,000 deaths), and the 2008 Myanmar cyclone (141,000 deaths) are examples of storm disasters that resulted in the most destruction and deaths in world history. According to the 2022 EMDAT report, the tropical storms Megi and Nalgae affected millions of people and caused destruction in the Philippines. Another hurricane is Hurricane Ian, which affected the USA (CRED, 2023).

In the study of Karaman et al., in the medium and long term, severe, she stated that rainfall and floods due to precipitation will increase, displacement will increase with the intensification of tropical cyclones, and the risk level of drought and increasing sea level rise will increase. Although storm events in Turkey appear as natural hazards, the potential for these hazards to turn into disaster risks and meteorological early warning systems should be evaluated within the framework of integrated disaster risk and crisis management (Karaman et al., 2022).

Extreme Temperatures

Another current climate-related hazard is extreme temperatures. In 2003, many people died from extreme temperatures in Europe. In 2010, thousands of people died as a result of extreme temperatures in Russia. In 2013, China (Southeast region) experienced extreme temperatures that exposed the population to illness and death. According to the 2019 World Disaster Report, 10 extreme temperatures have been recorded. In July-August 2019, 3,453 people were recorded to have died in Western Europe due to heat waves, affecting Belgium, France, Germany, Italy, the Netherlands, Spain, Switzerland, and the United Kingdom (IFRC, 2020:10). According to the EMDAT 2022 report, 12 extreme temperatures were recorded (CRED, 2023). Adverse weather conditions and climate change have serious impacts on public health and the ecological environment. Early warning systems and effective response methods for extreme temperatures can prevent extreme heat hazards from turning into disasters. These models are important approaches that should be included in disaster management.

Climatological Disasters

Disasters are caused by long-lived, medium- to macro-scale atmospheric processes. These are drought and wildfires.

Drought

Drought is another nature-induced disaster that causes great destruction in the world. AFAD defines drought as "a nature-induced disaster caused by air temperatures well above seasonal norms and annual precipitation averages falling below seasonal norms"(AFAD, 2014b).

The main factors affecting drought in Türkiye include atmospheric conditions, physical geography factors, and climatic conditions. Drought is characterized by multiple climatological and hydrological parameters and is classified as meteorological drought, agricultural drought, and hydrological drought (Wilhite and Glantz, 1985; T.C. Çevre, Şehircilik ve İklim Değişikliği Bakanlığı Meteoroloji Genel Müdürlüğü, 2024). In the 2022 natural disaster assessments, it was reported that 2465 people lost their lives in Uganda due to drought. On the other hand, the Democratic Republic of the Congo, Ethiopia, Nigeria, Sudan, Niger, Burkina Faso, and China have been the most affected countries due to drought. The countries that suffered the most economic losses due to the drought were the US, China, and Brazil (CRED, 2023).

Many factors, such as temperature differences, strong winds, low relative humidity, changes in rainy days, intensity, and duration of precipitation have an important role in the occurrence of drought. Droughts affect both groundwater and surface water. Due to the dependence of human activities on water resources, droughts have social, cultural, and economic impacts on society. Our country has an irregular rainfall regime and faces the risk of drought. With the integration of geographical information systems into drought analysis, together with technological developments, the developments are continuously evaluated. Although the factors that are effective in the transformation of a natural hazard into a disaster are evident in the drought event, the fact that it is a slowly developing hazard will be effective in reducing the risks and being prepared for the disaster (Zeybekoğlu, 2022). Technological advances include important approaches to monitoring drought hazards. In that way, precautions can be taken before natural hazards become disasters.

Wildfires

Partial or complete burning of forests for natural or human-induced reasons is defined as wildfire. According to EMDAT data, 15 wildfires have been reported in 2022 (CRED, 2023). Wildfires are encountered every year in many parts of the world and in Türkiye. Due to the effects of climate change and hot and dry summers, wildfire disasters occur in many geographical regions. Wildfires occur for human-related reasons as well as natural causes. The forests of subtropical climate zones and continental climate zones are at greater risk as they facilitate the spread of fires in summer. This is because in these regions, especially during the summer, the temperature values are high, and the relative humidity is low. Türkiye is geographically located in the Mediterranean basin, and a large part of the forests are under fire danger (Köçer and Aslan, 2023).

Extraterrestrial Dangers and Disasters

Disasters that cannot be prevented and originate from extraterrestrial dangers, such as meteorites, pose many risks to our living space.

Discussion and Conclusion

Dangers caused by nature have existed in every period of human life and, throughout history, these dangers have reached disaster levels. Today's technological developments provide successful approaches to disaster risk management. Having geographical information systems for natural hazards and early warning systems and national hazard maps can achieve success with the influence of technology and digitalization. Databases developed to monitor disasters and collect data provide accurate and rational data to society and governments with technological developments. The use of existing databases in disaster management processes is an important approach to reducing disaster risks and losses. Societies can become more resilient with the use of existing data and the development of technologies on the components that are effective in turning nature-related hazards and dangers into disasters. Thanks to databases, early warning systems, artificial intelligence, blockchain, and data-oriented applications, the effectiveness of disaster management can be increased and guided by governments. Using technologically based data sources and early warning systems to reduce nature-related disaster risks is necessary for successful disaster management.

Disclaimer

None.

References

AFAD. (2014a). Afet Türleri. *Afet ve Acil Durum Yönetimi Başkanlığı.* https://www.afad.gov.tr/afet-turleri (Accesed:03.03.2024).

AFAD. (2014b). Açıklamalı Afet Yönetimi Terimleri Sözlüğü. *Afet ve Acil Durum Yönetimi Başkanlığı.* https://www.afad.gov.tr/aciklamali-afet-yonetimiterimleri-sozlugu (Accesed:03.03.2024).

Afanasiev, G. (2021). The Russian and Ottoman Empires in the 19th - Early 20th Centuries. *Rusya Araştırmaları Dergisi,* 5, 20-34. https://doi.org/10.48068/rusad.881118.

Aleebrahim-Dehkordi, E., Soveyzi, F., Deravi, N., Rabbani, Z., Saghazadeh, A., & Rezaei, N. (2021). Human Coronaviruses SARS-CoV, MERS-CoV, and SARS-CoV-2 in Children. *Journal of Pediatric Nursing,* 56, 70-79.

Alptekin, A., & Yakar, M. (2020). Heyelan Bölgesinin İHA Kullanarak Modellenmesi. Türkiye İnsansız Hava Araçları Dergisi, 2(1), 17-21.

Aslan, R., & Şahinöz, S. (2023). The Experiences of People With Disabilities in the 2020 Izmir Earthquake: A Phenomenological Research. *International Journal of Disaster Risk Reduction,* 95, 103868, https://doi.org/10.1016/j.ijdrr.2023.103868.

Atalay, E., & Kaya, N. (2023). Afetlerde Exoskeleton Kullanımı (190-199). Afet Yönetimi ve Teknoloji Ed. Saime Şahinöz. İstanbul Tıp Kitabevi.

Çan, T., Duman, T. Y., Olgun, Ş., Çörekçioğlu, Ş., Karakaya-Gülmez, F., Elmacı, H. & Emre, Ö. (2013). Türkiye Heyelan Veri Tabani. TMMOB Coğrafi Bilgi Sistemleri Kongresi.

Çevre, T. C., Şehircilik ve İklim Değişikliği Bakanlığı Meteoroloji Genel Müdürlüğü. (2024). *Kuraklık analizi.* https://www.mgm.gov.tr/veridegerlendirme/kuraklik-analizi.aspx?d=yontemsinif (Accesed:03.03.2024).

CRED. (2023). "2022 Disasters in numbers," Brussels: CRED, 2023. The full report is available at https://cred.be/sites/default/files/2022_EMDAT_report.pdf (Accesed: 03.03.2024).

Dinçer, S. (2023). Afetlerde Coğrafi Bilgi Sistemleri (176-188). Afet Yönetimi ve Teknoloji Ed. Saime Şahinöz. İstanbul Tıp Kitabevi.

Ekici T. (2023). Yeryüzündeki Volkanik Aktivitenin Dağılımı ve Türkiye'den Örnekler. CUJAST, 2(1), 5-15.

EM-DAT (t.y.a). EM-DAT – Data Entry – Field Description/Definition. https://www.emdat.be/guidelines (Accesed:03.03.2024).

EM-DAT (t.y.b). General Classification. https://www.emdat.be/classification (Accesed:03.03.2024).

EM-DAT. (n.d.). Classification. Emergency Events Database. https://www.emdat.be/classification (Accesed:03.03.2024).

Gökçeoğlu, C., & Ercanoğlu, M. (2001). Heyelan Duyarlılık Haritalarının Hazırlanmasında Kullanılan Parametrelere İlişkin Belirsizlikler. Yerbilimleri, 22(23), 189-206.

Gülsoy, A., & Uyan, Y. (2023). Afetlere Genel Bakış (1-19). Afet Yönetimi ve Teknoloji Ed. Saime Şahinöz. İstanbul Tıp Kitabevi.

Hui, D. S., & Zumla, A. (2019). Severe Acute Respiratory Syndrome: Historical, Epidemiologic, and Clinical Features. *Infectious Disease Clinics,* 33(4), 869-889.

IFRC. (2020). World Disasters Report 2020 Come Heat Or High Water. https://media.ifrc.org/ifrc/world-disaster-report-2020 (Accesed:03.03.2024).

Kandilli Rasathanesi ve Deprem Araştırma Enstitüsü, 2024 http://www.koeri.boun.edu.tr/sismo/2/deprem-bilgileri/genel-bilgiler/(Accesed:03.03.2024).

Karaman, Z. T. (2016). Afet Yönetimine Giriş ve Türkiye'de Örgütlenme. Bütünleşik Afet Yönetimi (ss. 1-39). Editörler Karaman Z. T. ve Altay A. İzmir: Birleşik Matbaacılık.

Karaman, Z. T., Çakır, Ö., & Berge, M. A. (2022). Şiddetli Rüzgârlar, Fırtınalar İklim Değişikliğinin Neresinde?. Çevre Şehir Ve İklim Dergisi, 1(2), 225-250.

Kaya, N., Atalay, E., & Demir Doğan, M. (2023). Determining Levels of Search and Rescue Team's Knowledge on Avalanche Search and Rescue. Afet Ve Risk Dergisi, 6(2), 434-447. https://doi.org/10.35341/afet.1178842.

Köçer, M. S., & Aslan, R. (2023). Gönüllü Arama Kurtarma Ekiplerinin Orman Yangınlarındaki Tahliye Deneyimleri: 2021 Akdeniz Orman Yangınları. Afet Ve Risk Dergisi, 6(3), 829-851. https://doi.org/10.35341/afet.1236021.

Machhi, J., Herskovitz, J., Senan, A. M., Dutta, D., Nath, B., Oleynikov, M. D., & Kevadiya, B. D. (2020). The Natural History, Pathobiology, and Clinical Manifestations of SARS-CoV-2 infections. *Journal of Neuroimmune Pharmacology,* 15, 359-386.

Maden Tetkik ve Arama Genel Müdürlüğü, (2024). Türkiye'nin Volkanları. https://www.mta.gov.tr/turkvolc/tr/turkiyenin-volkanlari (Accesed:03.03.2024).

Our World in Data (2024). Natural Disasters. https://ourworldindata.org/natural-disasters (Accesed:03.03.2024).

Özer, Y. E. (2017). Afet Konusundaki Algı ve Yerel Aktörlerin Sorumlulukları. Sayıştay Dergisi 106, 1-34.

Peiris, J. S., Guan, Y., & Yuen, K. (2004). Severe Acute Respiratory Syndrome. *Nature Medicine,* 10(12), 88-97.

Sussman, G. D. (2011). Was the black death in India and China?. *Bulletin of the History of Medicine,* 319-355.

Taşdemiroğlu, M. (1970). Türkiyede Kütle Hareketleri. *Türkiye Jeoloji Bülteni,* 13(2), 26-35.

Turan, M., & Hamza Çelikyay, H. (2020). Türkiye'de KOVİD-19 ile Mücadele: Politikalar ve Aktörler. Uluslararası Yönetim Akademisi Dergisi, 3(1), 1-25. https://doi.org/10.33712/mana.733482.

UNDRR (2024). https://www.preventionweb.net/knowledge-base/hazards/insect-infestation (Accesed:03.03.2024).

Usta, G. (2021). Türkiye'de Meydana Gelen Deprem Afetlerinde Sunulan Hizmetlerin Değerlendirilmesi (1999-2020): Nitel bir araştırma. (Publication No. 688294) Doctoral dissertation, Gümüşhane University.

Usta, G., Aslan, R., Gündüz, F., Atalay, E., & Pala, İ. T. (2023). On-site Evaluation of Emergency Response Carried out After Earthquakes with Epicentral Base Kahramanmaraş/Pazarcık-Elbistan in Türkiye: An Observational Study. *Türk Coğrafya Dergisi*, (83), 145-154. https://doi.org/10.17211/tcd.1360428.

WHO (2024). Coronavirus Disease (COVID-19) Pandemic. https://www.who.int/europe/emergencies/situations/covid-19 (Accesed:03.03.2024).

Wilhite, D. A., & Glantz, M. H. (1985). Understanding: the Drought Phenomenon: The Role of Definitions. *Water International,* 10(3), 111-120.

Zeybekoglu, U. (2022). Spatiotemporal Analysis of Droughts in Hirfanli Dam Basin, Turkey By The Standardised Precipitation Evapotranspiration Index (SPEI). *Acta Geophysica,* 70(1), 361-371. https://doi.org/10.1007/s11600-021-00719-x.

Chapter 2

General Logistics in Disaster Management

Özlem Bölükbaş*, **MA**
Institute of Social Sciences, Marmara University, Kadıköy, İstanbul, Türkiye

Abstract

The response to disasters can vary depending on the type, scale, and affected area of the event, making intervention challenging. This study provides a general overview of disaster management, focusing on the systems used by three selected countries and the associated technological infrastructure. The objective is to underscore the significance of effective disaster management and the systems used to streamline response efforts. To achieve this, international and national reports, along with relevant literature and documents, were thoroughly analyzed. Successful disaster management necessitates a comprehensive and systematic approach that addresses pre-disaster preparedness, the management of the event itself, and post-disaster recovery. Through our examination, it became evident that the systems and technological innovations utilized in disaster management significantly enhance response capabilities.

Keywords: disaster, disaster logistics, disaster management, disaster technologies

---

* Corresponding Author's Email: ozlembolukbas87@hotmail.com.

In: Disaster Management in the Digital Age and Emerging Technologies
Editors: Ayşe Asiltürk and Ayşegül Sarıoğlu Kemer
ISBN: 979-8-89530-213-2

Introduction

It is known that disasters lead to economic problems on both the regional and national levels, resulting in social and psychological distress among people (Tanyaş et al., 2013). Therefore, it is crucial to focus on the social, economic, cultural content, and dimensions of disasters to develop policies accordingly. While the destructive power of some types of disasters, which can be defined as low-probability but high-impact risks, can be minimized with preventive measures, complete elimination is not possible (Giddens, 1998). Consequently, humanity has been compelled to learn to live with disasters, leading to the field of disaster management.

Countries tend to undertake significant efforts in disaster management following major and devastating disasters with widespread negative consequences. In Türkiye, the Marmara Earthquake marked a turning point in disaster management. Following the Marmara Earthquake, a new and serious restructuring process began in the disaster management organization. In the scope of emergency and disaster preparedness, Law No. 5902 on "Disaster and Emergency Situations" was enacted in 2009, establishing the Presidency of Disaster and Emergency Management Authority (AFAD), merging the Directorate General of Disaster Affairs, the General Directorate of Civil Defense, and the General Directorate of Public Works under this umbrella. Additionally, the AFAD directorates were established under the authority and command of the provincial governors (Önsüz and Atalay, 2015). Similar examples exist for the United States and Japan.

The effectiveness of disaster risk reduction efforts in disaster management is directly proportional to the level of development of societies. This is because as societies become more socioeconomically advanced, the importance placed on human life increases, leading to a greater emphasis on preventing disasters, one of the most common events posing a threat to human life, and allocating more resources to mitigate their impacts (Schram, 1993).

The Disaster Concept and Disaster Types

A disaster is defined as "an event that causes physical, economic, and social losses for the whole or certain segments of society, disrupting or halting normal life and human activities, and where the affected community's coping capacity is insufficient, whether it is of natural, technological, or human origin" (AFAD, 2014). According to The International Federation of Red

Cross (IFRC), disasters are significant disruptions that exceed a society's ability to cope using its own resources. In addition to natural, human-made, and technological hazards, various factors affecting a society's exposure and vulnerability can also lead to disasters (IFRC, 2024).

It is possible to classify disasters according to their types. Here are the types of disasters observed worldwide (Tanyaş et al., 2013):

1. Geological Disasters: Earthquakes, Landslides, Mudflows, and Rockfalls, Volcanic Eruptions, Land Subsidence, Tsunamis, and Meteor Showers.
2. Climatic Disasters: Extreme Heat Waves, Extreme Cold Waves, Droughts, Hailstorms, Sandstorms, Floods, Avalanches, Blizzards, Extreme Snowfall, and Ice Storms, Acid Rains, Fog, Rise in Sea and Lake Levels, and Air Pollution.
3. Biological Disasters: Erosion and Desertification, Forest Fires, Epidemics (Plague, Influenza, AIDS, etc.), and Insect Infestations.
4. Social Disasters: Fires, Wars, Terrorist Attacks, and Migrations.
5. Technological Disasters: Mining Accidents, Biological, Radiological, Nuclear, Chemical Attacks and Accidents, Industrial Accidents, and Transportation Accidents.

Human-Made Disasters

Disasters caused by humans resulting in significant loss of life and property, such as fires, major accidents, explosions, uncontrollable nuclear activities, technological disasters, wars, and uncontrollable social events, are referred to as Human-Made Disasters. Although Human-Made Disasters are anthropogenic in nature, they often involve the influence of natural factors in reaching disaster proportions. For instance, a fire caused by human activity may become uncontrollable due to strong winds, spreading over a wider area (Ankara University, 2024).

According to the Global Natural Disaster Assessment Report (2022), a total of 388 disasters occurred globally in 2022. Among the most devastating disasters of 2022 were floods in Pakistan, droughts in Africa, typhoons, and hurricanes in the Philippines. While floods were the most frequent disaster in 2022, extreme heat caused the highest number of deaths (16,416 deaths), accounting for more than half of all disaster-related deaths. However, earthquakes continued to cause the highest number of deaths per disaster event, with an average of 990 deaths per event from 1992 to 2021, while only 52 deaths per disaster event occurred in 2022. The damage caused by disasters in 2022 (223.84 billion USD) exceeded the annual average disaster damage of the last 30 years (annual average of 121.46 billion USD) (Global Natural Disaster Assessment Report, 2022).

Given that thousands of earthquakes occur worldwide every year and have a significant impact, earthquakes naturally hold a distinct place among disasters. Although the majority of these earthquakes may go unnoticed by humans, some of them have observable effects and can become disasters (Ceren, 2021). In 2023, a total of 1,712 earthquakes with a magnitude of five or greater were recorded worldwide as of December of that year (Dyvik, 2023).

Logistics

Logistics is derived from the combination of the words "logic" and "statistics," implying logical statistics. Etymologically, the term logistics originates from the Greek word "Logistikos," meaning the science of calculation or the skill in calculation. The term logistics was first used in the military context (Tanyaş, 2004).

According to the Council of Supply Chain Management Professionals, logistics is defined as the planning, implementation, and control of the efficient and effective movement and storage of materials, supporting services, and information flows in both directions within the supply chain, from the point of origin of raw materials to the point of consumption of the end product, to meet customer requirements (Adıgüzel, 2009). Following the "7 Rs" principle, logistics can be defined as ensuring that the right material reaches the right place, in the right quantity, in the right condition, at the right time, to the right customer, at the right cost (İTO, 2006).

Logistics management encompasses the planning, implementation, and control of the forward and reverse flows of goods, services, and information between production and consumption points, as well as their storage, to meet customer needs effectively and efficiently within a supply chain process (Salam and Khan, 2020).

Disaster Logistics

Disaster logistics is a process that involves bringing together relevant resources, information, and capabilities to assist people affected by extraordinary circumstances and in need of assistance due to the impact of disasters (Van Wassenhove, 2006). In other words, disaster and emergency logistics comprise processes and systems that effectively mobilize people, resources, skills, and knowledge to assist disaster-affected individuals (Tanyaş et al., 2013). The primary objective of disaster logistics is to ensure the provision of resources tailored to the needs, the delivery to the disaster area, and the prioritization of life-saving activities within an adequate framework of safety and security, depending on the type of disaster (Jinhui, 2014; Raillani et al., 2020).

Disaster Management

Disaster management is a multidimensional, multi-actor, multi-disciplinary, and dynamic form of management with its unique rules and methods, using contemporary research, development, engineering, planning, and management technologies. Its aim is to prevent or minimize the physical, social, economic, and psychological losses suffered by human communities from natural, technological, or human-origin events, and to keep potential losses to the lowest possible level (Ergünay, 2008).

Disaster management refers to a comprehensive community-wide effort that must be undertaken to prevent disasters, reduce their impact, respond promptly, effectively, and efficiently to events leading to disasters, and create a safer and improved new living environment for the communities affected by disasters. In other words, it involves planning, directing, coordinating, supporting, and effectively implementing the measures and actions to be taken before, during, and after disasters by all institutions and organizations of society, using resources and capabilities in accordance with defined strategic objectives and priorities. This is a multidimensional, multi-disciplinary, and multi-actor management process. Within this process, activities can be categorized into main stages such as disaster risk reduction, preparedness, response, recovery, and reconstruction. The success of activities undertaken in one stage affects the success of activities to be undertaken in the next stage. This process is illustrated with a circle or nested rings, referred to as the Disaster Management Cycle or Chain. Therefore, terms such as "Integrated" or "Comprehensive Disaster Management" are also used (Özel İhtisas Komisyon Raporu, 2014).

Disaster management is a complex and comprehensive management model that requires the involvement and coordinated use of resources from all institutions and organizations within society to plan, direct, support, coordinate, and implement activities aimed at mitigating the impacts of disasters and preventing them, in the four main stages of disaster management: mitigation, preparedness, response, and recovery. This multi-actor, multi-disciplinary, and extensive approach is designed to serve the common goal of preventing disasters and reducing their adverse effects (T.C. Başbakanlık, 1997).

Regardless of how it is defined, disaster management encompasses all aspects of disasters along with the following:

a) Planning and implementation of preventive measures to reduce the adverse effects of potential harmful events, considering the economic aspect, and reducing the occurrence of disasters,
b) Preparation and organization of warning systems against the potential threat of disasters at any time, and organization of effective emergency responses both during alarm or warning periods and in the aftermath of disaster impacts,
c) Identification and monitoring of the situation of regions and communities in the event of collapses, floods, famines, slow-moving disasters, and timely implementation of solutions and measures,
d) Urgent implementation of emergency relief efforts for life-saving immediately after the occurrence of disasters,
e) Accelerating the implementation of recovery measures to facilitate recovery and promote ongoing post-disaster development. The impact of natural or human-origin disaster hazards manifests itself by destroying the social and economic activities of communities, striking at their development, and leading to the depletion or exhaustion of resources. As a result, large populations suffer. Therefore, disaster hazards should be considered only as a part of the development planning of all environments, along with natural and human resources (Sarp, 1999).

As is evident from the definitions provided above, "disaster management" is a dynamic form of management that requires continuous attention. It encompasses various fields ranging from healthcare to urban planning, involving numerous organizations and institutions. Therefore, it can be said that the concept of "disaster management" adapts to the specific circumstances of each disaster event, depending on its scale and type. It is natural for the management model created in response to a small-scale disaster event to differ from the model implemented in the case of a regional or national disaster event that has widespread effects (Akdağ, 2001).

Every year, millions of people are affected by natural disasters, resulting in significant loss of life, injuries, and billions of dollars in economic losses. Natural disasters disrupt infrastructure, transportation, and communication systems to varying degrees, leading to outbreaks of infectious diseases. They also disrupt education, affect public safety services, and create challenges related to shelter, food, and clothing. Furthermore, they contribute to unemployment and psychological issues. Hence, how disasters are managed becomes critically important (Tanyaş et al., 2013).

The Stages of Disaster Logistics

To prevent disasters and mitigate their impacts; before the disaster, during the disaster, and after the disaster, directing, coordinating, and implementing the necessary activities in line with this goal by all institutions and organizations of the society is the management of resources in this direction (Sarp, 1999). The stages of disaster logistics can be evaluated as follows (Tanyaş, et al., 2013):

1. Pre-Disaster Preparation
    a. Planning
    b. Procurement
    c. Transportation Management
    d. Warehouse Management
    e. Development of Human Resources and Exercises
2. Disaster Response Process
    a. Preliminary Assessment and Needs Identification
    b. Development and Implementation of Logistics Action Plan
    c. Monitoring, Evaluation, and Reporting of the Response Process
3. Post-Response:
    a. Planning
    b. Material Collection and Maintenance Activities
    c. Monitoring, Evaluation, and Reporting

Disaster Management in Practice: Used Systems and Technologies

In this section of the study, examples from Türkiye, the United States, and Japan are used to discuss how disaster management is conducted in these countries during disasters. Firstly, the most authoritative institutions and organizations nationwide during disasters in these three countries are provided. Subsequently, digital systems used to manage disasters are discussed.

Türkiye

A turning point in disaster management and coordination in Türkiye was the Marmara Earthquake on August 17, 1999. This earthquake, which caused significant loss of life and extensive damage, highlighted the need to revisit disaster management in the country. The need arose to redefine the responsibilities and authorities of the institutions that needed to be coordinated in dealing with disasters, necessitating the consolidation of authority and coordination under a single entity during disasters. Accordingly, the Directorate General of Civil Defense under the Ministry of Interior Affairs, the Directorate General of Disaster Affairs under the Ministry of Public Works and Settlement, and the Türkiye Emergency Management Directorate under the Prime Ministry were closed down, and the Disaster and Emergency Management Presidency (AFAD) was established under the Prime Ministry through Law No. 5902 enacted in 2009, bringing together the authority and responsibilities under one roof. With the adjustments made within the scope of the Presidential Government System, the Disaster and Emergency Management Presidency was transferred to the Ministry of Interior Affairs through Presidential Decree No. 4 issued on July 15, 2018 (AFAD, 2024a).

In order to monitor, manage, and provide decision support to managers in disaster and emergency processes electronically, the Disaster Management and Decision Support System (AYDES) was developed. AYDES is a system built on geographic information systems, equipped with decision support mechanisms capable of effectively managing national resources during disasters. AYDES, which also constitutes the IT infrastructure of the Turkish Disaster Response Plan (TAMP), operates to cover all phases of the planning, intervention, and recovery processes during disasters. It has been made available to the relevant Ministries and provincial organizations involved in disaster management to plan and improve disaster processes, with the aim of faster and more effective management (AFAD, 2024b).

The purpose of the Turkish Disaster Response Plan (TAMP) is to define the roles and responsibilities of disaster response groups and coordination units that will be involved in intervention efforts related to disasters and emergencies and to establish the basic principles of intervention planning before, during, and after disasters. The Presidency of Communications, the Ministry of Interior Affairs, the Ministry of Family and Social Services, the Ministry of Environment, Urbanization and Climate Change, the Ministry of Energy and Natural Resources, the Ministry of Treasury and Finance, the Ministry of Health, the Ministry of Agriculture and Forestry, the Ministry of Transport and Infrastructure, and the Turkish Red Crescent (Kızılay) are the main solution partners of TAMP. Other ministries, institutions, and organizations serve as support solution partners (AFAD, 2024c).

AYDES, built on Geographic Information Systems, is a web-based application with decision support mechanisms that effectively manage all resources in disaster and emergency situations. It is also the information infrastructure of the Turkish Disaster Response Plan (TAMP). AYDES fundamentally consists of three components (AFAD, 2024d):

Incident Command System

The Incident Command System enables the preparation, planning, and intervention processes of the service groups identified within TAMP to be managed on an integrated system. It provides preparation and intervention opportunities for local and national disaster and emergency situations through a software-based management model. The needs arising from disasters and emergencies can be managed flexibly and effectively through transportation, resource, and demand management processes. From the moment a national or local-level disaster occurs, incident notifications can be sent to teams via SMS and email. Service groups defined within TAMP can maintain continuous interaction and communication through instant messaging and email via the system (AFAD, 2024d).

Spatial Information System

The aim of the Spatial Information System is to create a sustainable disaster management system using Geographic Information System technologies. It is designed to provide rapid and accurate access to data before, during, and after disasters, to generate new information from data quickly, and to enable fast decision-making regarding spatial queries and analyses in disaster-affected and potentially affected areas. The application subcomponents and menus provide various base maps, allowing for real-time updating, editing, querying, displaying, and reporting of spatial data and results (AFAD, 2024d).

Recovery System

The Recovery System aims to conduct post-disaster recovery efforts supported by the Geographic Information System in the information environment. This will ensure the uninterrupted progress of interconnected processes such as damage assessment, ownership determination, site selection, etc. Additionally, the system's support with mobile applications facilitates data collection from fieldwork, as well as obtaining location-based information (AFAD, 2024d).

At the local level, coordination is ensured through Provincial Emergency Coordination Boards and Provincial-District AFAD Centers. The Provincial Emergency Coordination Board convenes under the chairmanship of the relevant governor. In cases of disaster and emergency situations, it convenes at the Provincial AFAD Center without waiting for a call or instruction (AFAD, 2024c).

The United States

In 1979, the Federal Emergency Management Agency (FEMA) was established in the United States. The primary missions of FEMA were redefined to include preparedness for all types of crisis situations, protection, response, recovery, and mitigation activities. In addition to the earthquake division within FEMA, there are other important national agencies contributing to earthquake disaster relief efforts. For example, agencies such as the US Geological Survey, Centers for Earthquake Engineering Research, and the National Institute of Standards and Technology provide vital information such as geographic data, field data, earthquake probabilities, land use plans, building types and standards, suitable material types, engineering information, and emergency preparedness data (Çeper, 2014).

In the United States, preparations for natural disasters, including those considered to be at the highest level of emergency such as war situations, are carried out within the framework of the "Integrated Emergency Management System" developed by FEMA. Within this system, an all-hazards approach has been adopted for both minor and limited incidents and events affecting the entire country. The terrorist attacks of September 11, 2001, were a turning point for the entire disaster and emergency management system of the country, particularly in terms of national security and preparedness capabilities (Çeper, 2014).

FEMA views itself as the nation's leading emergency management and preparedness agency and emphasizes that disaster and emergency management are not matters that can be addressed by a single agency alone. FEMA seeks to organize a disaster management system in which all national and local agencies, as well as all citizens, come together, collaborate, and work in coordination. This system can be referred to as the "Integrated Emergency Management System (IEMS)." Instead of creating separate plans for each emergency situation, this approach involves addressing all disasters under a single overarching plan that can be shaped according to the specific event (Çeper, 2014).

With this understanding, FEMA sets goals and objectives for disaster management and provides all kinds of support, including financial and equipment assistance, to state and local governments as well as civil society sectors at the primary responsibility level. Intensive training, expert development, and professionalization efforts are also implemented. Disaster management is approached from the perspective that it begins with individuals and families. Various plans, projects, and programs such as the National Incident Management System (NIMS), the National Response Framework (NRF), and the National Emergency Alert System are utilized to operate the system and continuously improve it (Çeper, 2014).

Japan

In Japan, the organization responsible for disaster management and coordination mechanisms is the Disaster Prevention Bureau within the National Land Agency (NLA) (AFAD, 2002). This bureau is responsible for determining and planning preventive or mitigating policies for all kinds of disasters in Japan, as well as implementing the Basic Law on Measures Against Natural Disasters and the Measures Law for extensive earthquake disasters. The NLA, which is part of the Ministry of Land, Infrastructure, and Transport, is an agency under the Prime Minister's Office, and its president holds the position of a minister of state. Its main activities revolve around land, infrastructure, and transportation, but one of its key tasks is also to establish and improve disaster management (Akdağ, 2002).

During normal times, the NLA continuously inputs maps, settlement plans, building structures and locations, and detailed information about the residents of each house into a computer to create a database. Therefore, in the event of a disaster, detailed information about buildings and facilities affected by the disaster in that area, as well as the people living in each building, can be immediately searched and transmitted to search, rescue, and relief teams (Akdağ, 2002). Additionally, Japan has a national research institute dedicated to disaster prevention and earthquake research, known as the National Research Institute for Earth Science and Disaster Prevention (NIED, 2024).

The Disaster Management Basic Information Distribution System (SIPD) serves as a link between disaster-related information systems operated by government ministries, municipalities, and relevant organizations. It plays a vital role as a pipeline allowing the integrated use and implementation of information (NIED, 2024a).

Discussion and Conclusion

In conclusion, different countries experience various types of disasters based on their geographical characteristics. Each disaster has its own unique features. This can be exemplified through the three countries discussed: Türkiye is mainly affected by earthquake disasters, the United States is subject to devastating hurricanes similar to typhoons, and Japan is prone to earthquakes and tsunamis. Therefore, when countries approach disaster management, prioritizing measures and planning specific to the prevalent disasters in their regions would be appropriate. Türkiye, for instance, addresses earthquake preparedness measures in its Disaster Management Plan (TAMP) for provinces such as Kahramanmaraş, while implementing measures against floods, landslides, and soil erosion in regions such as Rize and Trabzon, where rainfall is abundant.

The main objective of disaster management is to minimize loss of life and property and to overcome disasters with minimal damage. However, disaster management is a comprehensive, multi-actor, and complex process. Effective and comprehensive disaster management can be achieved through pre-disaster, during-disaster, and post-disaster actions. Pre-planned assignment of duties and responsibilities among relevant institutions eliminates complexity, ensuring efficient use of financial and human resources and facilitating disaster response. With advancing technology, countries employ various applications for disaster management, such as the AYDES system used in

Türkiye. Integrating these technological infrastructures, systems, and applications into disaster plans makes disaster management more systematic and successful.

Disclaimer

None.

References

Adıgüzel, B. (2009). Depo ve Lojistik Elemanı Eğitimleri Kitabı. Kadıköy Belediyesi.

AFAD. (2014). Açıklamalı afet yönetimi terimleri sözlüğü, Ankara.

AFAD. (2024a). AFAD ve Tarihçesi. (Accesed: 11.03.2024) https://www.afad.gov.tr/afad-hakkinda.

AFAD. (2024c). Bütünleşik Afet Yönetim Sistemi. (Accesed: 11.03.2024) https://www.afad.gov.tr/butunlesik-afet-yonetim-sistemi.

AFAD. (2024c). Türkiye Afet Müdahale Planı. (Accesed: 11.03.2024) https://www.afad.gov.tr/turkiye-afet-mudahale-plani.

AFAD. (2024d). Afet Yönetim ve Karar Destek Sistemi Projesi (AYDES). (Accesed: 12.03.2024) https://www.afad.gov.tr/afet-yonetim-ve-karar-destek-sistemi-projesi-aydes21.

Akdağ, S. E. (2001). Afet yönetimi sistemi ve marmara depremi sonrasında yaşanan sorunlar. Ankara, s.6-7.

Akdağ, S. E. (2002). Mali Yapı ve Denetim Boyutlarıyla Afet Yönetimi. Sayıştay İnceleme Raporu. (Araştırma/İnceleme/Çeviri Dizisi: 20). Ankara, s. 16-18.

Ankara Üniversitesi. (2024). Afet ve Afet Yönetimi. (Accesed: 06.03.2024) https://acikders.ankara.edu.tr/pluginfile.php/190603/mod_resource/content/0/1.%20 HAFTA.pdf.

Başbakanlık, T. C. (1997). Doğal Afetler Genel Raporu, Ankara, Mayıs, s.16.

Çeper, K. (2014). ABD'de eyaletten devlete acil durum yönetimi, Türkiye ile mukayesesi. İçişleri Bakanlığı İnceleme-Araştırma Raporu. Mülkiye Müfettişliği Sayı: 64/. Ankara s.4,10,38.

Ceren, A. (2021). Afetler ve İnsani Yardım Faaliyetleri Üzerine Bir Değerlendirme. *Ankara Hacı Bayram Veli Üniversitesi İktisadi ve İdari Bilimler Fakültesi Dergisi, 23*(2), 289-312.

Dyvik, E. H. (2023). Number of earthquakes worldwide 2000-2023. (Accesed: 06.03.2024) https://www.statista.com/statistics/263105/development-of-the-number-of-earthquakes-worldwide-since-2000/.

Ergünay, O. (2008). Afet Yönetiminde Kurumsal Yapılanma ve Mevzuat Nedir? Nasıl Olmalıdır?, İstanbul Depremini Beklerken Sorunlar ve Çözümler Bildiriler Kitabı, 20 Eylül 2008 CHP İstanbul Deprem Sempozyumu, İstanbul, s. 97-108.

Giddens, A. (1998). Modern Law Review. Cambridge: Cambridge University Press.

Global Natural Disaster Assessment Report. (2022). (Accesed:06.03.2024) https://www.preventionweb.net/publication/2022-global-natural-disaster-assessment-report.

IFRC. (2024). What is a disaster? (Accesed: 06.03.2024) https://www.ifrc.org/our-work/disasters-climate-and-crises/what-disaster.

İTO. (2006). "Türkiye Lojistik Sektörü Altyapı Analizi," İstanbul, İTO Yayın 14.

Jinhui, D. (2014). Earthquake Disaster Emergency Logistics of Transport Route Optimization Research. International Conference on Logistics Engineering, Management and Computer Science, 1165–1170. (Accesed:12.03.2024) https://www.researchgate.net/publication/266650920_Earthquake_Disaster_Emergency_Logistics_of_Transport_Route_Optimization_Research.

NIED. (2024). 'About NIED'. (Accesed: 11.03.2024) https://www.bosai.go.jp/e/about/.

NIED. (2024a). 'Center for Comprehensive Management Disaster Information'. (Accesed: 12.03.2024) https://www.bosai.go.jp/e/research/center/jouhou.html.

Önsüz, M. & Atalay, B. (2016). Afet Lojistiği. *Osmangazi Tıp Dergisi, 37*(3): 1-6.

Özel İhtisas Komisyon Raporu. (2014). Onuncu Kalkınma Planı 2014-2018. Afet Yönetiminde Etkinlik. Kalkınma Bakanlığı. Ankara, s. 102.

Raillani, H., Hammadi, L., Altimari Samed, M. M., El Ballouti, A. & Barbu, V. S. (2020). Humanitarian logistics in the disaster relief supply chain: State of the art. WIT Transactions on Engineering Sciences, 129, 181–193.

Salam, M. A. & Khan, S. A. (2020). Lessons from the humanitarian disaster logistics management: A case study of the earthquake in Haiti. *Benchmarking An International Journal, 27*(4), 14551473.

Sarp, N. (1999). Disaster Management in Healthcare. *Deprem Araştırma Bülteni*. ss.14.

Schram A. (1993). *Public good provision and public bad prevention: The effect of framing.* Netherlands: University of Amsterdam, Faculty of Economics and Econometrics, Department of Micro-economics/CREED.

Tanyaş, M. (2004). Lojistik ve Tedarik Zinciri Yönetimi, Ders Notları. İstanbul: İ. T. Ü. Endüstri Mühendisliği Bölümü.

Tanyaş, M., Günalay, Y., Aksoy, L. & Küçük, B. (2013). İstanbul İli Afet Lojistik Planı Kılavuzu. İstanbul Kalkınma Ajansı Raporu DFD-39. İstanbul.

Van Wassenhove, L. N. (2006). Blackett memorial lecture. Humanitarian aid logistics: Supply chain management in high gear. *Journal of the Operational Research Society, 57*(5), 475-489.

Chapter 3

Technological Innovations in Disaster Shelter Services: A Comprehensive Exploration

Ramazan Aslan*, PhD

Department of Emergency Aid and Disaster Management, Ardahan University Faculty Health Sciences, Ardahan, Türkiye

Abstract

One of the biggest impacts of disasters is making dwellings uninhabitable, forcing hundreds of thousands of people to abandon their homes every year. Providing shelter, a fundamental human right, to displaced people is a challenging process that both governments and humanitarian organizations must deal with. Shelter initiatives can range from immediate emergency shelters following a disaster to long-term efforts spanning years. To successfully fulfill such a lengthy and comprehensive service, decisions must consider numerous standards and variables. A thorough analysis of these variables allows for appropriate solutions to be offered to those in need. Incorporating technological innovations undoubtedly facilitates the provision of such complex services. In this context, technologies such as artificial intelligence, mathematical algorithms, multi-criteria decision-making techniques, geographic information systems, and others offer significant advantages. Newly developed shelter designs and materials may also initiate changes in sheltering activities. Ensuring that all these technological advancements and changes are environmentally, economically, and socially sustainable is seen as the most crucial threshold for replacing traditional techniques.

* Corresponding Author's Email: ramazanaslan@ardahan.edu.tr.

In: Disaster Management in the Digital Age and Emerging Technologies
Editors: Ayşe Asiltürk and Ayşegül Sarıoğlu Kemer
ISBN: 979-8-89530-213-2

Keywords: shelter, response, recovery, housing, temporary

Introduction

In 2023, due to disasters, 86,473 people lost their lives, and 93.1 million people were affected. The earthquakes that struck Turkey and the Syrian Arab Republic in 2023 resulted in 56,683 deaths and $42.9 billion in economic losses, making them the most devastating events of the year in terms of both loss of life and economic damage (Center for Research on the Epidemiology of Disasters, 2024). One of the most apparent effects of disasters is the damage or complete destruction of buildings in affected areas, leading to homelessness for many (Félix et al., 2013; Hadafi and Fallahi, 2010). Generally, housing structures bear the largest share of economic losses resulting from disasters (Johnson et al., 2006). For example, because of the earthquake in Türkiye, the total housing damage amounted to ₺1,031.9 billion, equivalent to $56.9 billion. Additionally, 2,273,551 people required emergency shelter. The necessary economic resources to meet this need are approximately ₺40 billion (approximately $2.1 billion) (Republic of Türkiye, Presidency of Strategy and Budget, 2023, p. 36). Furthermore, conflicts are also among the primary causes of homelessness. Due to persecution, conflict, violence, human rights violations, or events that seriously disrupt public order, 108.4 million people worldwide were forcibly displaced in 2022 (United Nations High Commissioner for Refugees [UNHCR,] 2023).

After any natural or man-made disaster, providing temporary shelter for affected families has always been a primary concern (Hadafi and Fallahi, 2010). The provision of temporary housing can be crucial for the affected population to quickly recover and rebuild, thus saving time (Johnson, 2007a). Post-disaster reconstruction is a complex process, and providing temporary shelter is one of the essential tasks that must be fulfilled as soon as possible, as it provides conditions for people to gradually return to normal life (Félix et al., 2013). The provision of housing services is not only a lifesaving measure but also a fundamental human right to have a place to live (Office of the United Nations High Commissioner for Human Rights [OHCHR], 2010, p. 1; UNHCR, 2007, p. 219). Despite many efforts in this direction, meeting people's housing needs remains a significant challenge for governments, humanitarian organizations, and, most importantly, the affected populations (International Federation of Red Cross and Red Crescent Societies [IFRC], 2013, p. 4).

Stages of Sheltering in Disasters

Following a disaster or similar situation, there are various approaches and perspectives regarding meeting the need for emergency shelter. A widely accepted approach, focusing on post-disaster needs, is the shelter approach proposed by Quarantelli (1995). This approach addresses shelter needs in four different stages. The first stage is an emergency shelter, where people seek refuge immediately after disasters and stay for a few hours to several days. Following these emergency shelters, disaster victims transition to temporary shelters where they can stay for a longer period but still cannot resume their regular lives. Subsequently, the stages of temporary and permanent housing follow. In these stages, people have returned to their daily routines. Although these stages may appear sharply delineated, given the complex nature of the reconstruction process, it should be remembered that these stages may run parallel to each other, transition into one another, or occur without one stage necessarily preceding the other (Da Silva, 2010, p. 8; IFRC, 2013, p. 9; Johnson, 2007b; Mukherji, 2017).

IFRC uses the terms "emergency shelter," "temporary shelters," "transitional shelters," "progressive shelters," and "core shelters." Temporary shelters or "transitional" shelters are typically designed to be relocatable and reusable. Progressive shelters and core shelters, on the other hand, are constructed to become part of a permanent solution (IFRC, 2013, p. 8). The general approach advocates for the use of different types of housing, but the housing stages do not necessarily require different types of buildings. One structure can serve multiple purposes (Lizarralde et al., 2010, p. 7).

Types of Shelter in Disasters

Post-disaster housing needs can be addressed in various ways, which can be classified into two main groups: dispersed settlement and grouped settlement. Dispersed settlements can be further divided into host families, rural self-settlement, and urban self-settlement. Similarly, grouped settlements can be categorized into collective centers, self-settled camps, and planned camps (Corsellis and Vitale, 2005, p. 7). Each of these approaches has its own advantages and disadvantages. For example, staying with relatives may offer a quick solution, but it may later burden the hosts excessively. While planned camps can serve large populations, high population density can lead to serious health issues. Using schools as collective shelters can provide access to good infrastructure, but it can disrupt education. Therefore, after a thorough assessment of the situation, the most appropriate method should be chosen (UNHCR, 2007, p. 219; UNHCR and the International Organization for Migration [IOM], 2010, p. 62).

The sole purpose of housing is not merely to provide needy individuals with four walls and a roof. Adequate housing entails safe, culturally appropriate, economically affordable, accessible to services, close to livelihood sources, with sufficient living space, adequate water, sanitation, and hygiene facilities, and areas that offer food preparation and storage facilities (OHCHR, 2010, p. 4; Sphere Association, 2018, p. 244). According to the priority order, the functions that a shelter needs to fulfill are as follows (United Nations Office for the Coordination of Humanitarian Affairs [OCHA], IFRC, 2015, p. 18):

- Providing protection against cold, heat, wind, and rain.
- Storage of belongings and protection of property.
- Establishment of land rights.
- Setting up a preparation point for future actions such as reconstruction.
- Emotional security and privacy needs.
- An address for receiving medical assistance, food distribution, etc.
- Housing is located close to employment.

Considerations and Technological Developments in Providing Adequate Shelter

Ensuring adequate shelter for those in need is a crucial issue, and there are several points to consider in the provision of this service. This section addresses the considerations that need to be considered in shelter provision and the technological innovations related to these considerations.

Planning and Needs Assessment

Making decisions regarding the provision of shelter services raises a nested series of complex questions. Decisions need to be made about what types of temporary housing will be provided, where and how payments will be made, where the shelters will be located, and how long they will be used. Additionally, plans should be made for what will happen to shelters after the end of the usage period (Johnson et al., 2006). For the shelter service to be successful, the planning stage must also consider local standards of living, local industries, and local politics (Johnson, 2007a). Planning is one of the cornerstones of a successful shelter service (Félix et al., 2013). Despite its obvious importance, the planning process is often the most unsuccessful aspect of providing shelter services (Global Shelter Cluster, 2023, p. XX).

Any effort regarding housing or shelter provision should begin with an analysis of what constitutes "normal housing" in a particular community. This stage should not be considered a "low-priority concern" (Davis, 1978, p. 16). Following this stage, it should be determined who needs or will need housing. Rural communities often have the ability to create their own shelters. However, in industrialized societies where standard living conditions are generally higher, people are not accustomed to providing their own housing. This situation makes shelter provision inevitable in urban areas affected by disasters (Johnson, 2007a). It should not be overlooked that not only those whose houses are destroyed or damaged by earthquake-like disasters but also those who fear aftershocks will need shelter, and planning should take them into account. Similarly, extended family structures should be considered (Boğaziçi University and Istanbul Metropolitan Municipality, 2019, p. 177; OCHA, IFRC, 2015, p. 54). It should also be considered that the depletion of the existing housing stock by the disaster will exacerbate the housing need due to potential rent increases in the remaining houses (Johnson, 2007a).

The dimensions of potential housing needs can be better understood with the example of the 2023 earthquakes in Türkiye. As a result of the earthquakes, 2,273,551 people were directly faced with housing problems. Emergency shelter was provided to 1,440,668 people with 360,167 tents set up in the disaster area (Republic of Türkiye, Presidency of Strategy and Budget, 2023, p. 30). After an earthquake of magnitude Mw=7.5 in Istanbul, approximately 640,000 households, equivalent to about 2 million people (assuming an average household size of 3), are estimated to require emergency shelter (Boğaziçi University and Istanbul Metropolitan Municipality, 2019, vii).

Assessing the opportunities offered by the advancement of technology is important for increasing planning success and providing the right response. In the planning process, various prediction models can be used to determine potential shelter needs, and various mathematical algorithms can be used to allocate shelters to the right people. Examples of these developments include the use of the Internet of Things, unmanned aerial vehicles, and remote sensing technologies for rapid damage assessment and needs analysis (Bail et al., 2021; Félix et al., 2020; Gharib et al., 2022; Vecere et al., 2017; Yin et al., 2022).

Site Selection

The location of the shelter is critically important for life to return to normal during the extended process (Kim et al., 2021). Research has shown that land ownership, security, proximity to employment, and access to services are crucial to the effectiveness of shelter activities (OCHA, IFRC, 2015, p. 54). In one study, it was indicated that ten criteria significantly influence site selection, including accessibility, social facilities, health services, area capacity, soil type, terrain slope, security, infrastructure, ownership, climate, and vegetation (Ömürgönülşen and Menten, 2021). The Disaster and Emergency Management Presidency (AFAD, 2015) recommends paying attention to the following points for the selection of shelter center locations:

- Proximity to easily protected and controlled areas against threats.
- Preference for areas suitable for electricity, water, and sewerage connections.
- Planning for expansion in case of population growth.
- Conducting soil surveys rapidly.
- Selecting areas unsuitable for agriculture and free from ponding risks, and paying attention to the prevailing wind direction.

As mentioned above, site selection is a complex process that requires the consideration of many different variables, and traditional methods may overlook many criteria. Therefore, the use of mathematical models and multi-criteria decision-making techniques is recommended to ensure proper site selection (Chen et al., 2013; Dayanır et al., 2022; Kılcı et al., 2015; Ma et al., 2019; Omidvar et al., 2013). Furthermore, active utilization of geographic information systems creates significant opportunities for site selection (Uddin and Matin, 2021).

Shelter Design

Shelter design can vary depending on various factors such as geographical conditions, economic situation, technological capabilities, cultural structure, belief system, and many others in the area affected by the disaster.

When considering the construction process, there are generally two main approaches to providing temporary shelter: either it is produced and transported to the target area, or it is built locally using materials found locally by displaced populations or local workers (Albadra et al., 2018). Another approach to housing involves construction kits instead of prefabrication. These kits use small components assembled on-site, facilitating participation in the local community and overcoming transportation challenges (Félix et al., 2013).

Standardizing shelters in certain sizes or forms for ease of production and packaging is a common practice. However, this approach is not realistic, as it overlooks factors such as the high cost associated with the affected country, social consequences, cultural and climatic differences, family size, and local technological capacity (OCHA, IFRC, 2015, p. 53). Additionally, such aid activities can undermine the community's ability to cope (Barakat, 2003, p. 16). Imported industrialized designs in the 70s and 80s did not meet local needs, leading to cultural mismatches and material shortages for repairs. This has increased skepticism about prefabrication for post-disaster housing (Lizarralde et al., 2010, p. 13). Failure to involve the community in developing shelter design criteria, as seen here, has significantly reduced success in meeting needs (OCHA, IFRC, 2015, p. 51).

In emergencies, how and where shelter assistance is provided significantly affects people's situations. Therefore, active community participation in processes is of critical importance (Global Shelter Cluster, 2023, p. xv). Community participation in design processes helps to develop social bonds and promotes solidarity (Opdyke et al., 2019). Engaging beneficiaries, involving them in the process, and allowing them to make decisions about their own living space will help them better adapt to the process (Hadafi and Fallahi, 2010). It has been found that the lack of participation in decision-making processes or delayed participation leads to high levels of dissatisfaction with the design and causes some beneficiaries to refuse to move to new settlements (Davidson et al., 2007). In addition to social participation, cultural approaches and empowering the local community are important in all improvement and development processes (El-Masri and Kellett, 2001).

Various building materials are available for emergency shelter and housing reconstruction. These include unused inventories, local commercial and non-commercial materials, and salvaged materials from debris (OCHA, IFRC, 2015, p. 16). Local materials offer low environmental impacts and costs, but structural design requires more effort, while global materials can provide better technical performance (Zea Escamilla and Habert, 2015).

Shelter design should be adaptable and contextually appropriate due to the potential suitability of different designs for different regions (IFRC, 2013, p. 14). It is important for shelters to be convertible according to people's needs and usable for other purposes (Arslan, 2007). When the design of the shelter does not allow for adaptation, people can take steps to expand or transform these shelters using their own methods. This situation can lead to problems such as health issues, flammability, lack of privacy, and security concerns, while also making settlements irregular. Strictly preventing changes made by people can maintain order but significantly reduce their satisfaction (Alshawawreh et al., 2017).

The design of the shelter follows national and international minimum standards. In Türkiye, temporary shelters aim for a living area of 3.5-4.5 square meters per person in tents. Tents should allow airflow and sunlight protection, with insulation for cold climates and heat-resistant materials for hot climates. They should have a power capacity of 5-7 kilowatts per hour, and communal tents should have lockable doors (AFAD, 2015). These standards are parallel to international standards (Sphere Association, 2018, 238–287).

One of the most sensitive issues in constructed shelters is undoubtedly ventilation and temperature-related measures. Inappropriate shelter materials can lead people to despair and prompt them to develop their own risky methods, such as making a fire inside tents (Albadra et al., 2018). Another important issue is the accessibility of shelters. Necessary accessibility measures should be taken for the elderly, disabled individuals, and other vulnerable individuals. Studies have found that inaccessible shelter areas cause significant difficulties for vulnerable individuals (Aslan and Şahinöz, 2023; Kim et al., 2021).

Innovative shelter design methods include diversifying materials, altering form, and changing production. In one study, the use of recycled plastics was proposed in the design of shelters, emphasizing the positive physical and chemical properties of recycled plastics and their ease of accessibility (Moreno-Sierra et al., 2020). In another study, a shelter design inspired by quilts was proposed, demonstrating that these shelters have higher structural efficiency compared to existing solid-walled military shelters (Tumbeva et al., 2016). As an alternative method to accommodate displaced individuals in Syria, shelter production using 3D printing technology with sand has been realized (Labuda et al., 2024). Some studies have focused on the digitization of the design process, highlighting that the use of computer technologies and calculation programs accelerates construction processes and provides advantages to practitioners (Kuchai et al., 2020). Computer-aided design is seen to have strong potential for modularity, mass customization, and automatic production generation (Karaoğlan and Alaçam, 2019). The use of social network analysis as a decision-making tool to understand material and supplier networks related to shelters is also proposed as another suggestion (Copping et al., 2022).

Sustainability of Shelters

A significant portion of temporary shelter programs are unsustainable in terms of environmental, economic, and social dimensions due to failed strategies, urgency, misunderstandings of users' real needs, and misconceptions about local conditions and resources (Félix et al., 2013; Lizarralde et al., 2010, p. 20; Pomponi et al., 2019). Temporary shelter centers can remain functional for much longer than expected and can even transition to permanent settlements over time. The social, environmental, and economic impacts of this transformation are inevitable. Since the existing temporary shelter area is expected to be more affordable economically compared to other areas, it is expected that people will not evacuate from there, which is an anticipated outcome (Johnson et al., 2006). The potential political consequences of forcibly evacuating people from these areas can influence the decisions of authorities (Barakat, 2003).

From an environmental sustainability perspective, shelters often inflict difficult-to-reverse damage to the surrounding area due to materials used, infrastructure works, unplanned settlements, and other factors. Concrete slabs placed under shelters can be cited as an example. These slabs are often left in place without being removed, leading to significant environmental pollution (Johnson, 2007a, p. 49). In a study, emphasis was placed on the high energy consumption of temporary shelters (Song et al., 2016). To enhance environmental sustainability, renewable energy sources should be used, and passive heating and cooling techniques from local architecture should be adopted. Shelters should be made reusable or recyclable. Natural and recyclable materials should be preferred, and if possible, local materials should be used. A balance should be struck between local materials and prefabrication (Alshawawreh et al., 2020).

Economic sustainability is another crucial dimension. The provision of shelter services involves many stages, such as procurement, transportation, assembly, infrastructure, and services, making it a costly service. It is essential to complete the reconstruction process as soon as possible to reduce these costs (OCHA, IFRC, 2015, p. 13). To illustrate these costs with an example, it is estimated that expenditures for tent procurement, tent interior materials, and container purchases for the temporary shelters of earthquake victims in the 2023 Türkiye earthquakes will reach approximately ₺25 billion (approximately $1.3 billion) (Republic of Türkiye, Presidency of Strategy and Budget, 2023, p. 37). An important aspect at this point is the financing system of these programs. Direct cash assistance is effective in the short term but can create a dependency relationship between survivors and the assisting groups. It is much more advantageous for both individuals and society to participate in financing their own shelter programs, especially when permanent reconstruction is involved (OCHA, IFRC, 2015, p. 13). Additionally, it is important to take measures to save money and make choices. Encouraging the use of local resources whenever possible can be beneficial, as importing goods can create serious financial problems (Félix et al., 2013).

Considering social sustainability, the shortcomings in community participation and cultural adaptation are evident. For social sustainability, the involvement of residents in the design of housing units is important. In addition to individual facilities, dimensions suitable for age, gender, and cultural needs should be provided, and areas for animals should also be considered (Alshawawreh et al., 2020).

It is observed that most of the technological developments previously mentioned focus on sustainability, and the majority of new methodological approaches aim to enhance environmental, economic, and social sustainability. The post-disaster recovery period creates an environment where technological advancements can compete with traditional technologies in the field, and sustainability is the most crucial criterion that will determine the winner in this regard (Jigyasu, 2010, p. 49).

Discussion and Conclusion

Disaster shelter service is multidimensional and multi-actor, ranging from planning to implementation. In this section, shelter services have been examined broadly, and technological innovations that could bring advancements in the provision of these services have been highlighted. From the analyses conducted, it can be said that some of the major technological developments that could affect the services provided during disasters include the Internet of Things, Geographic Information Systems, Artificial Intelligence, Big Data, Advanced Robotics, Cloud Computing, Wireless Sensor Networks, Unmanned Aerial Vehicles, smartphones, and various mathematical models (Munawar et al., 2022; Niyazi and Behnamian, 2023). While these new technologies have the potential to make significant contributions to activities, addressing challenges in shelter services requires a systematic approach and a change in perspective. It is necessary to evaluate temporary housing from the users' perspective and focus more on people because a home is more than just a physical structure; it is a space where people live (Félix et al., 2013). Any construction is an integral part of the living and breathing infrastructure of the community with which it interacts. Therefore, for new technologies to be able to substitute existing practices, they must meet certain criteria, such as being economically, culturally, and climatically suitable, being usable with local resources, and having aesthetic sensitivity (Jigyasu, 2010, p. 49, 68).

Disclaimer

None.

References

Albadra, D., Coley, D., & Hart, J. (2018). Toward healthy housing for the displaced. *The Journal of Architecture, 23*(1), 115–136. https://doi.org/10.1080/13602365.2018.1424227.

Alshawawreh, L., Pomponi, F., D'Amico, B., Snaddon, S., & Guthrie, P. (2020). Qualifying the sustainability of novel designs and existing solutions for post-disaster and post-conflict sheltering. *Sustainability, 12*(3), 890. https://doi.org/10.3390/su12030890.

Alshawawreh, L., Smith, R. S., & Wood, J. B. (2017). Assessing the sheltering response in the middle east: studying Syrian camps in Jordan. *International Scholarly and Scientific Research & Innovation, 11*(8).

Arslan, H. (2007). Re-design, re-use and recycle of temporary houses. *Building and Environment, 42*, 400–406. https://doi.org/10.1016/j.buildenv.2005.07.032.

Aslan, R., & Şahinöz, S. (2023). The experiences of people with disabilities in the 2020 Izmir earthquake: A phenomenological research. *International Journal of Disaster Risk Reduction, 95*, 103868. https://doi.org/10.1016/j.ijdrr.2023.103868.

Bail, R. d. F., Kovaleski, J. L., Da Silva, V. L., Pagani, R. N., & Chiroli, D. M. d. G. (2021). Internet of things in disaster management: technologies and uses. *Environmental Hazards, 20*(5), 493–513. https://doi.org/10.1080/17477891.2020.1867493.

Barakat, S. (2003). Housing Reconstruction After Conflict and Disaster. London, United Kingdom. Retrieved from Humanitarian Practice Network website: https://www.files.ethz.ch/isn/95619/networkpaper043.pdf.

Boğaziçi University, & Istanbul Metropolitan Municipality (2019). Updating Project of Possible Earthquake Loss Estimates in Istanbul Province. İstanbul. Retrieved from https://depremzemin.mncdn.com/wp-content/uploads/2020/02/dezim_kandilli_deprem-hasar-tahmin_raporu.pdf.

Centre for Research on the Epidemiology of Disasters (2024). 2023 Disasters in Numbers. Retrieved from https://files.emdat.be/ reports/2023 EMDAT report.pdf.

Chen, Z., Chen, X., Li, Q., & Chen, J. (2013). The temporal hierarchy of shelters: a hierarchical location model for earthquake-shelter planning. *International Journal of Geographical Information Science, 27*(8), 1612–1630. https://doi.org/10.1080/13658816.2013.763944.

Copping, A. [Alex], Kuchai, N., Hattam, L., Paszkiewicz, N., Albadra, D. [Dima], Shepherd, P., . . . Coley, D. (2022). Understanding material and supplier networks in the construction of disaster-relief shelters: the feasibility of using social network analysis as a decision-making tool. *Journal of Humanitarian Logistics and Supply Chain Management, 12*(1), 78–105. https://doi.org/10.1108/JHLSCM-01-2020-0007.

Corsellis, T., & Vitale, A. (2005). Transitional settlement: Displaced populations. Oxford: Oxfam GB in association with University of Cambridge shelterproject.

Da Silva, J. (2010). Lessons from Aceh: Key considerations in post-disaster reconstruction. Rugby, Warwickshire, UK: Practical Action Publishing Ltd. Retrieved from http://www.loc.gov/catdir/enhancements/fy1305/2012361233-b.html.

Davidson, C. H., Johnson, C., Lizarralde, G., Dikmen, N., & Sliwinski, A. (2007). Truths and myths about community participation in post-disaster housing projects. *Habitat International, 31*(1), 100–115. https://doi.org/10.1016/j.habitatint.2006.08.003.

Davis, I. (1978). Shelter After Disaster. Oxford, UK: Oxford Polytechnic Press.

Dayanır, H., Çınar, A. K., Akgün, Y., & Çorumluoğlu, Ö. (2022). Post-Disaster Temporary Shelter Area Selection and Planning by Using Delphi Method: Izmir/Seferihisar Case. *Journal of Natural Hazards and Environment, 8*(1), 87–102. https://doi.org/10.21324/dacd.936585.

Disaster and Emergency Management Presidency (2015). Instructions on the establishment, management and operation of temporary accommodation centers.

El-Masri, S., & Kellett, P. (2001). Post-war reconstruction. Participatory approaches to rebuilding the damaged villages of Lebanon: a case study of al-Burjain. *Habitat International, 25*(4), 535–557. https://doi.org/10.1016/S0197-3975(01)00023-6.

Félix, D., Branco, J. M., & Feio, A. (2013). Temporary housing after disasters: A state of the art survey. *Habitat International, 40*, 136–141. https://doi.org/10.1016/j.habitatint.2013.03.006.

Félix, D., Monteiro, D., & Feio, A. (2020). Estimating the needs for temporary accommodation units to improve pre-disaster urban planning in seismic risk cities. *Sustainable Cities and Society, 61*, 102276. https://doi.org/10.1016/j.scs.2020.102276.

Gharib, Z., Tavakkoli-Moghaddam, R., Bozorgi-Amiri, A., & Yazdani, M. (2022). Post-disaster temporary shelters distribution after a large-scale disaster: An integrated model. *Buildings, 12*, 414. https://doi.org/10.3390/buildings12040414.

Global Shelter Cluster (2023). Shelter Projects: Case Studies of Humanitarian Shelter and Settlement Responses 2021-2022 (9th edition).

Hadafi, F., & Fallahi, A. (2010). Temporary housing respond to disasters in developing countries - Case study: Iran-Ardabil and Lorestan Province Earthquakes. *International Journal of Social, Behavioral, Educational, Economic, Business and Industrial Engineering, 4*(6), 1326–1332.

International Federation of Red Cross and Red Crescent Societies (2013). Post-disaster shelter: Ten designs. Geneva, Switzerland. Retrieved from https://www.shelterprojects.org/tshelter-8designs/10designs2013/2013-10-28-Post-disaster-shelter-ten-designs-IFRC-lores.pdf.

Jigyasu, R. (2010). Appropriate technology for post-disaster reconstruction. In G. Lizarralde, C. Johnson, & C. H. Davidson (Eds.), Rebuilding after disasters: From emergency to sustainability (pp. 49–69). London, New York: Spon Press.

Johnson, C. (2007a). Impacts of prefabricated temporary housing after disasters: 1999 earthquakes in Turkey. *Habitat International, 31*(1), 36–52. https://doi.org/10.1016/j.habitatint.2006.03.002.

Johnson, C. (2007b). Strategic planning for post-disaster temporary housing. *Disasters, 31*(4), 435–458. https://doi.org/10.1111/j.1467-7717.2007.01018.x.

Johnson, C., Lizarralde, G., & Davidson, C. H. (2006). A systems view of temporary housing projects in post-disaster reconstruction. *Construction Management and Economics, 24*(4), 367–378. https://doi.org/10.1080/01446190600567977.

Karaoğlan, F. C., & Alaçam, S. (2019). Design of a post-disaster shelter through soft computing. International *Journal of Architectural Computing, 17*(2), 185–205. https://doi.org/10.1177/1478077119849694.

Kılcı, F., Kara, B. Y., & Bozkaya, B. (2015). Locating temporary shelter areas after an earthquake: A case for Turkey. *European Journal of Operational Research, 243*(1), 323–332. https://doi.org/10.1016/j.ejor.2014.11.035.

Kim, M., Kim, K., & Kim, E. (2021). Problems and implications of shelter planning focusing on habitability: a case study of a temporary disaster shelter after the pohang earthquake in South Korea. International *Journal of Environmental Research and Public Health, 18*(6), 2868. https://doi.org/10.3390/ijerph18062868.

Kuchai, N., Shepherd, P., Calabria-Holley, J., Copping, A. [Alexander], Matard, A., & Coley, D. (2020). The potential for computational IT tools in disaster relief and shelter design. *Journal of International Humanitarian Action,* *5*(1). https://doi.org/10.1186/s41018-020-00069-1.

Labuda, I., Pugliese, F., & Dzwierzynska, J. (2024). An innovative concept for 3D sand-printed sustainable refugee shelters in a sandy desert in a hot and dry climate. *Sustainability, 16*(6), 2294. https://doi.org/10.3390/su16062294.

Lizarralde, G., Johnson, C., & Davidson, C. (2010). Rebuilding after disasters: From emergency to sustainability. In G. Lizarralde, C. Johnson, & C. H. Davidson (Eds.), Rebuilding after disasters: From emergency to sustainability (pp. 1–24). London, New York: Spon Press.

Ma, Y., Xu, W., Qin, L., & Zhao, X. [Xiujuan] (2019). Site Selection Models in Natural Disaster Shelters: A Review. *Sustainability, 11*(2), 399. https://doi.org/10.3390/su11020399.

Moreno-Sierra, A., Pieschacón, M., & Khan, A. (2020). The use of recycled plastics for the design of a thermal resilient emergency shelter prototype. *International Journal of Disaster Risk Reduction, 50*, 101885. https://doi.org/10.1016/j.ijdrr.2020.101885.

Mukherji, A. (2017). Post-Disaster Housing Recovery. In A. Mukherji (Ed.), Oxford Research Encyclopedia of Natural Hazard Science. Oxford University Press. https://doi.org/10.1093/acrefore/9780199389407.013.82.

Munawar, H. S., Mojtahedi, M., Hammad, A. W. A., Kouzani, A., & Mahmud, M. A. P. (2022). Disruptive technologies as a solution for disaster risk management: A review. *The Science of the Total Environment, 806, 151351.* https://doi.org/10.1016/j.scitotenv.2021.151351.

Niyazi, M., & Behnamian, J. (2023). Application of emerging digital technologies in disaster relief operations: a systematic review. *Archives of Computational Methods in Engineering, 30*(3), 1579–1599. https://doi.org/10.1007/s11831-022-09835-3.

Office of the United Nations High Commissioner for Human Rights (2010). The Right to Adequate Housing: Fact Sheet No. 21 (Rev. 1). Geneva: United Nations Centre for Human Rights. Retrieved from https://unhabitat.org/the-right-to-adequate-housing-fact-sheet-no-21rev1.

Omidvar, B., Baradaran-Shoraka, M., & Nojavan, M. (2013). Temporary site selection and decision-making methods: A case study of Tehran, Iran. *Disasters, 37*(3), 536–553. https://doi.org/10.1111/disa.12007.

Ömürgönülşen, M., & Menten, C. (2021). The selection of temporary shelter areas with fuzzy TOPSIS method for the Ankara Province. *Journal of Natural Hazards and Environment, 7*(1), 159–175. https://doi.org/10.21324/dacd.691088.

Opdyke, A., Javernick-Will, A., & Koschmann, M. (2019). Assessing the impact of household participation on satisfaction and safe design in humanitarian shelter projects. *Disasters, 43*(4), 926–953. https://doi.org/10.1111/disa.12405.

Pomponi, F., Moghayedi, A., Alshawawreh, L., D'Amico, B., & Windapo, A. (2019). Sustainability of post-disaster and post-conflict sheltering in Africa: What matters? *Sustainable Production and Consumption, 20*, 140–150. https://doi.org/10.1016/j.spc.2019.06.007.

Quarantelli, E. L. (1995). Patterns of sheltering and housing in US disasters. *Disaster Prevention and Management, 4*(3), 43–53. https://doi.org/10.1108/09653569510088069.

Republic of Türkiye, Presidency of Strategy and Budget (2023). 2023 Kahramanmaraş and Hatay Earthquakes Report. Retrieved from https://www.sbb.gov.tr/2023-kahramanmaras-ve-hatay-depremleri-raporu.

Scriven, K. (2010). ALNAP Case Study No. 5 - Transitional shelter: Understanding shelter from the emergency through reconstruction and beyond. Retrieved from https://reliefweb.int/report/world/alnap-case-study-no-5-transitional-shelter-understanding-shelter-emergency-through.

Song, Y., Mithraratne, N., & Zhang, H. (2016). Life-time performance of post-disaster temporary housing: A case study in Nanjing. *Energy and Buildings, 128*, 394–404. https://doi.org/10.1016/j.enbuild.2016.07.019.

Sphere Association (2018). The Sphere handbook: Humanitarian charter and minimum standarts in humanitarian response (Fourth edition). Geneva: Sphere. Retrieved from www.spherestandards.org/handbook.

Tumbeva, M. D., Wang, Y., Sowar, M. M., Dascanio, A. J., & Thrall, A. P. (2016). Quilt pattern inspired engineering: Efficient manufacturing of shelter topologies. *Automation in Construction, 63*, 57–65. https://doi.org/10.1016/j.autcon.2015.12.005.

Uddin, K., & Matin, M. A. (2021). Potential flood hazard zonation and flood shelter suitability mapping for disaster risk mitigation in Bangladesh using geospatial technology. *Progress in Disaster Science, 11*, 100185. https://doi.org/10.1016/j.pdisas.2021.100185.

United Nations High Commissioner for Refugees (2007). Handbook for Emergencies (Third Edition). Geneva, Switzerland. Retrieved from https://www.refworld.org/policy/opguidance/unhcr/2007/en/110283.

United Nations High Commissioner for Refugees (2023). Global Trends: Forced Displacement in 2022. Retrieved from https://www.unhcr.org/global-trends-report-2022.

United Nations High Commissioner for Refugees; International Organization for Migration (2010). Collective Centres Guidelines. Retrieved from https://emergency.unhcr.org/sites/default/files/IOM%2C%20Collective%20Centres%20Guidelines%2C%202010.pdf.

United Nations Office for the Coordination of Humanitarian Affairs, International Federation of Red Cross and Red Crescent Societies (2015). Shelter After Disaster (Second Edition). Genève (Suisse): International federation of red cross and red crescent societies. Retrieved from https://sheltercluster.s3.eu-central-1.amazonaws.com/public/docs/shelter_after_disaster.pdf.

Vecere, A., Monteiro, R., Ammann, W. J., Giovinazzi, S., & Melo Santos, R. H. (2017). Predictive models for post disaster shelter needs assessment. *International Journal of Disaster Risk Reduction, 21*, 44–62. https://doi.org/10.1016/j.ijdrr.2016.11.010.

Yin, Y., Zhao, X. [Xiangcheng], & Lv, W. (2022). Emergency shelter allocation planning technology for large-scale evacuation based on quantum genetic algorithm. *Frontiers in Public Health, 10,* 1098675. https://doi.org/10.3389/fpubh.2022.1098675.

Zea Escamilla, E., & Habert, G. (2015). Global or local construction materials for post-disaster reconstruction? Sustainability assessment of twenty post-disaster shelter designs. *Building and Environment, 92,* 692–702. https://doi.org/10.1016/j.buildenv.2015.05.036.

Chapter 4

The Management of Health Services in Disasters

Asuman Şener[*], **PhD**
Vocational School of Health Services, Ondokuz Mayıs University, Samsun, Türkiye

Abstract

In today's increasingly unpredictable world, the occurrence of natural and human-caused disasters poses a constant threat to communities and organizations. It is very important to organize health teams and ensure cooperation and coordination among all relevant institutions and organizations in the planning and execution of health services in disasters. In such challenging times, preparedness, response, and recovery efforts can be enhanced by leveraging the power of technological advances in healthcare management. From early warning systems to data management, communication networks, remote monitoring, and recovery planning, technological solutions can provide invaluable support to reduce the impact of disasters.

Keywords: disaster management, health services, technology

Introduction

Disasters are natural, technological, or human-induced events that can cause social, economic, and physical losses for the whole or certain segments of the society, interrupt or stop normal life and human activities, and also affect the society and the ability to cope inadequately (T.C. Resmî Gazete, 2021).

[*] Corresponding Author's Email: asuman.sener@omu.edu.tr.

In: Disaster Management in the Digital Age and Emerging Technologies
Editors: Ayşe Asiltürk and Ayşegül Sarıoğlu Kemer
ISBN: 979-8-89530-213-2

Disaster management aims to ensure that all institutions and organizations of the society use their means and resources in line with the determined strategic goals and priorities to plan, direct, coordinate, support, and effectively implement the measures to be taken and the work to be done before, during, and after the disaster to prevent disasters and reduce damages. It is a dynamic process that requires multi-faceted and discipline (Genç, 2021). Health services in disasters consist of many studies such as being prepared for disasters and emergencies, responding quickly and accurately, preventing and treating deaths, injuries, and disabilities, and preventing secondary comorbidities. It also covers the work carried out by all health institutions and organizations to access normal public health services as soon as possible and to fix the damaged healthcare infrastructure (T.C. Resmî Gazete, 2021). For disaster management to be successful, it is very important to determine hazards and risks in advance, to prevent possible damage, to ensure coordination and rapid response during a disaster, and to carry out post-disaster remedial activities with a holistic approach (Acar, 2019).

The Management of Health Services During a Disaster in Türkiye

In Türkiye, the Disaster and Emergency Management Department (AFAD) makes plans for disasters, extraordinary, and emergency situations within the scope of disaster and emergency management. It identifies the needs within the scope of this organization and ensures supply to the relevant units when necessary. In this context, a 24-hour flow of information is provided from all news sources and provincial command centers, and, when necessary, information is provided to managers through all kinds of communication tools. Services developed according to international service and quality standards are provided, when necessary, in cooperation with national and international organizations in the field of disaster (T.C. Sağlık Bakanlığı, 2024a).

Units

National Medical Rescue Team (UMKE)

UMKE was established to provide medical rescue and transportation services as soon as possible through appropriately equipped teams in the event of possible disasters and emergencies at home and abroad, and to establish and operate mobile hospitals. UMKE supports the medical aid provided by medical teams in disasters and emergencies with its specially trained and equipped personnel. UMKE takes an active role in health services within the scope of both disaster response and recovery activities during the establishment of a mobile hospital, emergency and surgical intervention unit, personnel accommodation unit, mobile command, and crime scene management center, depending on the size of the disaster (T.C. Sağlık Bakanlığı, 2024b).

Health Disaster and Coordination Center Unit (SAKOM)

It is the unit where disaster management is uninterruptedly monitored and information is collected in case of disasters and emergencies, crises that occur as a result of human health-related events that stop or interrupt normal life and activities, and also require urgent intervention. Furthermore, the information obtained is instantly shared with the necessary authorities and institutions (T.C. Sağlık Bakanlığı, 2024c).

Chemical, Biological, Radioactive, and Nuclear Hazardous Materials Unit (CBRN)

The CBRN unit makes plans for health services in case of accident, sabotage, or attack that may occur related to CBRN and industrial substances. The necessary cooperation and coordination of health services are ensured before, during, and after the event against all kinds of chemical, biological, radioactive, nuclear, and explosive threats and dangers that may arise at home and abroad and may ultimately affect our country (T.C. Sağlık Bakanlığı, 2024d).

Health Disaster Plans Unit

Within the scope of the Health Disaster Plans Unit, there is the Türkiye Disaster Response Plan (TAMP), Hospital Disaster and Emergency Plan (HAP), and National Earthquake Strategy Action Plan (UDSEP) (T.C. Sağlık Bakanlığı, 2024e).

Türkiye Disaster Response Plan (TAMP)

This unit ensures that field and mobile hospitals are ready in case of disaster, equipping them with emergency equipment that may be required after the disaster, sending a sufficient number of personnel, equipment, and materials to the region, and ensuring their dispatch and management.

It performs first aid, emergency medical aid, patient or injured evacuation and treatment in the disaster area, and quarantine and isolation services in the fight against infectious diseases when necessary. It ensures the coordination of relevant institutions to prevent environmental and water-related risk factors in the event of an epidemic. It determines reference region blood centers and develops their capacities, and ensures that all measures are taken regarding factors that may pose a risk in terms of environment and water sanitation. Additionally, it has many responsibilities, such as identifying reference hospitals and laboratories in Türkiye and increasing their capacities, taking precautions against dangerous substances and epidemics at border gates, and determining the official number of injured (T.C. Sağlık Bakanlığı, 2024f).

Hospital Disaster and Emergency Plan (HAP)

Due to HAP, all hospitals providing inpatient treatment services are enabled to take precautions to reduce the risks and damages that can occur due to disasters. In this way, necessary preparations are made for health services to be provided in disasters and emergencies. Furthermore, due to HAP, hospitals will be able to be self-sufficient without any help from outside the hospital during the first 72 hours after the disaster (T.C. Sağlık Bakanlığı, 2024g). Although the regulations try to standardize it in all hospitals, it can vary depending on the structural and institutional facilities of the hospitals (Koç, 2022).

National Earthquake Strategy and Action Plan (UDSEP)

This plan, which is carried out under the coordination of AFAD, it is aimed reducing the risk of earthquakes and creating a prepared and resilient society to cope with earthquakes, and establishing an institutional infrastructure for this purpose. Within the scope of UDSEP, there are a total of 13 Responsible Institutions, 3 Main Axes, 7 Targets, 29 Strategies, and 87 Actions (T.C. Sağlık Bakanlığı, 2024h).

Planning, Coordination, and Execution of Health Services in Disaster and Emergency Situations

Throughout the country, health services in disasters and emergencies are planned, coordinated, and conducted at the regional coordinatorship of the central organization of the Ministry of Health and at the provincial level. At the Ministry of Health level, the General Directorate of Emergency Health Services, at the regional level, the regional coordinator is the Provincial Health Director of the province, and at the provincial level, the Provincial Health Director is authorized and responsible. The health services to be provided in disasters and emergencies are carried out by the director at the provincial level, the president responsible for emergency health services, and the vice president responsible for emergency health services. In disasters and emergencies, the planning and coordination of health services at the country level are carried out by SAKOM in the Ministry and Provincial SAKOM in the provinces (T.C. Resmî Gazete, 2021).

Ministry Central Organization in Disaster and Emergency Situations

General Directorate of Emergency Health Services

It plans and conducts health services to be provided throughout the country in disasters and emergencies. It provides or ensures that all kinds of logistics materials, tools, and equipment are needed by UMKE. It organizes medical rescue and emergency health services teams in disasters and emergencies within the country and ensures cooperation and coordination with all relevant institutions and organizations. It participates in health and humanitarian aid activities in cooperation with national and international organizations and non-governmental organizations in disasters and emergencies that occur abroad. Provides the necessary coordination to develop preparedness and response to health threats related to CBRN events. It cooperates with relevant institutions and organizations to determine and procure the minimum standard materials to be used in healthcare services in CBRN incidents. Ensures that minimum standard materials are available in directorates. Ensures that CBRN training is provided to UMKE and 112 emergency health services personnel. UMKE organizes trainer training and cooperates with relevant institutions and organizations for advanced level training. Within the scope of TAMP, the national level health working group prepares the plan. It coordinates the preparation of health disaster and emergency plans prepared at the provincial level. HAP organizes trainer training. It organizes national and international drills and participates in the drills. It ensures that all kinds of communications and communication infrastructure of SAKOM operate uninterruptedly 24/7. It plans the mobile hospital services to be provided in disasters and emergencies and provides personnel planning and training regarding the supply and maintenance of the necessary hospital systems and the establishment and operation of these hospitals. It is conducted or ensures the planning, supply, distribution, and storage activities for communication, medicine, medical, and technical materials that will be needed in the provision of health services for disasters and emergencies (T.C. Resmî Gazete, 2021).

Provincial Organizations of Ministries and Affiliates in Disaster and Emergency Situations

Provincial Health Directorate

In disasters and emergencies, it coordinates all provincial organizations of the organizations and service units affiliated with the Ministry throughout the province, ensures that the work is performed, and cooperates with other institutions and organizations. It plans and implements all activities carried out within the scope of the mitigation, preparation, response, and recovery phases for disasters and emergencies at the provincial level. It creates UMKE teams to respond to disasters and emergencies, provides the necessary equipment and materials, plans, performs, and reports training, exercises, and activities. SAKOM ensures an uninterrupted flow of information between the provincial organization of the General Directorate of Border and Coastal Health of Türkiye and other relevant institutions. It prepares, coordinates, reviews, and sends health disaster and emergency plans at the provincial level to the approval authority. It establishes and operates mobile hospitals and provides personnel, logistics, and similar support. It ensures that the evaluation reports prepared following the end of disasters and emergencies are sent to SAKOM. It plans, coordinates, and ensures the provision of communications, medicine, medical, and technical materials that will be needed to provide health services in disasters and emergencies. It enables the use of the Triage Card prepared by the General Directorate during interventions in disasters and emergencies (T.C. Resmî Gazete, 2021).

Regional Coordinator Province

It coordinates the health service activities to be provided regarding the preparation, intervention, and recovery phases for disasters and emergencies in the provinces in the area of responsibility located in the emergency health services regions determined by the Ministry of Health. The regional coordinator coordinates, organizes, and reports the planning and reporting of training, exercises, and other activities of the province and its affiliated provinces during the year and for the following year and reports them to the Ministry. It ensures the provision of all personnel, logistics, and technical materials requested by the Ministry during the intervention process and controls the stock level, expiration, and availability status (T.C. Resmî Gazete, 2021).

Türkiye Borders and Coasts General Directorate of Health

The provincial organization of this directorate provides support services to the directorate in the preparation of health disaster and emergency plans prepared at the provincial level. It takes precautions against dangerous goods and epidemic diseases at national entry points. It prepares and implements training programs in cooperation with relevant institutions and organizations to train personnel related to its field of duty and responsibility, organizes exercises, and participates in national and international exercises. It ensures that the evaluation reports prepared following the end of disasters and emergencies are delivered to the directorate (T.C. Resmî Gazete, 2021).

The Role of Universities and Private Hospitals in Disaster and Emergency Situations

It provides support services to the directorate in the preparation of health disaster and emergency plans prepared at the provincial level. It prepares the HAP and sends the plans to the university rectorate for university hospitals, and to the directorate responsible for health services for private hospitals, for review. It performs planning, supply, and storage activities for communication, medicine, medical, and technical materials that may be needed. It carries out the necessary work to ensure that routine hospital services continue without disruption. It provides personnel and logistic support, when needed, to health facilities established or operated by the Ministry. It delivers the evaluation reports prepared following the end of disasters and emergencies to the directorate (T.C. Resmî Gazete, 2021).

Health Services Unit in Disaster

It plans, coordinates, and performs activities to be carried out within the scope of the Provincial Health Directorate's mitigation, preparation, intervention, and recovery phases for disasters and emergencies. It coordinates the preparation and updating of prepared health disaster and emergency plans at the provincial level, forwards them to the relevant units, and ensures their implementation. Plans drill organizations at the regional and provincial levels. Participates in, prepares, and executes national and international organizations and projects related to disasters and emergencies, with the knowledge and approval of the Ministry. It evaluates the HAPs sent by the university rectorate, the mayor's office, the directorate's public hospitals, public health, and health services departments, according to the HAP standard inspection form, and submits them for approval. It manages the planning and preparation processes for health threats related to chemical, biological, radiological, and nuclear hazards. UMKE creates teams, ensures their dispatch and management in disasters and emergencies, and coordinates them. It performs the provision of medical rescue and health services in disasters and emergencies in cooperation and coordination with all relevant parties. Participates in, performs, and reports on national and international exercises with the permission of the Ministry. It provides the necessary organization to respond to health threats caused by chemical, biological, radiological, and nuclear materials. It determines the needs of all healthcare teams providing healthcare services in the field and coordinates the provision of these needs. It ensures the recruitment of personnel for the assignments to be done by the Ministry. It ensures that the UMKE personnel in the province are ready for duty at all times and creates a communication chain. It performs and coordinates the plans, supply, distribution, and storage activities for the communication, medicine, medical and technical materials, and other tools and equipment that will be needed (T.C. Resmî Gazete, 2021).

It ensures that communication, medicine, medical and technical materials, and similar logistics materials, mobile hospital systems, personnel accommodation units, and hospital equipment to be used in disasters and emergencies that may occur at home or abroad are delivered to the scene of the incident as soon as possible. The training plans and performs determined by the Ministry for the personnel to be assigned. With the approval of the Ministry, it prepares and implements training programs in cooperation with relevant national and international institutions and organizations to train healthcare personnel who will work in disasters and emergencies and participate in organized training. It is planned, performed, and reported the training of UMKE personnel. Coordinates HAP practitioner training. Prepares the data inventory of the province regarding disasters and emergencies. Keeps records, archives, reports, and reports all information regarding the unit to the Ministry (T.C. Resmî Gazete, 2021).

The Use of Technology in Health Services Management During Disasters

Considering that technology has become the focal point of our lifestyles in recent years, disaster management also benefits from the benefits of technology. As promising technological developments for reducing risk and effective intervention in disasters in the disaster management process; there are early warning systems, crowd resource management, geographic information systems (GIS), global positioning systems (GPS), computer-aided programs, social media, and artificial intelligence technology (Demiröz, 2023).

In disaster management, it is seen that disaster-related information is frequently shared on social media during the disaster as well as in the pre-disaster phase (Çakı, 2020). Social media is also very important in terms of accessing instant information, as well as during disasters, search, and rescue, in-kind and cash donations, communication, coordination, organization, problem-solving, decision-making, resource management, ensuring social unity, social networking, it is also used for information dissemination and collection purposes (Şahinsoy, 2017). The September 11, 2001, Twin Towers Attack was the first disaster in which social media was used. In this attack, the social media platform was used to obtain information about missing persons.

The WHO also announced information on the bird flu epidemic that started in China in 2011 via social media (Demiröz, 2023).

It seems that it has become easier to access the desired information about disasters thanks to applications that can be downloaded from mobile phones. In this regard, many organizations such as AFAD, Türk Kızılay, General Directorate of Meteorology, Kandilli Observatory, and Non-Governmental Organizations have developed many mobile applications to predict disasters, deliver early warning messages, provide training, and create calls (Karahisar, 2016). These practices have been used in many cases in our country, such as the 2020 Elazığ and İzmir Earthquake, the Covid-19 epidemic, and the Ankara Kızılay explosion (Deniz Öztürk, 2023).

Disaster Management and Decision Support System (AYDES) uses web/mobile-based, geographic information technologies. It is an integrated decision support platform that brings together solution and support solution partners, including all institutions or organizations, the private sector, and NGOs, and provides effective use in pre-disaster preparation, intervention during disaster, and recovery processes (AYDES, 2024). Due to this platform, the decision-making process can be accelerated and administrative problems can be eliminated. Due to the AYDES database, building damage assessment can be done by expert teams in the field. "Hayat Eve Sığar" mobile application, developed by the Ministry of Health in Türkiye during the COVID-19 period, has personal add-ons such as disease risk control, infection prevention, quarantine period tracking, a risk map, and the creation of personal barcodes (Deniz Öztürk, 2023). The national software AFAD-RED System, created with our own resources and people, has the ability to make automatic and manual solutions after a real earthquake and can calculate estimated damage and loss of life. The scenario has the ability to create earthquakes. The system is flexible and can be constantly improved. The analyses take place in approximately 35-40 seconds. It works integrated with earthquake observation stations and AYDES (AFAD, 2024).

Unmanned Aerial Vehicles (UAV), popularly called drones, can reach hard-to-reach places (such as mountainous, deserts, volcanic, earthquakes, floods, and accidents) that manned aerial vehicles cannot reach (Kanat, 2023). CBRN is able to collect information in contaminated areas and does not endanger human life (Erdil, 2021). UAVs are used in disaster management for reconnaissance and mapping, structural assessment, meeting temporary infrastructure needs, detecting and extinguishing forest fires, and responding to high-rise building fires and CBRN incidents. It is also used in early warning systems, emergency communication, logistics, information gathering, and search and rescue (Kanat, 2023).

Another technology used in disaster management is robot technology. For example, small tactical robots can easily move in areas where humans or dogs cannot physically enter. Robots are used to extinguish fires, rescue people under rubble, transport rescued people, and clean roads. In general, it is seen that robots are used in disasters for search and rescue, communication, mapping, hazardous material cleaning, infrastructure inspection, evacuation, area or damage assessment, and disaster victim detection (Usta ve Şahinöz, 2023).

Artificial intelligence has recently been preferred by health authorities, especially in the management of infectious diseases. Corona Prevention Practice was implemented by the Republic of Türkiye immediately after the World Health Organization declared the Coronavirus as a global pandemic and the first case was observed in Türkiye. It is an artificial intelligence application launched by the Ministry of Health and made available to all citizens. This application was developed to make a preliminary assessment based on coronavirus (COVID-19) symptoms and to that recommend citizens visit a health facility based on this preliminary assessment. Due to this application, citizens are directed to detailed questions about the disease, and with algorithms approved by the Scientific Board, people are enabled to evaluate themselves based on risky diseases and symptoms (Tarcan et al., 2024).

Discussion and Conclusion

Disasters are one of the biggest problems of the 21st century. While countries all over the world face natural disasters such as earthquakes, floods, landslides, hurricanes, droughts, and fires, some countries face disasters caused by human beings, such as war and terrorist attacks. In the globalizing world, it is seen that disasters are not only a problem of countries, but their consequences affect the whole world (Akgün, 2022a). However, disasters do not respect borders, and in the globalizing world, the consequences of someone else's disaster can easily affect the entire global system. As a final result, although disaster management is, above all, the duty of the national authority, it must begin to be integrated into international systems (Hecker and Domres, 2018). The Türkiye Disaster and Emergency Management Presidency (AFAD) operates under the Ministry of Internal Affairs. The institutional structure and duties of the AFAD are determined by law. The institution does not have a hierarchical structure among its departments. When necessary, it can create working groups to which it can assign personnel from outside the institution. AFAD can also cooperate with universities, public institutions, the private sector, and NGOs. Türkiye has the second-largest earthquake monitoring network in Europe. AFAD monitors seismic activity in Türkiye and its surroundings 24/7 with 1,143 active earthquake observation stations. The Türkiye Disaster Response Plan (TAMP) covers institutions and organizations involved in the effective response to disasters and emergencies of all types and sizes occurring in the country. This national plan defines roles and responsibilities through an integrated planning approach and identifies the key elements of response planning. TAMP is flexible, modular, and can be adapted to disasters of all types and sizes. It is responsible for the planning and coordination of public institutions and non-governmental organizations in case of a disaster at the local/national level. It aims to minimize loss of life and property through effective resource management. TAMP gives the personnel of ministries, public institutions, and non-governmental organizations the responsibility to work under 28 working groups after a disaster at the local or national level. There are over 100 support partners, with a total of 10 ministries and NGOs appointed as main partners (European Union, 2024a).

Germany has a decentralized emergency management system. This means that the state delegates EMS authority to its provinces, and states delegate EMS authority to their counties, cities, and communities. The main tool of medical civil defense in disasters is Regional Emergency Health Systems. To cope with mass casualties and disasters, these forces must have basic capabilities to assist the population beyond regular medical assistance. This situation varies from city to city, region to region, and state to state. Sometimes, when local authorities have difficulties, the organizational structure is reversed. In other words, the larger the disaster, the command structure may shift from region to state and then to the federal state (European Union, 2024b).

The Directorate General for Civil Protection and Crisis Management (DGSCGC), within the French Ministry of Internal Affairs, is responsible for anticipating and monitoring crises affecting internal and civilian security. The region is divided into 12 defense and security regions, and the governor has expanded powers in terms of coordination and mobilization of resources in the event of an exceptional situation or a department exceeding its limits or capabilities. It has an inter-ministerial headquarters and an operations center for this purpose. When disasters occur, the head of the ministry ensures the coherence and unity of public action by coordinating public, private, and integrative actors and local authorities (European Union, 2024c).

The US Federal Emergency Management Agency (FEMA) is responsible for emergency and disaster management in the United States of America (USA). National emergency aid and rescue efforts are carried out with a participatory approach, at the local and federal levels, with the participation of public and private institutions and organizations. How institutions will work together is regulated according to the Federal Response Plan. In this plan, there are 12 separate emergency aid interventions, from fire extinguishing to providing medical services, from removing debris to providing nutrients (Akgün, 2022b). FEMA's Mitigation Directorate is responsible for identifying pre-disaster risks, preventing injuries or deaths, and saving time. The institution has basic analysis programs for hurricanes, earthquakes, and floods (Bullock et al., 2013).

After the 1995 Kobe earthquake in Japan, the Japanese National Disaster Medical System (NDMS) was developed. It basically consists of a disaster base hospital, an emergency medical information system, Disaster Key Hospitals, Disaster Medical Assistance Team (DMAT), and air ambulances (Homma, 2015). The 2nd World Conference on Disaster Reduction, held in Kobe, Japan in 2005, aimed to reduce the damages caused by disasters between 2005 and 2015. For this purpose, the HYOGO Framework Action Plan (HÇEP) was prepared and adopted by 168 countries that are members of the United Nations. HÇEP is an action plan to reduce disaster risks in the above-mentioned period. In 2015, the HYOGO Framework ended and was replaced by the Sendai Framework (Delshad et al., 2020). To support the purpose of the Sendai Framework and the assessment of global development in achieving the outcome, 7 global targets have been adopted. It has created a legally based structure for comprehensive disaster risk reduction (Mizutori, 2020).

As a result, human beings have faced many situations that can be defined as disasters since the formation of the Earth (Gülsoy and Uyan, 2023). Disasters are among the most difficult situations that people have to deal with (Öztaş, 2022). The common point in disaster management systems around the world is that the disaster and emergency management structures of countries have a hierarchical structure from the central government to the local government. It is seen that in cases other than CBRN that require special expertise, the first response is made by local search and rescue teams. When the disaster area grows, the organizational structure also grows, and command and control can reach the government level (Akgün, 2022).

The deficiencies in emergency aid plans made for disasters are observed when disasters occur. The reasons for these include the fact that those who prepared the plans did not consider bad scenarios sufficiently, numerical measurements are not easy, the plans are not reinforced sufficiently with drills, not enough awareness is provided, and disaster response teams are not given enough information. For this reason, plans can generally change after major disasters. Plans should be shared with the public electronically or through training, criticisms of the plans should be considered, and plans should be constantly updated. While preparing disaster plans, non-governmental organizations and volunteers should be involved (Baskak and Büyüksaraç, 2021).

Disclaimer

None.

References

Acar, N. (2019). *Afet Yönetiminde Görev Alması Beklenen Acil Sağlık Hizmetleri Unsurları ve Görev Tanımları.* Afet Yönetimi ve Tıbbi Uygulamalar-Temel Başvuru Kitabı, chief edited by. Eroğlu, S. E. edited by Yılmaz, S., Dursun, R., Karakayalı, O., 7-26. İstanbul: EMA Tıp Kitapevi Yayıncılık.

Afet ve Acil Durum Yönetimi Başkanlığı (AFAD). 2024. Accessed April 12. https://www.afad.gov.tr/on-hasar-ve-kayip-tahmin-sistemi.

Afet Yönetimi ve Karar destek Hizmeti (AYDES). 2024. Accessed April 12. aydes.gov.tr.

Bullock, J. A., Haddow, G. D., & Coppola, D. P. (2013). Mitigation, prevention, and preparedness. Introduction to Homeland Security, 435-94.

Çakı, F. (2020). Afet Yönetiminde Sivil Toplum ve Güven Algıları. *Sosyal ve Kültürel Araştırmalar Dergisi, 6*(12),1-45.

Coşkun, D., & Büyüksaraç, A. (2021). Afetlerde acil yardım planlaması. *Paramedik ve Acil Sağlık Hizmetleri Dergisi, 2*(2), 68-83.

Delshad, V., Pourvakhshoori, N., Rajabi, E., Bazyar, J., Ahmadi, S., & Khankeh, H. R. (2020). International Agreements on Disaster Risk Management Based on World Conferences, Successful or Not: A Review Study. *Health in Emergencies and Disasters Quarterly, 6*(1), 1-8.

Demiröz, K. (2023). *Afetlerde Sosyal Medya.* Afet Yönetimi ve Teknoloji, Edited by Şahinöz, s. 20-38. İstanbul: İstanbul Tıp Kitapevi.

Erdil, B. (2021). İnsansız Hava Araçlarının Kullanım Alanları ile Bu Araçların Türkiye'nin Yurtdışı Operasyonlarındaki Yeri ve Önemi. *Bölgesel Araştırmalar Dergisi, 5*(2), 581-607.

European Union, (2024a). "European Civil Protection and Humanitarian Aid Operations." Accessed April 12. https://civil-protection-humanitarian-aid.ec.europa.eu/what/civil-protection/national-disaster-management-system/turkiye_en.

European Union, (2024b). "European Civil Protection and Humanitarian Aid Operations." Accessed April 12. https://civil-protection-humanitarian-aid.ec.europa.eu/what/civil-protection/national-disaster-management-system/germany_en.

European Union, (2024c). "European Civil Protection and Humanitarian Aid Operations." Accessed April 12. https://civil-protection-humanitarian-aid.ec.europa.eu/what/civil-protection/national-disaster-management-system/france_en.

Genç, F. N. (2021). Afet Yönetimi. Ankara: Nobel Akademik Yayıncılık.

Gülsoy, A., Uyan, Y. (2023). *Afetlere Genel Bakış*. Afet Yönetimi ve Teknoloji, edited by Şahinöz, s. 1-19. İstanbul: İstanbul Tıp Kitapevi.

Hecker, N., & Domres, B. D. (2018). The German emergency and disaster medicine and management system—history and present. *Chinese Journal of Traumatology, 21* (02), 64-72.

Homma, M. (2015). Development of the Japanese national disaster medical system and experiences during the Great East Japan Earthquake. *Yonago Acta Medica*, 58 (2), 53-61.

Kanat, M. (2023). *Afetlerde Drone ve İHA Uygulamaları*. Afet Yönetimi ve Teknoloji, edited by Şahinöz, S. 55-68. İstanbul: İstanbul Tıp Kitapevi.

Karahisar, T. (2016). Olası Bir Depremde Koordinasyonun Sağlanması ve Sosyal Medyanın Rolü. *Beykoz Akademi Dergisi, 4* (2), 43-64.

Koç, A. (2022). *Hastane Afet ve Acil Durum Planı (HAP)*. Afet Hemşireliği, edited by Atasoy, E., Özpulat, F. 197-214. Ankara: Nobel Akademik Yayıncılık.

Mizutori, M. (2020). Reflections on the Sendai Framework for disaster risk reduction: five years since its adoption. "(2), 147-151.

Öztaş, D. (2022). *Afet Tanımı ve Türleri*. Afet Hemşireliği, edited by Esma Atasoy, Funda Özpulat, 1-18. Ankara: Nobel Akademik Yayıncılık.

Öztürk, Y. D. (2023). *Afetlerde Mobil Uygulamalar*. Afet Yönetimi ve Teknoloji, edited by Şahinöz, S. 39-54. İstanbul: İstanbul Tıp Kitapevi.

Şahinsoy, K. (2017). Kriz Yönetimi Açısından Geleneksel ve Sosyal Medya. *İstanbul Aydın Üniversitesi Dergisi, 36*, 1-19.

Şenay, A. (2022a). *Dünya'da ve Türkiye'de Afetler*. Afet Hemşireliği, edited by Atasoy, E., Özpulat, F. 33-50. Ankara: Nobel Akademik Yayıncılık.

Şenay, A. (2022b). *Dünya'da ve Türkiye'de Afet Yönetimi*. Afet Hemşireliği, edited by Atasoy, E., Özpulat, F. 51-83. Ankara: Nobel Akademik Yayıncılık.

T.C. Resmî Gazete. (2021). Accessed April 12. https://www.resmigazete.gov.tr/eskiler/2021/05/20210525-3.htm.

T.C. Sağlık Bakanlığı. (2024a). "Görev ve Yetkiler." Accessed April 12. https://ashgmafetacildb.saglik.gov.tr/TR-73730/gorev-ve-yetkiler.html.

T.C. Sağlık Bakanlığı. (2024b). "Ulusal Medikal Kurtarma Ekibi Birimi (UMKE)." Accessed April 12. https://ashgmafetacildb.saglik.gov.tr/TR-80255/ulusal-medikal-kurtarma-ekibi-birimi-umke.html.

T.C. Sağlık Bakanlığı. (2024c). "Sağlık Afet ve Koordinasyon Merkezi Birimi (SAKOM)." Accessed April 12. https://ashgmafetacildb.saglik.gov.tr/TR-80256/saglik-afet-ve-koordinasyon-merkezi-birimi-sakom.html.

T.C. Sağlık Bakanlığı. (2024d). "Kimyasal, Biyolojik, Radyoaktif ve Nükleer Tehlikeli Maddeler Birimi (KBRN)." Accessed April 12. https://ashgmafetacildb.saglik.gov.tr/TR-80257/kimyasal-biyolojik-radyoaktif-ve-nukleer-tehlikeli-maddeler-birimi-kbrn.html.

T.C. Sağlık Bakanlığı. (2024e). "Sağlık Afet Planları Birimi." Accessed April 12. https://ashgmafetacildb.saglik.gov.tr/TR-80258/saglik-afet-planlari-birimi.html.

T.C. Sağlık Bakanlığı. (2024f). "Türkiye Afet Müdahale Planı (TAMP)." Accessed April 12. https://ashgmafetacildb.saglik.gov.tr/TR-81376/1-turkiye-afet-mudahale-plani-tamp.html.

T.C. Sağlık Bakanlığı. (2024g). "Hastane Afet ve Acil Durum Planı (HAP)." Accessed April 12. https://ashgmafetacildb.saglik.gov.tr/TR-81377/2-hastane-afet-ve-acil-durum-plani-hap.html.

T.C. Sağlık Bakanlığı. (2024h). "Ulusal Deprem Strateji ve Eylem Planı (UDSEP 2012 – 2023)." Accessed April 12. https://ashgmafetacildb.saglik.gov.tr/TR-81379/3-ulusal-deprem-strateji-ve-eylem-plani-udsep-2012--2023.html.

Tarcan, G. Y., Balçık, P. Y., & Sebik, N. B. (2024). Türkiye ve Dünyada Sağlık Hizmetlerinde Yapay Zekâ. *Mersin Üniversitesi Tıp Fakültesi Lokman Hekim Tıp Tarihi ve Folklorik Tıp Dergisi*, 14(1), 50-60.

Usta, G., Şahinöz. S. (2023). "Afetlerde Robotik Sistemler." Afet Yönetimi ve Teknoloji, Edited by Şahinöz, S. 130-142. İstanbul: İstanbul Tıp Kitapevi.

Chapter 5

The Management of Infectious Diseases in Disasters

İlknur Başaran*, RN, MsC

Trabzon Public Hospital, Trabzon, Türkiye

Abstract

Natural disasters result in significant destruction, loss of many lives, and exposure to infectious diseases referred to as a significant threat to those who need to continue their lives. The main disasters seen in Türkiye include earthquakes, floods, storms, fires, droughts, and avalanches. Following these disasters, infectious diseases such as measles, malaria, diarrhea, acute respiratory infections, tetanus, typhoid fever, dengue fever, meningitis, and leptospirosis are observed due to the inability to provide basic needs such as air, water, food, and shelter necessary for human life. Disaster management is one of the most critical processes before and after a disaster in a country. In this context, one of the fundamental tasks of disaster management could be to prevent or reduce loss of life and property damage. Moreover, it should be prepared for possible disasters, take necessary measures to reduce damage, and support the population. Furthermore, it should use technological means effectively and appropriately to inform and educate the people of the country after the disaster. The purpose of this study is the management of infectious diseases following disasters. It is prepared as a qualitative study using a large number of literature reviews. As a result of the literature review, it has been observed that infectious diseases such as measles, malaria, diarrhea, acute respiratory infections, tetanus, typhoid fever, dengue fever, meningitis, and leptospirosis occur in Türkiye after

* Corresponding Author's Email: nurbasaran35@gmail.com.

In: Disaster Management in the Digital Age and Emerging Technologies
Editors: Ayşe Asiltürk and Ayşegül Sarıoğlu Kemer
ISBN: 979-8-89530-213-2

natural disasters. The importance of disaster management before and after disasters has been understood. Recommendations can include adhering to disaster regulations to prevent disasters and minimize damage, ensuring that individuals and institutions involved in disaster management receive adequate training, possessing the necessary skills to implement measures during disasters, and paying attention to the selection of settlement areas.

Keywords: disaster, disaster management, infectious disease, technology

Introduction

Looking at the records of organizations like the Red Cross and Red Crescent worldwide, it is observed that more than 2 million people are affected both mentally and physically by disasters, with around 200,000 people losing their lives. The annual cost of disasters to the global economy exceeds $20 billion (Ekşi, 2016:16). Generally, disasters include droughts, earthquakes, volcanoes, storms, refugee issues, floods, tsunamis, migrations, and cyclones (Kılıç, 2017:8). In Türkiye, floods and earthquakes are the most common natural disasters. Additionally, landslides, fires, droughts, avalanches, chemical, biological, nuclear, radioactive, human, and technology-induced disasters are observed. Natural disasters can occur in various forms such as earthquakes, outbreaks, droughts, heatwaves, floods, and landslides. Depending on the type of disaster, the likelihood of infectious diseases and epidemic risks varies.

When looking at their effects on individuals, natural disasters with a more destructive nature compared to others are climatic events known as storms and floods rather than geological events such as volcanic eruptions and earthquakes. Geological disasters are less frequent compared to climatic disasters. When evaluated in terms of both loss of life and economic losses, Asia is the continent where natural disasters are most common (Güler, 1994:11). Disasters are among the leading causes of sudden deaths. They lead to a decrease in quality of life and a deterioration in healthcare services. In general, disasters are known as extraordinary events that cause significant social destruction.

Disaster Management

According to the definition of the World Health Organization, a disaster is defined as an extraordinary event of such magnitude that it requires external assistance and requires preventive measures. Disasters disrupt the existing way of life of a community, leading to loss of life and property. The environmental destruction caused by disasters results in the disruption of social structures and economies, as well as negative effects on the mental, physical, and social health of individuals. Without taking the necessary measures to mitigate the damage following disasters, numerous deaths and illnesses occur due to post-disaster factors (Kınıklı & Cesur, 2020:15-23).

The primary goal of disaster management should be to prevent loss of life and property. Based on this, disaster management is defined as the management approach that involves planning, implementing, supporting, directing, and coordinating all of society's institutions, organizations, and resources for the prevention and mitigation of disasters, both before and after they occur (Gökçe & Tetik, 2012:8). One of the most important aspects of disaster management is the provision of emergency health services and rescue operations in the first few days following a disaster. Rapid intervention for victims and injured individuals according to the type of disaster, immediate provision of the necessary first aid and emergency assistance on-site, and swift transfer to healthcare facilities are crucial. Additionally, one of the significant factors threatening the lives of healthy individuals remaining in disaster-stricken areas is the environmental risks arising after the disaster. These risks include the inability to provide clean drinking water, the lack of suitable living conditions for shelter, and various infectious diseases. Factors such as inadequate healthcare services and disruptions in disaster areas contribute to the inability to maintain the necessary hygienic conditions for health. Without implementing the necessary health measures for adverse environmental conditions following disasters and taking measures to sustain the lives of the remaining individuals, numerous casualties and infectious diseases can arise. Therefore, immediate action is required for emergency treatment, rescue, shelter, nutrition, first aid, the provision of clean water, and resolving infrastructure issues in temporary settlements following natural disasters.

The initial stage of disaster management involves planning living areas, providing tents, and ensuring their proper setup. Conducting this stage systematically helps prevent an increase in infectious diseases following disasters. Swift provision of clean water and the establishment of the necessary system for waste disposal are crucial (Canbulat, 2023:59). Effective disaster management and communication measures can minimize potential damages. Communication led by the highest-ranking disaster manager during both pre-disaster and post-disaster periods is considered the most effective element during disasters. For instance, organizing a press conference with the participation of all media outlets to inform the public and build trust is crucial. Similarly, the importance of social media in communication during disasters is recognized both globally and in Türkiye, as exemplified by the use of Twitter by the Ministry of Health during the COVID-19 pandemic (Alay & Eke, 2023: 318,320). Communication during disasters should be accurate, fast, and honest, providing affected individuals with information about the events, potential risks, and possible consequences.

The Management of Infectious Diseases in Disasters

One of the crucial issues to address following natural disasters and emergencies is infectious diseases. During disasters, trauma-related skin and soft tissue infections are observed. In the days following the disaster, water and foodborne infections, respiratory tract infections, and wound and vector-borne infections can occur (Kınıklı & Cesur, 2020:15-23). In sudden natural disasters like hurricanes and earthquakes, injuries and deaths occur immediately due to trauma. However, infectious diseases begin to emerge as people are displaced, clean drinking water becomes inaccessible, and healthcare services are not effectively accessible after the disaster. For example, malaria following floods in Ecuador, cases of infectious diarrhea following cyclones and floods in Bangladesh, and leptospirosis after floods in Brazil are observed (Uçku, 2001:651).

In slower-onset natural disasters such as wars, famine, drought, and forced migration, serious deaths occur due to infectious diseases. If the scale of the outbreak expands, measures such as blocking public transportation, implementing quarantine measures to protect public health, and closing schools and workplaces until the epidemic subsides can prevent further spread of infectious diseases (Ünlü & Çiçek, 2021:104).

The risk of outbreaks varies depending on the type of natural disaster. These differences are observed in three periods. In the first period, infectious diseases are not observed within (0-3) days based on the impact of the natural disaster. The second period, between 4 days and 4 weeks, can see infectious diseases caused by endemic pathogens in the region. The third period encompasses more than 4 weeks, during which infectious diseases occur due to prolonged incubation periods (Uçku, 2001:651).

During a disaster, health services should be provided urgently in certain areas. Vaccinations for diseases with available vaccines should be administered to prevent potential outbreak diseases following the disaster. It is essential to ensure that the teams arriving at the disaster site do not carry infectious diseases. Factors that could lead to an outbreak should be identified, and necessary measures should be taken to prevent transmission. Priority should be given to vaccination of the elderly, pregnant women, and children in vaccination efforts. Regular vaccinations for diseases such as pertussis, hepatitis B, tetanus, and diphtheria in the country where the disaster occurred provide protection even after the disaster, preventing transmission.

An important aspect of controlling infectious diseases after disasters is ito identify contacts early and implement isolation measures. Moreover, timely contact tracing efforts are crucial (Canbulat, 2023:59). Health services should be prepared to initiate interventions to prevent and alleviate infectious diseases after a disaster. Healthcare facilities should ensure the capability to operate with limited personnel and resources and plan adequately for external assistance (Rathore, 2020:507). Search and rescue teams and healthcare workers involved in disaster response should be well-versed in infection control measures and possess the skills to implement them.

When determining settlement areas, access to clean water should be facilitated, essential materials for personal hygiene should be provided, and tents erected for shelter should not be placed in areas that could serve as sources of infection. Proper disposal of the deceased, controlled disposal of waste, and vector control measures are necessary. Adequate data collection, evaluation, and efficient utilization are essential for post-disaster areas (Kınıklı & Cesur, 2020:15-23).

In 2006, the World Health Organization provided some recommendations to prevent infectious diseases after natural disasters. These recommendations include "do not use floodwaters for direct food preparation, keep hands clean, separate cooked and raw foods, thoroughly cook food, store food at appropriate temperatures, wear protective clothing during rescue operations, and prioritize vaccinations" (Canbulat, 2023:58-59).

Potential Issues Following Disasters

Below are mentioned several factors determining the sustenance of the lives of disaster survivors after disasters. These include providing shelter, ensuring access to clean water suitable for consumption, remediation of human-generated hazardous waste, ensuring control of infectious diseases, and ensuring control of specific pollutants.

Shelter Issues

After natural disasters, one of the most significant problems is the shelter issue for the remaining individuals. As seen in the earthquakes that occurred in Kahramanmaraş and the surrounding provinces of Türkiye (2023), the problem of shelter also causes infectious diseases. Individuals, having lost their loved ones, homes, and material possessions in disasters are reluctant to leave the disaster areas. Consequently, inappropriate living conditions bring along life-threatening factors such as "infectious diseases, injuries, climatic events, cold, heat, and rain." Furthermore, inadequate nutrition and failure to provide hygiene conditions also facilitate the spread of infectious diseases. If proper attention is not given to unsuitable environmental conditions and the accompanying environmental health services after disasters, and precautions are not taken, an increase in the number of deaths due to infectious diseases can be observed (Ekşi, 2016:17).

Potable Clean Water Provisions

Ensuring access to clean potable water is of vital importance for disaster survivors after natural disasters. In the aftermath of disasters, survivors often have to use whatever water they can find without checking whether it is clean or not. This situation leads to various health problems, primarily infectious diseases. As a result of such situations, deaths due to waterborne infectious diseases that negatively impact human life occur. For instance, in 1994 in Zaire, more than 50,000 Rwandans contracted Vibrio Cholera from contaminated water sources, resulting in loss of life. On average, after disasters, each person in shelter areas requires 15-40 liters of clean water per day (Ekşi, 2016:19).

One of the most common challenges immediately after a disaster is finding clean water. Disasters such as nuclear incidents, earthquakes, and floods can significantly damage water supply sources and lead to contamination. In such cases, it is recommended to consume bottled water if possible. After a disaster, the sewage and water systems of settlements should be repaired, and clean water suitable for consumption should be provided from the networks (Ekşi, 2016:19). Until tap water can be used in disaster areas, the water tanks in these areas should be easy to clean, portable, and have a capacity of at least 200 liters. It is crucial to remember that there is a possibility of water contamination even at this point. Therefore, the water used in disaster areas must be disinfected.

Various methods are available for disinfection processes. However, in emergency situations, chlorination and boiling, which are the least costly and easiest to implement, are prioritized. Chlorine tablets are commonly used for disinfection with chlorine. For disinfection, 160 mg tablets are used for 40 liters of water. In cases where water needs to be used urgently, Sodium Hypochlorite Solution containing 4-6% chlorine, also known as bleach, can be used. For disinfection, 2 drops of clean water and 4 drops of cloudy water per liter of water are used. Boiling water for 5-10 minutes at 100 degrees Celsius is also considered sufficient for disinfection. In case droppers are not available for emergency disinfection, bleach soaked in cotton can be applied. It is recommended to let chlorinated water rest for 30 minutes before use. If the disinfected water is still cloudy and is sourced from the city network, the amount of chlorine can be doubled. The water transported by tankers should be chlorinated at the source and if chlorination was not done at the source, it should be chlorinated before use. Disaster survivors living in disaster areas should receive the necessary training on the disinfection of water (Ekşi, 2016:20).

The Remediation of Human-Generated Hazardous Waste

Harmful microorganisms are present in human feces. Therefore, in disasters, emergency toilets should be promptly constructed in residential areas for people to use, and sufficient water and soap should be provided to ensure hand hygiene. The most important factor to consider here is preventing the accumulation of human waste and water, as this can promote the proliferation of microorganisms and the emergence of infectious diseases. Adequate education should be provided to disaster victims in residential areas regarding this issue. Additionally, each family should be provided with a toilet. If this is not possible, one toilet should be provided for every 20 people. Furthermore, garbage in disaster areas should be collected regularly.

Deaths immediately following disasters generally occur after injuries. Therefore, such deaths pave the way for the emergence of infectious diseases. Hence, corpses should not be buried near water sources. Animal carcasses should be buried immediately without delay and should be prevented from being consumed by the surviving animals. Animal carcasses should be buried at least 1 km from residential areas or incinerated. Carcasses that pose a risk of infection should be incinerated in special furnaces or buried at a depth of 2 meters. If an animal carcass is found in a water source, its use by humans should be prevented (Ekşi, 2016:21).

Ensuring the Control of Infectious Diseases

After disasters, the rate of transmission and spread of infectious diseases tends to increase. Staying in a disaster area for more than 4 weeks increases the risk of infectious diseases. Several factors contribute to the occurrence of infectious diseases, including the lack of clean water in disaster areas, environmental risk factors, the presence of pre-existing microorganisms causing infectious diseases in the disaster area, the inability to neutralize human waste, the inability to control animals like dogs and cats after a disaster, and the conditions reducing individuals' protection against infectious diseases.

To prevent infectious diseases that may occur after disasters, it is essential to keep records regularly, take all necessary precautions promptly, and intervene immediately in cases of detected infectious diseases. Otherwise, if the outbreak grows, it can result in an uncontrollable situation, leading to the loss of many lives. Due to water contamination and the inability to ensure food safety, infectious diseases such as Cholera and Typhoid, Hepatitis A transmitted through contact, Malaria, and Plague due to the inability to control transmitting vectors, and Measles transmitted through the respiratory route are observed. Historically, in temporary shelter areas after disasters, infectious diseases such as respiratory tract infections, Malaria, Measles, and Diarrheal diseases are commonly seen (Ekşi, 2016:21).

Ensuring the Control of Specific Contaminants

The environmental risk factors that occur after disasters are typically associated with the new environmental conditions that emerge. Factors such as the presence of workplaces containing flammable and explosive materials, petroleum products, radioactive substances, and waste, and factories of toxic chemical production always pose dangers and risks in disaster areas. For example, the gas leak from an insecticide factory in the city of Bhopal, India, on December 2, 1984, resulted in thousands of injuries and deaths, while the triggering of the TÜPRAŞ Fire after the Izmit Earthquake on November 17, 1999, occurred after a natural disaster such as an earthquake.

Radiation-related accidents occur, and these accidents cause serious damage to the environment. Major accidents such as Fukushima, Chernobyl, and Three Mile Island can occur. The Chernobyl nuclear accident, which occurred on April 26, 1986, resulting in the death of 29 people, can be cited as an example. Prolonged exposure to radiation damages the brain membranes, severely affects the cardiovascular system and leads to sudden deaths. After nuclear incidents, the cleanliness of food and water should be monitored, necessary precautions should be taken, and the people should be provided with the necessary education (Ekşi, 2016:22-23).

Diseases That Lead to Epidemics

An outbreak is defined as the occurrence of an infectious disease in a specific region, community, or season with a frequency higher than the expected normal rate (Nişancı, 2020:88). The main diseases that cause outbreaks worldwide are listed and explained below.

Coronavirus (MERS-CoV)

Coronavirus (MERS-CoV) is a novel coronavirus within the betacoronavirus family that causes a viral respiratory disease. Common symptoms among infected individuals include shortness of breath, fever, acute respiratory disease, and cough. According to global data from the World Health Organization, there have been 2122 confirmed cases and 740 deaths attributed to MERS up to the present day (Çalışkan, 2018:58).

Salmonella Agona

Salmonella is the most significant bacterium worldwide responsible for foodborne diseases. Symptoms of infection typically appear 12-72 hours after ingestion and include nausea, fever, vomiting, abdominal pain, and diarrhea. In France, in December 2017, 35 confirmed cases of Salmonella infection were reported in infants under 6 months old, with no reported fatalities among these cases (Çalışkan, 2018:58).

Yellow Fever

The causative agent of this disease is an Arbovirus of the Flavivirus family. It is prevalent in the subtropical and tropical regions of Africa and South America. It is transmitted to humans through the bite of infected mosquitoes. This disease causes liver failure and febrile disease, leading to death. To prevent infection, vaccination should be administered, protective clothing should be worn, and insect repellents should be used. In Nigeria, 341 suspected cases were reported between July and December 2017.

Monkeypox

Monkeypox is a lethal zoonotic disease that is known sporadically in the forested areas of Central Africa and West Africa. The virus responsible for monkeypox belongs to the Orthopox species of the Poxviridae family. Symptoms of the disease include headache, rash spreading from the face to the body, and fever. Between 4 September and 9 December, 172 suspected cases were reported in various regions of Nigeria.

Diphtheria

The causative agent of diphtheria is Corynebacterium diphtheriae. Diphtheria is a disease that can cause paralysis in individuals. Symptoms of the disease include shortness of breath, sore throat, fever, difficulty swallowing, and swelling of the lymph nodes of the neck. The disease is fatal in 5-10% of affected individuals. In Yemen, between August 13 and December 21, 2017, there were reported 333 suspected cases and 35 fatal cases. Similarly, in

Bangladesh, between November 3 and December 12, 2017, there were reported 804 suspected cases and 15 fatal cases (Çalışkan, 2018:59).

Cholera

The causative agent of cholera is known to be Vibrio cholerae, a deadly disease characterized by acute watery diarrhea resulting from the consumption of food and water contaminated with the bacterium. Although the disease often presents without symptoms, it can sometimes lead to life-threatening complications. Between September 28 and December 7, 2017, a total of 547 cases were reported, including 15 fatal cases. Following the cholera epidemic in Yemen that began in October 2016 and continued until June 27, 2017, Kenya reported 3,967 cases, including 76 fatal cases, between January 1 and November 29, 2017 (Çalışkan, 2018:60).

The Plague

The causative agent of the plague disease is Yersinia pestis, and it is a zoonotic bacterial infectious disease. The transmission route of the disease generally occurs through direct contact with infected tissues, bites of infected fleas, and inhalation of infected respiratory droplets. The mortality rate of plague disease is reported to be between 30% and 60%, but if left untreated, it is said to be 100% fatal. Countries where plague disease is observed include Madagascar, the Democratic Republic of the Congo, and Peru. In Madagascar, between August 1 and November 22, 2017, there were reported 2348 cases and 202 deaths (Çalışkan,2018:60).

Marburg Fever

Marburg virus disease (Marburg hemorrhagic fever) is described as a highly lethal disease in both primates and humans. Symptoms of the disease include headache, sudden onset of fever, myalgia, and tremors. The mortality rate among affected individuals is around 50%. It has been noted that reliable laboratory practices, community mobilization, close monitoring of infection control and prevention measures, and safe burial of infected animals facilitate the control of disease outbreaks. The most recent occurrence of Marburg fever was reported in the Kween region of Uganda.

Dengue Fever

The dengue virus is prevalent in tropical and subtropical regions where more than a third of the world's population resides, leading to the occurrence of the disease. The Aedes aegypti mosquito species, which carry the virus, also pose a risk for other diseases such as chikungunya, Zika infection, and yellow fever. While the symptoms of the disease can resemble those of the flu, it can be more severe and even fatal in some cases. It has been increasingly reported in recent years, with half of the world's population considered at risk. In Burkina Faso, located in the western part of the African continent, there were reported 9,029 cases and 18 fatal cases of the disease as of November 3, 2017 (Çalışkan, 2018:61).

Influenza-H7N9

The influenza virus causes seasonal flu outbreaks every year. It is indicated that influenza spreads through direct contact with contaminated environments or infected animals. Continuous human-to-human transmission occurs. The Influenza virus can lead to various diseases such as sepsis, pneumonia, mild upper respiratory tract infections, acute respiratory distress, and severe shock. Influenza A (H5N1) and A (H7N9) virus diseases are usually transmitted through indirect or direct contact with dead poultry or infected animals. As of 2013, the number of cases related to the disease was reported at 1564 (Çalışkan, 2018:62).

Avian Influenza

Avian influenza is a viral infection that affects birds such as chickens, geese, and ducks, but it has also been observed to affect other animal species such as tigers and pigs. Although rare, this disease can cause serious infections in humans. Avian influenza is a pandemic that can cause seasonal outbreaks worldwide (Gökçekuş et al., 2018:12).

The Impact of Technology Utilization in Disasters on the Management of Infectious Diseases

Various technologies used in different stages of disaster management have direct and indirect effects on the management of infectious diseases, as detailed in the examples below.

In the pre-disaster period, technology is utilized in many areas, such as ensuring communication in disaster areas, detecting meteorological events and taking necessary precautions accordingly, conducting data analysis, sharing data between institutions, determining land use, water sources, natural gas lines, food warehouses, identifying risks and hazardous areas (Memiş & Babaoğlu, 2020:781). During the preparation phase, data management plays a significant role in identifying shelter locations, site identification, education and evacuation, emergency simulation studies, communication systems, establishment of personnel lists and material inventory, and coordination systems among stakeholders. The mapping of an aircraft is used to obtain accurate and up-to-date data. Early warning systems provide the necessary time to reduce material damage, loss of life, and potential damage to the environment before disasters occur. Various technological applications, such as early detection of warnings through earthquake recording stations, are available (Memiş & Babaoğlu, 2020:782). The use of smart emergency management systems in supply chain management facilitates emergency management and accelerates information sharing. In the United States and Japan, where severe hurricanes occur, rapid and smart supply chain information systems technology is used to prevent energy, time, and material waste (Adıgüzel, 2022:48).

In disasters, the rapid and effective planning of interventions, logistics, and other practices is achieved successfully through technological applications. The cause of the disaster, its geographical location, intensity, management of logistic systems, planning of supply processes, and storage of materials are swiftly carried out using technological methods. The effective and efficient use of communication technologies in disaster management allows stored data to be used for simulation studies aimed at disaster prediction. The Integrated Disaster Management System developed through the use of information systems enables all data to be stored, processed, and transmitted through computer and mobile systems. RFID (Radio Frequency Identification) and barcode technologies facilitate the transfer of information to the system, such as the treatment and assistance provided to disaster victims, ensuring consistent aid delivery and preventing aid misuse. Technological devices such as portable RFID or barcode technology and fingerprint readers are used to identify deceased victims of disasters, enabling their identification to be accurately recorded in the system (Adıgüzel, 2022:49-50).

Discussion and Conclusion

It can be argued that there is a formulation in disaster situations of "needs > resources" or "required intervention > intervention made." Disaster management is about expecting the unexpected and managing the worst. In the digitalized world, new technologies and innovative solutions accelerate the discovery and implementation of infectious diseases from a scientific, clinical, and public health perspective by enhancing disease management strategies (Fitch, 2015). The use of information technology-based systems (e.g., artificial intelligence, remote sensors, surveillance systems, computational methods) can assist in epidemic management in the intervention phase of disasters (Asadzadeh et al., 2020). Indeed, it is noted that VR and AR technologies have potential applications in the emergency management of infectious diseases (Asadzadeh et al., 2021). New technologies such as the next-generation wireless Internet and geographic positioning play an important role in improving communication, information management, general disaster response, and emergency medical care in disasters (Chan et al., 2004). By using bio-pressure in the creation of virus disease models, potentially infectious diseases in disasters can be responded to more efficiently and effectively (Zimmerling & Chen, 2020). Similarly, the use of new technologies such as web-based surveillance tools and epidemic intelligence methods, is expected to predict and prevent future infectious disease threats (Christaki, 2015). It is emphasized that wearable technologies for physiological monitoring can improve health service delivery during outbreaks (Ming et al., 2020). It is seen that technologies aiming to manage infectious diseases within the scope of disaster management hold promise, and there is a need for new studies to integrate these two areas.

Disclaimer

None.

References

Adıgüzel, S. (2022). Afet durumlarında yapay zekâ teknolojisi ile lojistik yönetimi örnekleri. *Akademik İzdüşüm Dergisi. 7*(1). 47-70.

Alay, D., Eke, E. (2023). Sağlık Bakanlığı'nın kriz iletişimi üzerine bir araştırma. *Süleyman Demirel Üniversitesi Sosyal Bilimler Enstitüsü Dergisi, 2* (46), 317-337.

Asadzadeh, A., Pakkhoo, S., Saeidabad, M., Khezri, H., & Ferdousi, R. (2020). Information technology in emergency management of Covid-19 outbreak. *Informatics in Medicine Unlocked, 21,* 100475. https://doi.org/10.1016/j.imu.2020.100475.

Asadzadeh, A., Samad-Soltani, T., & Rezaei-Hachesu, P. (2021). Applications of virtual and augmented reality in infectious disease epidemics with a focus on the COVID-19 outbreak. *Informatics in Medicine Unlocked, 24,* 100579. https://doi.org/10.1016/j.imu.2021.100579.

Canbulat, Ş. (2023). Doğal afetler ve olağanüstü durumlarda görülen salgın hastalıklar. *Tarsus Üniversitesi Uygulamalı Bilimler Fakültesi Dergisi, 3,* (1), 47-64. https://dergipark.org.tr/en/download/article-file/2871996.

Chan, T., Killeen, J., Griswold, W., & Lenert, L. (2004). Information technology and emergency medical care during disasters. *Academic Emergency Medicine, 11* (11), 1229-36. https://doi.org/10.1197/J.AEM.2004.08.018.

Christaki, E. (2015). New technologies in predicting, preventing and controlling emerging infectious diseases. *Virulence, 6,* 558-565. https://doi.org/10.1080/21505594.2015.1040975.

Çalışkan S. (2018). Salgın haberleri. *Türk Dünyası Uygulama ve Araştırma Merkezi Halk Sağlığı Dergisi, 3*(1), 56-64.

Ekşi, A. (2016). Afetlerden sonra ortaya çıkabilecek çevresel risklerin yönetimi. *Hastane Öncesi Dergisi, 2*(1), 15-25. https://dergipark.org.tr/tr/pub/hod/issue/25839.

Fitch, J. (2015). Engineering a global response to infectious diseases. *Proceedings of the IEEE, 103*, 263-272. https://doi.org/10.1109/JPROC.2015.2389146.

Gökçe, O., Tetik, Ç. (2012) Teoride ve Pratikte Afet Sonrası İyileştirme Çalışmaları. T.C. Başbakanlık Afet ve Acil Durum Yönetimi Başkanlığı. 8. https://www.afad.gov.tr/kurumlar/afad.gov.tr/3479/xfiles/afet_sonrasi_iyilestirme_calismalari-1.pdf.

Gökçekuş, H., Barlas, C., Almuhisen, M. & Eyni, N. (2018). Doğal ve insan kaynaklı afetler, sonuçları ve afet yönetimi. Yakın Doğu Üniversitesi. http://www.iusarc.org/wp-content/uploads/2020/11/Dogal-ve-Insan-Kaynakli-Felaketler-Sonuclari-ve-Afet-Yonetimi.pdf.

Güler, Ç, Çobanoglu, Z., Arıkan, N., Bunsuz, O., Açıkgöz, B., Yılmaz, H., Sakarya, S., Şentürk, Ş., Kale, Ş., Keçeci, M. (1994). Çevre Sağlığı Bilimsel İmece Eğitim Toplantısı Uygulama Kitabı. Çevre Sağlığı Temel Kaynak Dizisi. http://cevresagligi.org/saglik-bakanliginca-personel-ve-toplumun-cevre-sapligi-egitimine-yonelik-ozel-bir-donem-1994-1997-zakir-cobanoglu/.

Kılıç S. (2017). Afetlerde Bulaşıcı Hastalıklar Ve Salgınlar. Ortadoğu Afet Ve Hastane Öncesi Yönetim Kongresi. 8-11 Ekim 2017. İstanbul. https://dergipark.org.tr/tr/pub/ijchs/issue/53784/682592.

Kınıklı, S., Cesur, S. (2020). Afetlerde enfeksiyon kontrol önlemleri. *Int J Cont Health, 1,* 15-23. https://dergipark.org.tr/en/download/article-file/1064481.

Memiş, L., Babaoğlu, C. (2020). Acil durum ve afet yönetiminde süreç yaklaşımı ve teknoloji. *Ömer Halisdemir Üniversitesi İktisadi ve İdari Bilimler Fakültesi Dergisi, 13 (4),* 776-791.

Ming, D., Sangkaew, S., Chanh, H., Nhat, P., Yacoub, S., Georgiou, P., & Holmes, A. (2020). Continuous physiological monitoring using wearable technology to inform individual management of infectious diseases, public health and outbreak responses. *International Journal of Infectious Diseases, 96*, 648-654. https://doi.org/10.1016/j.ijid.2020.05.086.

Nişancı, D. (2020). Salgın hastalıklar ve salgın hastalıklar özelinde sağlık hakkına Avrupa İnsan Hakları Mahkemesi'nin bakış açısı ile ulusal mevzuatın Covid-19 özelinde değerlendirilmesi. *Türkiye Barolar Birliği Dergisi, 150*, 85-122.

Rathore, M. H. (2020). Infections after natural disasters. *Pediatrics in Review, 41*(10), 501-510.

Uçku, R. (2001). Doğal afetler ve olağanüstü durumlarda görülen infeksiyon hastalıkları: kontrol önlemler. *Ankem Derg.15*(3). 651-656.

Ünlü, H., Çiçek, E. (2021). Salgın hastalıklardan koruma ve kontrol önlemleri, Covid-19 pandemisi ile mücadele ve yaşanan güçlükler. *Hacettepe Üniversitesi Hemşirelik Fakültesi Dergisi, 8*(1), 101-107. https://doi.org/10.31125/hunhemsire.907978.

Chapter 6

Psychological First Aid in Disasters

Mükerrem Kabataş Yıldız*, PhD
and Tuğba Çınarlı, PhD
Vocational School of Health Services, Ondokuz Mayıs University, Samsun, Türkiye

Abstract

Psychological first aid in disasters includes psychosocial interventions applied instantly or later in events that cause various adverse effects, such as natural disasters and accidents. Its main purpose is to identify the most basic physical or mental needs of the affected individuals and aim to solve the problems. Psychological first aid is of crucial importance not only for individuals affected by the disaster or adverse situation but also for aid workers. All individuals, families, support teams, and search, rescue, and intervention units facing a traumatic or emergency situation, especially children and the elderly, should benefit from psychological support. Psychological first aid aims to help disaster victims express their feelings, and relax with basic psychological information, and understand what they have experienced and felt. Psychological first aid is applied at the moment contact is made with people under stress. Early intervention in psychological first aid makes it possible for individuals to cope with the situation.

Keywords: coping, disaster, psychological first aid, psychological interventions, well-being

* Corresponding Author's Email: mukerrem.kabatas@omu.edu.tr.

In: Disaster Management in the Digital Age and Emerging Technologies
Editors: Ayşe Asiltürk and Ayşegül Sarıoğlu Kemer
ISBN: 979-8-89530-213-2

Introduction

It is reported that the demand for mental health services worldwide greatly exceeds the availability of mental health resources. Additionally, it is noted that this "gap" between demand and mental health resources expands during local mental health emergencies and larger-scale disasters (Everly and Lating, 2021). Many people on Earth are quite vulnerable in the face of events with significant life impacts. Disaster, known as an unexpected and merciless event, causes serious social disorder and significant psychological distress (Azzolini et al., 2018). Psychological First Aid (PFA) is a tool widely used by trained providers or first responders to apply early interventions to people experiencing disaster or traumatic experiences during disasters, epidemics, or wars (Ni et al., 2023). The PFA protocol determines the actions to be applied to victims in the event area. Through these actions, a non-interventional protocol is applied to victims, and they are helped to stay calm but not passive (Azzolini et al., 2018). At the same time, an important goal of PFA is to enhance people's recovery capacity. PFA supports recovery by helping people identify their urgent needs and their strengths and abilities to meet these needs (Australian Red Cross, 2020).

The Concept of Psychological First Aid

The PFA is based on the understanding that people affected by disasters will experience a series of early responses (physical, psychological, emotional, and behavioral) (Australian Red Cross, 2020). The impact of the disaster is experienced by many people directly affected by the event, witnesses, or indirectly affected. Disasters create both physical and psychological stress because people perceive their lives to be threatened or under duress. Therefore, it is common for people to respond with behavioral and emotional stress responses and re-adaptation issues (Unicef, 2021). In this respect, PFA is an early intervention to reduce acute distress and facilitate adaptation after exposure to trauma (Wang et al., 2024a). This is a humane and supportive response to someone who is suffering and needs support (Australian Red Cross, 2020). PFA is a combination of two conceptual and practical foundations, physical first aid and psychological crisis intervention (Everly and Lating, 2022). In other words, PFA can be considered a mental health analog of physical first aid (Everly and Lating, 2021). A small portion of the population affected by disasters will show more complex reactions and will need more mental health support to help recovery. However, most disaster victims will be able to recover on their own or with the support of compassionate and helpful disaster workers, family, and friends (Australian Red Cross, 2020).

The Development of Psychological First Aid

PFA has a long history (Australian Red Cross, 2020). This concept was first mentioned in a guide prepared by Drayer et al. in 1954 on behalf of the American Psychiatric Association (Drayer et al. 1954). Various methods related to critical incidents were used in the 1970s and 1980s. Research was conducted in the 1990s on these methods, and the research results determined that some applications re-traumatized individuals and showed various symptoms of mental problems such as post-traumatic stress disorder, and major depression in the following processes (Kılıç and Şimşek, 2018). Since the emergence of research showing the dangers of critical incident stress briefings, it has become more popular, and since 2002, psychological first aid has been recommended as an important part of providing psychosocial support after disasters (Australian Red Cross, 2020).

Methods of Applying Psychological First Aid

Information on where, how, and by whom PFA applications can be applied is specified in manuals published by various organizations. In these guides, PFA is implemented with the basic principles of "Look, listen, connect, and direct." PFA is not designed as a method that can only be applied only by mental health professionals because it does not include treatment elements such as psychological counseling or psychotherapy (World Health Organization [WHO], 2011). In the PFA application guides, it is stated that anyone who receives training can apply psychological first aid (Sargın and Demirelli, 2023; Gispen and Wu, 2018; WHO, 2011). PFA can be applied by mental health professionals, first responders, and even the general public to provide immediate psychological support to those in need after an emergency (Jung and Han, 2021). In disaster situations, a trained psychosocial work team can provide training to recognized and trusted local leaders, imams, nurses, or teachers in the affected area, enabling these people to support the work and sustain the applications through these local actors (Arslan Tomas and Şavur, 2018).

PFA, because it requires early intervention, can be applied during or immediately after an incident. The intervention time can sometimes last for a few minutes or even days or weeks. This diversity stems from the wide coverage of PFA; because any psychosocial support intervention made for people affected in disaster or crisis zones can be considered within this scope as long as it is compatible with PFA principles (Jung and Han, 2021; WHO, 2011).

PFA is usually applied on-site, in the disaster area, or in the affected area. However, when necessary, applications can also be made at various places such as treatment centers, schools, hospitals, homes, workplaces, and airports. The most important factor in choosing the application site is to provide a safe environment for individuals. In the disaster area, an environment where privacy and security are protected, and individuals can feel comfortable should be preferred (Sargın and Demirelli 2023; WHO, 2011).

The PFA application is usually carried out in three stages. These stages include preparation for the application, psychological triage to prioritize needs and demands, and the application of the 8-step PFA model (Jung and Han, 2021; Demircioğlu, 2017).

Preparation

The first stage of the PFA includes the preparation steps that need to be done before starting fieldwork. These steps are intended to maintain the psychological health of the PFA provider as well as to improve the quality of the service provided. The preparation process is divided into two categories: individual and event-specific (Everly et al., 2021; Demircioğlu et al., 2019).

Individual Preparation

It is helpful for individuals participating in the PFA process to self-assess in various ways before starting fieldwork, to prepare themselves mentally. Being prepared provides a protective effect against possible impacts of traumatic stress for PFA providers and alleviates potential challenges that may be encountered during work. It is important that the person who will work in this field ask themselves the following questions and assess their competence during the individual preparation process before the task (Everly et al., 2021; Demircioğlu et al., 2019; Stroud, 2012):

- Have I been involved in an incident of similar nature before?
- Could there be any aspect of the incident itself or working with the affected groups that could challenge me?
- Is my current health status appropriate for involving in this task?
- Are the working conditions in the task suitable for my current living conditions? (The necessity to stay in the field for a long time, travel restrictions, the presence of a person with caregiving obligations in the family, etc.)
- Do I know and can I apply what I need to pay attention to regarding psychological self-care when working with such incidents?

Event-Specific Preparation

Each disaster, crisis, or emergency has its own unique aspects. These aspects can affect the planning, implementation, and conclusion of post-event work. While the basic principles of approach to emergencies or PFA remain constant, the characteristics of the event can lead to various differences in applications. For PFA practitioners, it is thought that self-assessment before starting the studies would be beneficial, and the following questions are suggested for event-specific preparation (Demircioğlu et al., 2019; Demircioğlu 2017):

- Do I have enough information about the event that occurred? What happened? When did it happen? How many people were affected? Are there other important details that should be known about the event?
- What are the sociocultural characteristics of the individuals or groups affected by the event?
- Which institution coordinates the psychosocial services to be carried out after the event? Which other institutions will be involved in this process? Where does my institution stand in this planning process and what kind of role does it take on? What is the scope of my duties and responsibilities?

Triage

After disasters, individuals at the scene can experience shock reactions in the acute phase. These symptoms can occur in different forms or intensities in everyone, and in some cases, it is extremely important that individuals with some symptoms are noticed and directed correctly (Sargın and Demirelli 2023; Everly and Lating 2021). Psychological triage is an attempt to prioritize and thus provide the fastest and most effective help to as many people as possible by identifying individuals who show severe psychological reactions and have a high level of impact in conditions where there are many affected individuals and limited resources (Demircioğlu, 2017; WHO, 2011). Situations such as confusion, orientation problems, excessive anger reactions, severe physical or verbal agitation, panic or excessive anxiety reactions, freezing, psychosis, dissociation, risk of self-harm, and alcohol or substance use disorders can be urgently evaluated in psychological triage (Everly Jr. and Lating, 2021). Especially in the acute periods, careful field observation provides the opportunity to determine and refer the individuals to be primarily identified in psychological triage for the necessary medical support without wasting time (Demircioğlu, 2017).

Eight-Step PFA Model

There are many guides prepared by different organizations to be used after a disaster. Although these guides show similarities in terms of basic steps, small differences are observed in their areas of use (Demircioğlu et al., 2019; WHO, 2011). In this section, the proven validity and reliability of the 8-step PFA model developed by Brymer et al. (2006) are included.

1. Building Connection and Contact: This is the first stage where communication begins between the PFA provider and the affected person. The PFA provider needs to introduce themselves and state why they are talking to the disaster victim. At this level, more importance should be given to listening and understanding rather than making them talk. Before moving on to an in-depth interview, it is assumed that disaster survivors may have some basic needs, so asking their needs; is the stage aimed at developing a sense of trust by meeting physical needs such as water, a blanket, clean clothes, or food.
2. Security and comfort: One of the most basic interventions applied to disaster victims is to support them to feel safe by ensuring their transition to safer areas where they will not be injured.
3. Balancing: Disaster victims may exhibit abnormal reactions in the early stages. The aim of this stage is to stimulate the senses of disaster

victims, helping them mentally and physically acclimate to their environment and confront their abnormal reactions.

4. Information Gathering: This stage includes the process of gathering information to determine psychosocial support services and the roadmap for planning interventions through interviews with disaster victims.
5. Meeting Basic Needs: Based on the information obtained, it is the stage of arranging the basic needs of individuals or the community according to urgency, and working in coordination with all units in the region to identify and meet needs.
6. Social support: After disasters, ensuring that individuals receive social support and strengthening their social ties can help them feel less alone and be able to receive support from their environment. At this stage, PFA providers can engage in a series of activities that will activate the existing social support systems for disaster victims. Sharing meetings, hobby groups, creating activity groups, and building sustainable social support systems can be beneficial.
7. Information: In crisis situations, chaos can be prevented by ensuring the flow of correct information. It is important for individuals experiencing a crisis to recognize situations such as acute stress responses and trauma responses. Additionally, being informed about the sources where basic food and hygiene supplies are obtained is of vital importance for maintaining order in emergency situations.
8. Contacting with the institutions to be collaborated with: This stage includes the management of services applied to individuals or society, directing, completion of individuals' duty periods, changes in duty, and such situations.

Psychological First Aid Applications in Sensitive Groups

Children and Adolescents

It is important to proceed carefully to meet the psychological needs of children and adolescents after a disaster. This can help them recover and adapt in a healthy way. The points to consider when applying PFA to children and adolescents in disasters are the following (Demircioğlu et al., 2019; Reveruzzi et al., 2016; WHO, 2011):

- A safe environment should be provided for children and adolescents after a disaster. Thus, they can be made to feel physically safe and psychologically good.
- It is important for them to be together and communicate with their families to increase feelings of safety and support after the disaster. During communication, it is important to be at eye level with the child and to use language appropriate to their level of development. Parents should be informed about possible responses to traumatic stress and behavioral changes in their children. Guiding mothers, fathers, and other adult caregivers on what they can do and how they can behave to cope with these situations they may observe in their children will be supportive of both children and their families.
- They should be allowed and supported to express their feelings (fear, anxiety, sadness, etc.).
- Children and adolescents often prefer to express their feelings through games and activities. Therefore, support should be provided to them through games and creative activities.
- They should be encouraged to return to their routines as much as possible. In this way, their feelings of security and well-being can be increased.
- When working with children and adolescents who have lost their caregivers, it is very important to show special sensitivity, ensure their safety, determine who will be in charge of them, and protect them from abuse.
- In necessary situations, access to professional help and resources that can provide psychological support should be ensured.

Individuals with Health Problems

This group includes the elderly, the disabled, and those with chronic diseases. Points to consider when applying PFA to individuals with health problems in disasters are as follows (Demircioğlu et al., 2019; WHO, 2011):

- First, physical health needs should be met. Providing their medication, medical intervention, or other medical needs in a timely and effective manner can help improve their health conditions and make them feel psychologically better.

- It should be checked whether they have a caregiving/supporting relative. If the individual needs care and no caregiver can be found, a referral to an institution where this service can be obtained or other solutions should be planned.
- Providing a safe environment can reduce their worries and help them feel psychologically more stable.
- It is important to understand and support the emotional difficulties experienced by individuals in this group. Listening, understanding, and showing empathy can support their emotional well-being.
- Individuals should receive the necessary information about the disaster, and communication channels should remain open. Thus, they can be informed and feel safe about their situation.
- Their access to the help and resources they may need (medical help, rehabilitation services, psychological support, etc.) should be ensured.
- In case of possible health problems, they should be informed about where they can apply and whom they can ask for help, and, if necessary, contact information should be shared with them in writing.
- A customized approach should be adopted according to the needs of each individual with a different health problem. PFA should be provided according to their specific health conditions, preferences, and needs.

Different Cultures

When applying PFA to individuals of different cultures, it is important to take into account cultural diversity and different needs. The points to consider when applying PFA to individuals from different cultures in disasters are as follows (Demircioğlu et al., 2019; WHO, 2011):

- Different cultures may have different beliefs, values, and practices. Therefore, cultural sensitivity should be shown when applying PFA. Understanding and respecting individuals' cultural differences enables trust and bonding.
- The style of communication and language differences may exist in different cultures. Communication should be established using individuals' mother tongue, or translation services should be provided

when necessary. Additionally, communication tools such as gestures, expressions, and body language should be taken into account.

- The beliefs and rituals of different cultures are important for psychological well-being after a disaster. Understanding and supporting individuals' religious or cultural beliefs can enhance their psychological well-being.
- Social support systems may vary in different cultures. Individuals' access to social support sources such as their families, communities, and religious leaders should be provided.
- The diversity between different cultures, handshaking and greeting rituals, norms or perceptions related to dress styles, attitudes in communication with the opposite sex, and religious features should be taken into account.
- Post-traumatic stress symptoms can vary in different cultures. Individuals' emotions should be recognized, and information about post-traumatic stress symptoms should be known.

Ethical Principles in Psychological First Aid Applications

- The ethical principles guiding PFA applications in disasters include the following (Korndörffer et al., 2023; Dieltjens et al., 2014; WHO, 2011):
- To preserve the well-being of individuals and provide support without causing harm,
- To ensure the physical and psychological safety of individuals,
- To keep the information shared by the disaster victims confidential and to respect their privacy,
- To provide calmness and emotional balance to help people cope after a disaster,
- As an implementer, avoid making promises that cannot be fulfilled, do not force individuals to tell their experiences, and do not accept money or any gift in return for the support provided,
- To increase individuals' confidence in their ability to cope with difficulties and strengthen the spirit of solidarity in the community,
- To facilitate their social connections and support networks to reduce the feeling of isolation and encourage resilience,
- To avoid judging during or after individuals' sharings,

- To instill a sense of hope and optimism for the future to aid the recovery process,
- To avoid misinforming disaster victims,
- It encompasses the principles of PFA practitioners not to exaggerate their own skills and capabilities and to act only within their authority and skills.

Technological Applications Related to Psychological First Aid

The way community members interact during disasters has changed significantly due to technological innovations and the development of new media (Tan et al., 2017). The widespread use of smartphones in the 21st century and its adoption as an integral part of everyday life have enabled the development of many mobile applications in different fields. In recent years, it has been observed that one of the areas of use of mobile applications, which have begun to be used in various fields, is psychological aid services (Kabadayı and Güven, 2023). In the following, mobile applications related to PFA are discussed.

Training for Life Skills (TLS)

This mobile application is a free-to-download app developed to support individuals affected by traumatic experiences. The TLS mobile application consists of three categories: information, psychological recovery, and mood change. The information category consists of self-assessment (Post Traumatic Stress Disorder, depression, anxiety, sleep disorder), post-disaster responses, coping mechanisms (anger, sleep disorder, addiction), and guidance on how to use information obtained from specialized institutions. The psychological recovery category includes exercises such as breathing, meditation, positive affirmation, butterfly embrace, and yoga. The mood change category includes activities such as matching picture games, coloring books, and Tetris (Choi et al., 2021).

SAMHSA Disaster

The Substance Abuse and Mental Health Services Administration (SAMHSA) provides comprehensive disaster behavioral health resources to assist disaster victims in both natural and human-caused disaster preparation, response, and recovery processes. The SAMHSA Behavioral Health Disaster Response App (SAMHSA Disaster App) is one of these applications. This mobile application enables behavioral health practitioners to respond to all kinds of traumatic events by providing easy access to disaster-specific information and other important materials directly with their mobile devices (Seligman et al., 2015).

Psychological First Aid

The PFA Mobile app was developed by the U.S. Department of Veterans Affairs National Center in partnership with the National Child Traumatic Stress Network, Defense Health Agency Connected Health, and the VA's Patient Care Services. The PFA Mobile app is designed to assist professionals who will provide psychological first aid to adults and children affected by disasters or emergencies. The purpose of this app is not to train intervention teams on PFA. This app allows individuals who will intervene with disaster victims to review the PFA guidelines and listen to mentoring tips for implementing PFA. It is also reported that this app is not designed to replace PFA training; it can be used as an addition to the resources that trained individuals use before, during, and after disaster intervention (U.S. Department of Veterans Affairs, 2023).

PTSD Coach

In response to the growing interest in using technology to manage symptoms of Posttraumatic Stress Disorder (PTSD), the PTSD Coach was developed by the U.S. Department of Veterans Affairs National Center for PTSD and the Department of Defense's National Center for Telehealth & Technology. This free app, for Apple or Android devices, includes evidence-based assessment, psycho-educational, and cognitive-behavioral strategies. Psycho-educational information includes materials about what PTSD is, how it develops, associated problems, treatment approaches, and when to seek professional help. In the self-management section of the app, users are asked to identify their symptoms, after which the app suggests symptom-specific evidence-based strategies such as planning enjoyable activities, functional analysis of traumatic triggers, and coping self-statements (Possemato et al., 2017).

Psychological First Aid Tutorial

This application provides on-the-spot information on PFA procedures for those who have received psychological first aid. The availability of this application for Apple and Android devices provides quick accessibility and ease of use while in the field after a traumatic event, natural disaster, emergency situations threatening public health, acts of terrorism, and individual crises. The application also includes appropriate self-care actions for PFA practitioners and intervention teams before, during, and after an emergency event (Olson, n.d.).

Discussion and Conclusion

Disasters are generally considered events that result in material damage and physical injuries that exceed the intervention capacity of local resources. History shows that disasters also lead to an increase in the demand for psychological support among survivors and those who intervene in the disaster. Therefore, the need for an effective acute mental health intervention is inevitable. However, the effectiveness of traditional multi-session counseling during and immediately after disasters has been investigated, and instead, the use of effective crisis-focused psychological interventions as an acute period alternative has been suggested. In this context, it has been suggested that PFA is a special mental health intervention in disaster that can be used during and after a disaster (Everly and Lating, 2021).

In a systematic review evaluating the effectiveness of PFA in improving the mental health and psychosocial well-being of individuals exposed to potentially traumatic events, it is reported that PFA has a positive effect, and in many studies, symptoms of anxiety, depression, post-traumatic stress, and distress have decreased, and there have been improvements in mood, safety experience, attachment, and sense of control in both adolescents and adults (Hermosilla et al., 2023). On the other hand, in a very recent study examining the evidence on the effectiveness of PFA interventions and how PFA interventions are designed, implemented, and experienced, it is reported that the PFA intervention performed after exposure to trauma showed a positive effect in terms of reducing anxiety in the short and medium term and facilitating adaptive functioning; however, evidence for reducing post-traumatic stress disorder/depressive symptoms was reported to be less convincing (Wang et al., 2024).

Mobile applications in psychological aid services contribute in a wide range such as providing an appropriate cost, ensuring accessibility, potential to strengthen therapeutic relationship, saving time, enabling users to self-assess, identifying the sources of psychological disorders, connecting with online support groups, and offering the chance of online psychological assistance (Kabadayı and Güven, 2023). In a systematic review related to mobile applications in crisis informatics, it is reported that it is important for future research to focus on user-centered studies that will enable citizens to gain more knowledge about the use of mobile applications to fully enhance resilience against disasters through technology (Tan et al., 2017).

Disclaimer

None.

References

Arslan Tomas, S., Şavur, E. (2018). Afet ve acil durumlarda psikososyal destek hizmetleri saha çalışanları için el kitabı. *Ankara, UNICEF Türkiye Temsilciliği.*

Australian Red Cross, 2020. Psychological First Aid: Supporting people affected by disaster in Australia.

Azzollini, S. C., Depaula, P. D., Cosentino, A. C., & Bail Pupko, V. (2018). Applications of psychological first aid in disaster and emergency situations: Its relationship with decision-making. *Athens Journal of Social Sciences*, *5*(2), 201-214.

Brymer, M., Layne, C., Jacobs, A., Pynoos, R., Ruzek, J., Steinberg, A., Eric, V. & Watson, P. (2006). *Psychological first aid field operations guide.* National Child Traumatic Stress Network.

Choi, Y. J., Ko, E. J., Choi, E. J., & Um, Y. J. (2021). Managing traumatic stress using a mental health care mobile app: A pilot study. *International Journal of Mental Health Promotion, 23*(3), 385-393.

Demircioğlu, M. (2017). Olağandışı koşullarda psikolojik ilk yardım. *Türkiye Psikiyatri Derneği*, 21.

Demircioğlu, M., Şeker, Z., & Aker, A. T. (2019). Psikolojik ilk yardım: Amaçları, uygulanışı, hassas gruplar ve uyulması gereken etik kurallar. *Psikiyatride Güncel Yaklaşımlar, 11*(3), 351-362.

Dieltjens, T., Moonens, I., Praet, K., Buck, E., & Vandekerckhove, P. (2014). A systematic literature search on psychological first aid: Lack of evidence to develop guidelines. *Plos One, 9*(12), e114714. https://doi.org/10.1371/journal.pone.0114714.

Drayer CS, Cameron DC, Woodward WD, Glass AJ. (1954). Psychological first aid in community disaster. *J Am Med Assoc. 156*(1):36-41. PMID: 13183803.

Everly Jr, G. S., & Lating, J. M. (2021). Psychological first aid (PFA) and disasters. *International Review of Psychiatry*, 33(8), 718-727.

Everly Jr, G. S., & Lating, J. M. (2022). The Johns Hopkins guide to psychological first aid. JHU Press.

Gispen, F., & Wu, A. W. (2018). Psychological first aid: CPR for mental health crises in healthcare. *Journal of Patient Safety and Risk Management, 23*(2), 51-53.

Hermosilla, S., Forthal, S., Sadowska, K., Magill, E. B., Watson, P., & Pike, K. M. (2023). We need to build the evidence: A systematic review of psychological first aid on mental health and well-being. *Journal of Traumatic Stress, 36*(1), 5-16.

Jung, H. O., & Han, S. W. (2021). Implementing psychological first aid ontology. *PLoS One,* 16(6), e0252891. https://doi.org/10.1371/journal.pone.0252891.

Kabadayı, F. & Güven, M. (2023). Psikolojik yardım hizmetlerindeki mobil uygulamalarla ilgili bir inceleme. *TEBD, 21*(3), 1660-1689. https://doi.org/ 10.37217/tebd.1291285.

Kılıç, N., & Şimşek, N. (2018). Psikolojik ilk yardım ve hemşirelik. *Psikiyatri Hemşireliği Dergisi, 9*(3), 212-218.

Korndörffer, L., White, C., & Mackelprang, J. (2023). Psychological first aid principles within a community-led arts initiative: Lessons from the blacksmiths' tree. *Disasters, 47*(3), 806-829. https://doi.org/10.1111/disa.12564.

Ni, C. F., Lundblad, R., Dykeman, C., Bolante, R., & Łabuński, W. (2023). Content analysis of psychological first aid training manuals via topic modelling. *European Journal of Psychotraumatology, 14*(2), 2230110.

Olson, D. (n.d.). *Psychological first aid tutorial for iPhone and android (20140244, Dr. Debra Olson).* University of Minnesota Driven to Discover. Retrieved from April 05, 2024 from(https://license.umn.edu/product/psychological-first-aid-tutorial-for-iphone-and-android).

Possemato, K., Kuhn, E., Johnson, E. M., Hoffman, J. E., & Brooks, E. (2017). Development and refinement of a clinician intervention to facilitate primary care patient use of the PTSD Coach app. *Translational Behavioral Medicine*, *7*(1), 116-126.

Sargın, N., & Demirelli, M. A. (2023). Psikososyal Destek ve Psikolojik İlk Yardım. *Current Research and Reviews in Psychology and Psychiatry*, *3*(2), 81-92.

Seligman, J., Felder, S. S., & Robinson, M. E. (2015). Substance Abuse and Mental Health Services Administration (SAMHSA) behavioral health disaster response app. *Disaster medicine and public health preparedness*, *9*(5), 516-518.

Stroud, M. L. (2012). Psychological triage in general practice. *Australian Family Physician, 41*(7), 508-512.

Tan, M. L., Prasanna, R., Stock, K., Hudson-Doyle, E., Leonard, G., & Johnston, D. (2017). Mobile applications in crisis informatics literature: A systematic review. *International journal of Disaster Risk Reduction*, *24*, 297-311.

Unicef, 2021. Trainers' guide: Psychological First Aid (PFA) and Mental Health and Psychosocial Support (MHPSS) Training Module. For sub-national social service workforce in Lao PDR. https://www.unicef.org/laos/media/5641/file/UNICEF%20and%20MOLSW_PFA%20and%20MHPSS%20Training%20Manual_Eng.pdf.

U.S Department of Veterans Affairs. (2023, December 6). *PTSD: National Center for PTSD.* https://www.ptsd.va.gov/appvid/mobile/pfa_app_pro.asp.

Wang, L., Norman, I., Xiao, T., Li, Y., Li, X., Liu, T., ... & Leamy, M. (2024a). Feasibility and acceptability of a culturally adapted psychological first aid training intervention (Preparing Me) to support the mental health and well-being of front-line healthcare workers in China: a feasibility randomized controlled trial. *European Journal of Psychotraumatology*, *15*(1), 2299195.

Wang, L., Norman, I., Edleston, V., Oyo, C., & Leamy, M. (2024b). The effectiveness and implementation of psychological first aid as a therapeutic intervention after trauma: an integrative review. *Trauma, Violence, & Abuse*, 15248380231221492.

World Health Organization, (2011). War Trauma Foundation and World Vision International. Psychological first aid: Guide for field workers.

Chapter 7

A Current Assessment of Technological Developments in Disaster Management and Marketing Perspectives

Müzeyyen Özhavzalı*, PhD
Department of Business Management, Fatma Şenses Social Science Vocational High School, Kırıkkale University, Bahşılı, Kırıkkale, Türkiye

Abstract

Disasters that disrupt the normal flow of life and have devastating effects have caused social, economic, physical, and psychological damage throughout human history. Today's technologies and human knowledge have not yet been able to produce solutions to prevent or control disasters. However, solutions are being sought to limit the effects of disasters and minimize the loss of life and property. Disaster management, which includes conducting research to minimize social damage, developing plans, determining strategies, developing effective disaster coordination programs, establishing protocols, and creating a human resources structure that can urgently implement all these preparations in case of a disaster, and taking measures to ensure coordination in advance, is versatile and comprehensive. It is a dynamic process in which all segments of society must assume a role and take responsibility. The integration and coordination of all academies of businesses, communities, citizens, and government are crucial for effective disaster management. With the changing technologies and digitalization, businesses have had to organize and develop their marketing strategies in a versatile way before, during, and after disasters. Especially the emergence of new communication tools along with digitalization in

* Corresponding Author's Email: thavzalimuzeyyen@hotmail.com.tr.

In: Disaster Management in the Digital Age and Emerging Technologies
Editors: Ayşe Asiltürk and Ayşegül Sarıoğlu Kemer
ISBN: 979-8-89530-213-2

technology and rapid and uninterrupted accessibility has made it necessary for businesses to update their marketing methods in disaster management. In this study, information will be given on disaster management with changing technological developments and how businesses carry out the adaptation process from the perspective of the marketing method.

Keywords: digitalization, disaster management, marketing, technological developments

Introduction

Disasters are events that negatively affect societies and individuals, halting or interrupting life. These disasters are often triggered by natural events or human-induced factors. Among natural disasters are earthquakes, floods, fires, landslides, hurricanes, and droughts, while examples of human-induced disasters include explosions, environmental pollution, traffic accidents, and fires. These disasters are often intertwined; when one occurs, others are frequently triggered. They do not occur in isolation; rather, one disaster often leads to the occurrence of others. For instance, the occurrence of an earthquake not only causes destruction but also leads to additional events such as floods, fires, or landslides (Limoncu and Bayülgen, 2005:19).

Present-day technologies and humanity's knowledge have not yet produced solutions for preventing or controlling disasters. However, it is possible to limit the effects of disasters and minimize loss of life and property. Conducting research to minimize the damage caused by disasters, developing plans, determining strategies, developing effective disaster coordination programs and protocols, establishing human resource structures capable of implementing all these preparations in emergency situations, and taking precautionary measures for coordination constitute a multi-faceted, comprehensive, dynamic process requiring the involvement and responsibility of all segments of society. The integration and coordination of businesses, communities, citizens, and the government are crucial to effective disaster management. With changing technologies and digitalization, businesses have been compelled to adapt and improve their marketing strategies before, during, and after disasters. Especially, digitalization in technology has led to the emergence of new communication tools and rapid, and uninterrupted accessibility, necessitating businesses to update their marketing methods in disaster management. This study aims to provide information about how changing technological developments have influenced the adaptation process of disaster management and marketing methods of businesses. Natural disasters not only reduce the Gross Domestic Product of countries but also lead to significant economic costs. Since these disasters usually occur unexpectedly, economic losses and damages affect the country's economy and lead to budget deficits. In addition to these financial liabilities, they also cause production losses in businesses. Therefore, addressing natural disasters as a research topic has a significant impact on restoring public order and balancing economic functioning (Partigöç and Tarhan, 2019).

The Role of Information and Communication Technologies (ICT) in Disaster Management

Resilience in disaster management is closely related to individual and environmental factors. The concept of resilience has been defined by Varol and Kaya (2018) through three different layers. Core resilience focuses on the fundamental physiological structure based on the psychological and physical characteristics of the individual, which remain relatively stable over time. Internal resilience refers to resilience based on internal factors such as personal characteristics, family status, social environment, and educational level. External resilience, on the other hand, represents resilience that develops depending on a series of external factors such as environmental geographical conditions, socio-economic status, infrastructure, and institutional status (Varol and Kaya, 2018). It can be said that a wide range of factors and disciplines, from individual characteristics to the characteristics of the environment in which one lives, are effective in determining the resilience of a community to disasters (Partigöç and Tarhan, 2019). As the resilience of a community or individual against disasters increases and vulnerability decreases, the capacity to cope with disasters also increases. Information technologies are often preferred and provide significant advantages in the planning of activities to reduce disaster damage and in ensuring an effective disaster management process involving various stakeholders.

The policies of the OECD countries to reduce digital inequality in disaster management can be listed as follows (Kılıç, 2011): widespread access to technology for individuals, households, and workplaces; initiation of state development and implementation projects; initiation of education and training initiatives; encouragement of international cooperation. Various projects have been developed in our country to enable the effective use of IT products and services in disaster management (AYDES, Radius, TAMP, etc.). However, there are problems in reaching users in the case of national- and urban-scale studies. Problems such as citizens' unawareness of the existence of disaster gathering areas in the urban area where they reside, and not knowing the location of the nearest gathering area to their residence or workplace after a disaster are some of the main indicators that IT is not sufficiently utilized (Partigöç and Tarhan, 2019).

In situations such as disasters, it may not be possible to effectively use Information and Communication Technologies (ICT) products and services due to reasons such as insufficient technological infrastructure, technical glitches, and inadequate information about social media usage. In such cases, various studies are conducted by institutions and organizations to guide citizens correctly. For example, during Hurricane Sandy in 2012, which affected America and Canada, efforts were made to control the spread of false news and rumors on social media (Çalışkan and Mencik, 2015). While the full-time integration of ICT into disaster management offers significant advantages in many ways, when not used correctly and effectively, it can create disadvantages for authorities, citizens, and social media users. Situations such as the uncontrolled flow of information, the rapid spread of false news and rumors, the acquisition of misinterpreted information, and the lack of quick feedback to citizens can lead to the emergence of new crises (Sayımer, 2008).

ICT plays a crucial role in disaster management, encompassing preparedness before a disaster, response during a disaster, and recovery efforts after a disaster. Here are some key points regarding the role of ICT in disaster management:

1. *Preparedness Before a Disaster:* ICT aids in information gathering, analysis, and sharing during the preparedness phase before a disaster. Tools such as disaster risk assessments, Geographic Information Systems (GIS), and remote sensing technologies help identify high-risk areas and create disaster scenarios. Additionally, ICT is used to develop emergency plans and establish communication networks.
2. *Response During a Disaster:* ICT provides essential communication and coordination tools during a disaster. In the cases where the communication infrastructure is damaged, alternative communication technologies such as satellite communication, radio communication, and mobile phone networks can be utilized. Furthermore, GPS and other location-based technologies are used to locate and rescue individuals in disaster-stricken areas.

3. *Information Management and Sharing:* ICT facilitates the management and sharing of information collected during a disaster. Data from sensor networks, weather forecasting models, and social media analysis are used to monitor the disaster situation. This information is crucial for guiding rescue operations and coordinating emergency services.
4. *Raising Community Awareness*: ICT helps to raise awareness about disaster risks within communities. Mobile applications, websites, and social media platforms are used to disseminate information and educational materials about disaster preparation. Additionally, emergency news and alerts are communicated through these channels.
5. *Recovery After a Disaster:* ICT continues to play a significant role in the recovery phase after a disaster. Data analysis and data mining techniques are used to assess the impact of the disaster and prioritize recovery efforts. Moreover, online fundraising campaigns and other support services are organized to aid disaster victims.

In summary, information and communication technologies are critical tools in disaster management, enabling effective preparedness, response, and recovery efforts. With technology advancing rapidly, the business world is undergoing profound changes. The concept of gig work has become significant in altering traditional business models and work arrangements, thanks to the new opportunities brought about by digital transformation. When viewed from the perspectives of the digital economy and entrepreneurship, gig work has emerged as a new and significant trend. While Knowledge-Based Entrepreneurship (KBE) remains a dynamic concept and globalization is not universally embraced, the principles of KBE embody resilient methodologies and tactics for socioeconomic advancement. Intrinsically, the phenomena of KBE and globalization can be viewed as interrelated but distinct avenues for leveraging capital for socioeconomic development. Research indicates that both KBE and globalization significantly influence a nation's wealth and are inherently interconnected (Kutnjak et al., 2007: 193).

Disasters and The Use of Artificial Intelligence (AI)

Artificial intelligence (AI) in disaster management is an innovative approach encompassing a range of methods and technologies. AI can be utilized in various areas, such as data analysis, prediction, decision support, risk management, and emergency response. Here are some contributions of AI usage to disaster management:

1. Prediction and Early Warning Systems: AI algorithms can analyze meteorological data, geological data, and other relevant information to predict the timing and location of disasters. This contributes to the development of early warning systems and helps to inform the public about potential hazards.
2. Risk Assessment and Planning: AI can process complex datasets to identify risk areas and predict the potential impacts of disasters. This information can be used to develop disaster management plans and strategically allocate resources.
3. Emergency Response and Rescue Operations: AI can assist in making rapid and effective decisions during emergency situations. For example, when combined with technologies such as drones and unmanned aerial vehicles, AI can be used for tasks such as damage assessment, search and rescue operations, and emergency medical assistance.
4. Big Data Analysis: Large datasets collected for disaster management can be analyzed using AI algorithms to uncover meaningful patterns and relationships. This enables more effective decision-making and guides post-disaster recovery efforts.
5. Communication and Crisis Management: AI can monitor large streams of data from social media and other online platforms to understand the emotional state and needs of the public during crises. This information allows for the updating of communication strategies and crisis management plans.
6. Education and Simulations: AI can create interactive simulations and educational materials to train disaster response teams. This enables teams to practice real-time scenarios and test their response to disaster situations in advance.
7. Fast Data Processing: In disaster situations, rapid data processing and analysis are crucial. AI can automatically process large amounts of data and reduce the need for human intervention to derive meaningful insights.

AI has the potential to provide significant benefits in disaster management in these and other areas. However, effective utilization of this technology requires addressing various challenges such as infrastructure, data security, and ethical issues.

With the advancement of technology today, artificial intelligence (AI) is being used in various fields, including disaster management. AI technologies are utilized in various applications before, during, and after disasters. These applications include disaster prediction, risk analysis, emergency response planning, damage assessment, and coordination of rescue operations.

Disasters pose serious emergencies worldwide, resulting in significant human and material losses. In addition to traditional methods, AI technologies are increasingly being used to cope with such situations. AI has made significant advancements in areas such as big data analytics, machine learning, and deep learning, providing opportunities for various applications in disaster management. There are many important studies in the literature regarding the use of AI in disaster management. For example, Smith and Johnson (2018) developed an AI-based model for pre-disaster risk analysis. This model uses large datasets to assess different risk factors and predict the probabilities of disasters. Similarly, Jones et al. (2020) introduced a new system for AI-supported emergency response planning. This system enables rescue teams to manage their resources more effectively.

AI technologies have significant potential in disaster management. However, various challenges need to be addressed for the effective utilization of these technologies. These include data security, algorithm accuracy, and ethical concerns. Therefore, more research and effort is needed in the future to develop and implement of AI-based disaster management systems.

Effects of Disasters on Businesses' Marketing

Disasters can have various effects on businesses' marketing:

1. *Supply Chain and Supplier Relationships:* Disasters can directly impact production and supply chains. Disruptions in the supply chain can hinder or delay the procurement of materials and products. This can adversely affect businesses' ability to provide timely service or products to customers. Businesses can mitigate such risks by strengthening supplier relationships or establishing partnerships with alternative suppliers.
2. *Marketing Communication and Advertising Strategies:* Disasters can alter consumer behaviors and perceptions. Businesses should reassess

their post-disaster marketing communication and advertising strategies and approach customers in an appropriate and sensitive manner. For example, engaging consumers through messages offering assistance to disaster victims or demonstrating sensitivity toward the disaster can positively impact businesses' reputation and brand image.

3. *Inventory Management and Product Diversity:* Disasters can lead to sudden and significant changes in demand for specific products. Businesses should review their inventory management strategies to be prepared for disasters and be able to respond quickly to changes in demand. Additionally, they can adapt to changes in demand by increasing product diversity or offering alternative products.
4. *Pricing Strategies:* Changes in supply and demand resulting from disasters can lead to fluctuations in prices. Businesses should follow fair and ethical pricing strategies during disaster periods and avoid opportunistic behavior. Moreover, adopting appropriate pricing policies to contribute to the post-disaster community recovery efforts can be beneficial.
5. *Corporate Social Responsibility and Reputation Management:* Disasters require businesses to consider their social responsibilities and reputations. Businesses can initiate social responsibility projects to help disaster victims or to support the community. This can help businesses build a positive image in society and increase customer loyalty.

In summary, businesses should be prepared for disasters and adjust their marketing strategies accordingly. Careful and strategic action in areas such as supply chain management, marketing communication, inventory management, pricing policies, and social responsibility can help businesses minimize the impact of disasters and recover faster from crises.

Disasters and Social Marketing

Social marketing is the systematic planning and execution of strategies to create awareness and behavior change on a particular issue with the aim of improving or enhancing social well-being (Coşkun, 2012:227). It involves designing and implementing programs and activities to ensure the acceptance of a social idea by the target audience (Torlak, 2022). Social marketers, like their counterparts in commercial marketing, use the marketing mix to promote purchases or the adoption of a behavior (Kotler and Lee, 2010:91). Therefore, the methods used in social marketing to achieve objectives are fundamentally based on those used in commercial marketing.

The relationship between social marketing and disasters has also begun to attract attention in the marketing literature. Social marketing can be used both before disasters to raise awareness in communities and after disasters in normalization efforts. The post-disaster social responsibilities of businesses have been discussed (Kuo 2011). The corporate social responsibilities assumed by businesses in natural disasters have been analyzed (Johnson, Connolly, and Carter, 2011). The legal dimension of the relationship between disasters and corporate social responsibility has been discussed (Cherry and Sneirson, 2010). The relationship between corporate social responsibility activities of companies and stock values after disasters has been examined (Malik, Chowdhury, and Alam, 2023). A case study has been conducted on how marketing can be used in post-disaster normalization efforts in tourism areas (Durocher, 1994).

Social marketing can be utilized before disasters to take preventive measures, develop disaster awareness in communities, and normalize life after disasters. Categorizing the needs of disaster-stricken areas, prioritizing the fulfillment of urgent and important needs, planning for other needs, and providing resources for them, organizing donation campaigns, and creating a trusting environment for donors, all fall under the responsibility of social marketing. Social marketing is communication-based, education-focused, and requires economic behavior. Like in commercial marketing, it involves consumer orientation, consideration of target audience expectations and needs, utilization of the marketing mix, strategic research, planning, implementation, and evaluation, and the preparation of strategic plans (Eser et al., 2006:10). Disaster communication and disaster crowdfunding are products of social marketing. Therefore, the design, implementation, responsibilities of the parties involved, campaign organization, achievement of targeted outcomes, delivery of donations, monitoring, and evaluation of processes are conveyed in other sections in line with the functioning of social marketing.

Marketing Strategies for Businesses in the Face of Disasters

Disasters can have a profound impact on businesses, disrupting normal operations, and posing significant challenges to marketing efforts. To effectively navigate through these crises, businesses need to adapt their marketing strategies accordingly. Below are some key considerations for businesses in developing marketing strategies to respond to disasters:

- *Communication Preparedness:* Establish robust communication channels to disseminate information to customers, employees, suppliers, and other stakeholders during a disaster. Utilize various platforms such as email, social media, websites, and mobile apps to provide updates, address concerns, and convey important messages.
- *Customer-Centric Approach:* Prioritize the needs and concerns of customers during a disaster. Tailor marketing messages and offerings to provide relevant solutions and support to customers affected by the crisis. Demonstrate empathy and understanding in all communications.
- *Adaptability and Flexibility:* Remain agile and adaptable when adjusting marketing strategies and tactics in response to changing circumstances during a disaster. Be prepared to modify campaigns, promotions, and product/service offerings to align with evolving customer needs and market conditions.
- *Brand Reputation Management:* Safeguard the reputation of the brand by maintaining transparency, honesty, and integrity in all communications and actions during a disaster. Demonstrate corporate social responsibility through meaningful contributions and initiatives that support relief efforts and community recovery.
- *Employee Engagement and Support:* Ensure the well-being and safety of employees while maintaining open lines of communication and providing the necessary support and resources. Engage employees as brand ambassadors to convey consistent and reassuring messages to customers and the public.
- *Crisis Preparedness Planning:* Develop comprehensive crisis management and response plans that outline roles, responsibilities, and protocols to address various scenarios. Conduct regular training and drills to ensure readiness to implement these plans effectively during a disaster.
- *Utilization of Technology:* Leverage technology and digital platforms to facilitate remote operations, customer engagement, and data-driven decision-making during a disaster. Explore innovative solutions such

as virtual events, online customer support, and e-commerce platforms to maintain business continuity.
- *Community Engagement and Solidarity:* Demonstrate solidarity with affected communities by actively engaging in relief efforts, supporting local initiatives, and contributing to rebuilding efforts. Engage with stakeholders and collaborate with government agencies, NGOs, and other organizations to coordinate collective responses to the disaster.

By incorporating these considerations into their marketing strategies, businesses can effectively navigate the challenges posed by disasters, protect their brand reputation, and demonstrate resilience and empathy in the face of adversity.

Discussion and Conclusion

Climate change and the disasters it causes lead to financial crises globally and in Türkiye, affecting individuals, businesses, governments, and the global community. Disasters such as earthquakes and floods affect local populations and businesses in the affected areas. Disasters occurring in countries experiencing disasters necessitate measures to reduce credit and collateral risks in the banking and insurance sectors. This calls particularly for the diversification of insurance products in the insurance sector. Solutions such as disaster insurance, risk pools, and micro-insurance are among these measures.

Among the sectors most affected by climate change and natural disasters are agriculture and industry. These sectors are closely related to the banking and insurance sectors and are significantly affected by adverse events in the financial system. With ongoing climate change, cost increases in the agriculture sector and consequent decreases in profit margins are inevitable. These developments may lead to changes in the trade structure of the agriculture sector.

Preventive measures need to be taken for potential future problems because this situation can make it difficult to maintain the diversity of agricultural products. Therefore, public authorities should provide the necessary incentives and initiatives to promote environmentally friendly agricultural practices. Preventive measures can have positive effects on controlling future costs in agricultural activities and product prices. In this context, two concrete proposals are made to address financial risks arising from climate change or natural disasters. First, it is proposed to establish a natural disaster fund by allocating a certain share of tax and similar revenues in terms of public finance. Second, it is recommended to make natural disaster insurance policies mandatory for businesses against potential future risks and to include these insurance expenses in the scope of legal expenses under tax legislation (Partigöç and Tarhan, 2019).

Although disasters generally occur over a short period of time and in specific geographies, their effects are long-lasting and widespread. Recently, it has been emphasized on various platforms that preventive measures taken before disasters are more important than intervention and recovery efforts after disasters. In this context, it should be noted that activities such as creating the necessary and adequate technological infrastructure to benefit from the opportunities provided by technology in disaster management, raising awareness among relevant institutions and organizations, providing training to citizens on the use of social media during and after disasters, and organizing various campaigns for more effective use of social media are vital to reducing potential loss of life and property.

Disclaimer

None.

References

Akar, S. (2013). The Impact of Natural Disasters on Public Finance: The Case of Türkiye, Marmara University, Institute of Social Sciences, Department of Finance, Fiscal Economics Branch, Istanbul. (https://katalog.marmara.edu.tr/eyayin/tez/eTez023422.pdf) (January 13, 2024).

Akdeniz, F. (2022). The Impact of Natural Disasters on Financial Reporting. In S. Cengiz (Ed.), *Natural Disasters and Their Financial Impacts* (pp. 102-130). Gazi Bookstore Co. Ltd.

Çalışkan, M. & Mencik, Y. (2015). The New Face of a Changing World: Social Media. *Academic Perspective Journal,* 50, 254-277.

Can, İ. & Elçi, S. (2021). An Evaluation on the Effects of Global Climate Change and Natural Disasters on the Financial System of Businesses and the Public Sector in the World and Türkiye. *Social Science Development Journal, SSD journal, 6* (25), 50-63. http://dx.doi.org/10.31567/ssd.385.

Careem, M., De Silva, C., De Silva, R., Raschid, L., & Weerawarana, S. (2006, December). Sahana: Overview of a disaster management system. *In 2006 International Conference on Information and Automation (pp. 361-366). IEEE.*

Carter, W. N. Disaster management: A disaster manager's handbook. Mandaluyong City, Phil.: Asian Development Bank. 2008. Philippines. ISBN 978-971-561-006-3.

Cherry, M. A., & Sneirson, J. F. (2010). The Legal Dimensions of Disaster and Social Responsibility. *Law and Policy, 32*(4), 446-467.

Coppola, D. P. (2006). Introduction to international disaster management. Elsevier.

Coşkun, R. (2012). Social Marketing. In H. Özkara, E. Günay, & Y. Genç (Eds.), *Principles of Marketing: Strategy, Planning, Implementation, and Control,* (p. 227). Ankara: Seçkin Yayıncılık.

Durocher, M. (1994). Marketing Strategies for Recovery in Tourism Areas after Disasters. *Journal of Travel Research, 32*(2), 41-45.

Eser, H., Akşit, Z., & Kınay, A. (2006). Social Marketing: Theory and Practice. *Marketing Science Journal, 25*(3), 5-20.

Garcia, M., Smith, T., & Nguyen, H. (2021). The Effects of Inventory Management and Product Variety on Disaster Marketing for Businesses. *Journal of Product Research, 18*(4), 207-225.

Gürcan, F. (2015). Digital Divide in Türkiye and Worldwide. Expertise Thesis, Ministry of Development, Directorate of Information Society, Ankara.

James, J. (2005). *The global digital divide in the Internet: developed countries constructs and Third World realities. Journal of Information Science, 31* (2), 114 - 123.

Johnson, S. R., Connolly, D. J., & Carter, W. S. (2011). The Corporate Social Responsibility of Disaster. *Corporate Governance, 11*(4), 399-414.

Jones, A., & Brown, K. (2019). The Effects of Marketing Communication and Advertising Strategies on Disaster Marketing for Businesses. *Journal of Marketing Research, 25*(2), 112-129.

Jones, A., Brown, K., & Williams, S. (2020). AI-supported emergency response planning: A case study in disaster management. *Journal of Emergency Management, 15*(3), 102-115.

Karaarslan, M. H., & Altuntaş, B. (2023). Disaster Crowd Funding: A Model Proposal from a Marketing Perspective. *Meriç International Journal of Social and Strategic Research, 7*(Special Issue), 1-34. https://doi.org/10.54707/meric.1323032.

Kılıç, Ç. (2011). The Problem of Digital Divide in the Globalizing World. *Erzincan University Journal of Education Faculty.* Volume-Issue: 13-1, Pages: 81-91.

Korea Agency (2004). *How to measure the digital divide*, Korea Agency For Digital Opportunity and Promotion.

Kuo, Y., & (2011). Corporate Social Responsibility in Disasters. *Journal of Business Ethics, 100*(1), 85-101.

Limoncu, S., & Bayülgen, C. (2005). Housing Problems After Disasters in Türkiye, YTU Faculty of Architecture e-Journal, Volume 1. https://jag.journalagent.com/megaron/pdfs/MEGARON-97720-ARTICLE-LIMONCU.pdf.

Malik, A., Chowdhury, J., & Alam, S. (2023). The Relationship between Corporate Social Responsibility Activities and Share Values after Disasters. *Journal of Financial Reporting and Accounting, 21*(3), 1-19.

OECD (2001). *Understanding the Digital Divide*, Paris. https://doi.org/10.1787/236405667766.

Öztürk, M. & Demir, Y. (2023). Bilgilendirme ve Kaos Arasında: Afet Yönetiminde Medyanın Rolüne Yönelik Bibliyometrik Bir Analiz. *TRT Akademi, 8* (18), 506-527. doi: 10.37679/trta.1270615.

Partıgöç, N. S., & Tarhan, Ç. (2019). Disaster Management in the Digital Age: A Review from the Perspective of Digital Divide. *Resilience, 3*(2), 301-306. https://doi.org/10.32569/resilience.619192.

Robinson, L. (2020). The Effects of Pricing Strategies on Disaster Marketing for Businesses. *Journal of Pricing Research, 12*(1), 36-52.

Sayımer, İ. (2008). Public Relations in the Virtual Environment, 1st Edition, Beta Publications, Istanbul.

Smith, J. (2020). The Effects of Supply Chain and Supplier Relationships on Disaster Marketing for Businesses. *Journal of Disaster Research, 15*(3), 45-60.

Smith, J., & Johnson, R. (2018). Artificial intelligence for pre-disaster risk analysis: A review of current practices. *Disaster Management Journal, 10*(2), 45-58.

Smith, J., & Johnson, R. (2018). The Effects of Corporate Social Responsibility and Image Management on Disaster Marketing for Businesses. *Journal of Corporate Responsibility Research, 8*(3), 145-162.

Smith, J., & Johnson, R. (2020). Bridging the digital divide: Strategies for inclusive technology access. *Journal of Digital Inclusion*, *5*(2), 78-92.

Torlak, Ö. (2022). Social Marketing. In A. İlter (Ed.), Marketing Strategies: Design, Implementation, and Activities (p. 91). Ankara: Palme Publisher.

Uluğ, A. (2009). What Kind of Disaster Management? *İzmir Urban Symposium Proceedings Book.* Retrieved from https://www.tmmob.org.tr/sites/default/files/dc4876f3f08201c_ek.pdf.

URL 1 (2019). Social Media and Disasters. Retrieved from https://www.acilci.net/sosyal-medya-ve-afet/, Access date: February 1, 2019.

URL 2 (2019). Information Society. Retrieved from https://www.bilgitoplumu.gov.tr/, Access date: February 1, 2019.

Varol, N., & Kaya, Ç. M. (2018). Transdisciplinary Approach in Disaster Risk Management. *Disaster and Risk Journal, 1*(1), Pages: 1-8.van Dijk, J. (2006). *Digital divide research, achievements and shortcomings.* Poetics, Volume 34, Issues 4–5, Pages 221-235.

Wang, Z., & Ye, X. (2018). Social Media Analytics for Natural Disaster Management. *International Journal of Geographical Information Science, 32*(1), 49-72.

Xiao, Y., Huang, Q., & Wu, K. (2015). Understanding Social Media Data for Disaster Management. *Natural Hazards, 79*, 1663-1679.

Chapter 8

Risk Management in Disasters

Ahmet Gülsoy*, **PhD**
and Yusuf Uyan, PhD
Department of Medical Services and Techniques, Vocational School of Taşköprü, Kastamonu University, Kastamonu, Türkiye

Abstract

Although countries' disaster risks vary, losses due to disasters increase significantly and pose an obstacle to economic sustainability. To minimize the vulnerability of societies, it is necessary to recognize and evaluate the disaster risks they face and produce solutions to these risks. This is only possible with an effective disaster risk management strategy. Involving society in disaster risk management is critical to establishing an effective policy. Additionally, risk management should be implemented proactively not only before the disaster but also at every stage of the disaster management cycle. The use of technology to manage disaster risks offers great convenience today. Integrating technology into disaster risk management will make significant contributions (such as early warning systems) at every stage of the disaster management cycle (mitigation, preparation, response, and recovery).

Keywords: disaster management, disaster risk management, use of technology in risk management

---

* Corresponding Author's Email: agulsoy@kastamonu.edu.tr.

In: Disaster Management in the Digital Age and Emerging Technologies
Editors: Ayşe Asiltürk and Ayşegül Sarıoğlu Kemer
ISBN: 979-8-89530-213-2

Introduction

Disaster risk management is a policy implemented to reduce the losses that disasters can cause to people and the environment in the future. In this context, it includes activities such as identifying, analyzing, and measuring losses to minimize the possibility of loss and measures to prevent, reduce, mitigate, and transfer risks. Societies need sustainable disaster risk management actions to achieve sustainable development goals (Mora, 2009). Disaster losses increase significantly day by day and constitute an obstacle to the development of communities. For example, in 2023, more than 200 billion dollars were lost due to disasters in the world. Only approximately 10% of these losses are insured losses. The top three disasters that affected the most people in 2023 occurred in Indonesia, India, and Türkiye, and more than 38 million people were affected by these three disasters (EMDAT, 2024). It is inevitable that disasters with such a high damage potential threaten the security and welfare of societies and pose a significant obstacle to the development of communities. Therefore, developing an effective disaster risk management strategy is vital to cope with disasters and minimize risks. Disaster risk management is a comprehensive approach to identifying, assessing, mitigating, and preparing for various disasters.

Disaster Types and an Identification of Threats

Events such as earthquakes, tsunamis, floods, forest fires, hurricanes, droughts, and heat waves have occurred at different magnitudes in every period of history. These events are part of the "risk environment" people learn to manage and live with (Rosselló et al., 2020). It is possible to classify disasters that pose risks to communities differently. Disasters: These can be classified according to the type of disaster, the type of hazard, the rate of development, the scale size, or the cause of the event that caused the disaster (Gülsoy and Uyan, 2023). In this section, disasters will be classified according to hazard classes. When the definition of hazard by the United Nations Office for Disaster Risk Reduction (UNDRR) is examined, it is understood that hazard is both a natural process and a phenomenon and can arise due to human activities. In this case, it is possible to examine disasters in two classes, natural and human-caused, according to the origin of the danger that may cause a disaster (CRED, 2016; Shi, 2019). Natural and technological disasters cause physical death, injury, material damage, business interruption, or economic halt. Therefore, sound studies on risk and risk impacts are vital for effective policies and practical actions before, during, and after natural hazards or human-caused disasters (Shen et al., 2023).

Natural Disasters

Natural disasters are "a combination of natural hazards and vulnerabilities that endanger vulnerable communities that cannot withstand the adversities resulting from them" (Yu et al., 2018). A natural hazard is "the emergence of a natural situation or phenomenon that threatens or acts dangerously at a specific place and time" (Alcántara-Ayala, 2002). Natural hazards arise from internal (below the Earth's surface), external (topographic), weather-related (meteorological/hydrological), and biological events (Mohamed Shaluf, 2007) (Table 1).

Table 1. Classification of natural disasters*

Geophysical	Hydrological	Meteorological	Climatological	Biological	Extra-Terrestrial
Earthquake	Flood	Extreme temperature	Drought	Animal incident	Impact
Mass movement (dry)	Mass movement (wet)	Fog	Glacial lake outburst flood	Epidemic	Space weather
Volcanic activity	Wave action	Storm	Wildfire	İnfestation	

* EMDAT (2024).

Human-Made (Technological) Disasters

It is possible to examine disasters that occur as a result of human decisions in two groups. These include sociotechnical disasters and war disasters. Sociotechnical disasters occur in at least four types. These: 1. Technological disasters (factory and factory failures); 2. Transport errors; 3. Stadium or other "public place" malfunctions; and 4. Production failures (Mohamed Shaluf, 2007). Human-caused disasters are shown in Figure 1.

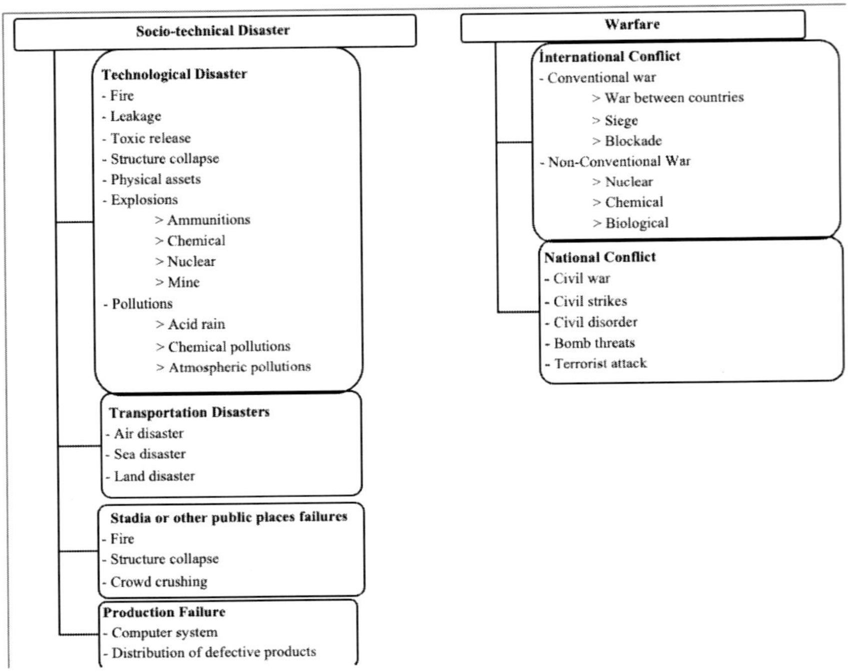

Figure 1. Classification of Human-Made Disasters.
Adapted from Mohamed Shaluf, 2007. (Disaster Tree, pp:699).

Risk Assessment

The risk assessment process is carried out in three steps. The first step in the risk assessment process is to identify the risk. It is not possible to manage something unknown. *Risk* can be defined as "the combination of the probability of an event and its negative consequences" (Papathoma-Köhle et al., 2016). Disaster risk is "the probability of losses that a certain hazard may cause to people, human settlements, and the natural environment in proportion to their damage or vulnerability if it occurs within a certain period in the future" (AFAD, 2014). As can be understood from the definition of disaster risk, much data is needed to define the risk. The most important is data about the event that can potentially cause a disaster (recurrence, size, spatial scope) and elements at risk (building inventory, inhabitants, values, etc.).

Regarding elements at risk, their vulnerability may have different dimensions (e.g., physical, economic, environmental, social, etc.). The extent of vulnerability considered is directly related to the purpose of the risk assessment and the nature of the elements at risk. Physical vulnerability is "the degree of loss of a specific item or set of items within the area affected by the hazard." It is expressed on a scale from 0 (no loss) to 1 (total loss) (Papathoma-Köhle et al., 2016). Calculation of disaster risk in a mathematically expressible way (AFAD, 2014);

Hazard (H) x Values (V) (elements that may be affected) x Vulnerability (V) (affect rate) = Disaster Risk (DR);

It can be formulated as DR=H x V x V

Disaster risk analysis is considered the essence of disaster management and control. Risk analysis is the primary research needed to prevent, reduce, and mitigate disaster risks (Jia et al., 2016). After the identification of the risk is made, the risk analysis studies are carried out in the second stage. Risk analysis is the most essential part of the risk assessment process in terms of time and effort and, simultaneously, the part where data is needed the most. Risk analysis is "the process of understanding the nature of risk and determining the level of risk." Risk analysis can be qualitative, where the risk is defined as high, medium, or low, or quantitative, where the risk is defined as a specific predefined risk metric (e.g., costs in €, number of potential victims, etc.) (Papathoma-Köhle et al., 2016). Today, there are various methods for natural disaster risk analysis. These are the probability and statistics method, the fuzzy mathematical method, the gray system theory method, the comentropy method, the qualitative analysis method, the geographic information system (GIS), modeling, the risk characterization method, ensemble forecast technology, and scenario analysis method (Sun et al., 2020).

Risk assessment usually begins with risk identification, continues with disaster risk analysis, and is performed simultaneously. Identified risks are classified into low, medium, and high categories. The acceptable risk level of society needs to be determined (Papathoma-Köhle et al., 2016). Acceptable risk is "the level of potential losses that a society or community considers acceptable given existing social, economic, political, cultural, technical, and environmental conditions" (UNISDR, 2009).

Emergency Preparedness and Response

Disaster preparedness is "the knowledge and capacity developed by governments, professional response and recovery organizations, communities, and individuals to effectively anticipate, respond to, and recover from disasters" (Granberg, 2013). The four stages of the disaster management cycle (mitigation, preparation, response, and recovery) are coordinated in disaster risk management. Studies to take measures to reduce the risk are carried out in the mitigation phase (Galloway, 2014). Developing the organizational capacity to respond to disasters and emergencies is very important. Preparation is critical to developing organizational capacity (Herstein et al., 2021). Disaster preparedness aims to establish the authorities and responsibilities of institutions and organizations operating in disasters and emergencies and to bring together the necessary resources for disaster response. Although mitigation efforts have been conducted, the hazard sensitivity must be addressed. In the response phase, when a disaster occurs, necessary efforts are carried out to reduce losses, save lives, protect property, and resume essential public services (Galloway, 2014).

One of the most essential components of disaster preparedness efforts is practical training. Because it significantly affects the perception of potential response teams regarding disaster preparedness (Hsu et al., 2013). Because the competence of personnel responding to disasters is gained through successful training programs, training programs need to be organized to improve disaster response teams' capacities. Adequately trained response teams provide good service and high operational efficiency (Jacquet et al., 2014).

Risk Reduction Strategies

Risk expresses uncertainty (i.e., something that may or may not occur) and implies the loss of some tangible or intangible asset. According to this idea, risk can be defined as a function between danger and vulnerability and can provide answers regarding the preparedness of an individual, community, or system (Morgan et al., 1990). Heijman mentioned that dangers are always a part of the world, and the important thing is to determine the risks that affect human life. He also emphasized the need for communities living in hazardous areas to create risk-reduction strategies using knowledge, skills, abilities, and technology (Heijmans, 2015).

Disaster Risk Management can generally be divided into two main approaches: Disaster Risk Reduction and Disaster Management. Disaster risk reduction is a proactive approach that aims to minimize the risks that may occur due to catastrophic events. It includes mitigation and preparedness measures based on the predictions from the data obtained. Therefore, it is essential to adopt proactive strategies to reduce future disaster risks and make society resilient (Mojtahedi and Oo, 2017). In this sense, the studies to be carried out are defined as all the structural and strategic studies that must be taken to prevent technological and human-induced hazards and environmental degradation from resulting in disasters and reduce their effects. These long-term studies require many institutions and organizations and a wide range of disciplines to work toward a specific goal.

Structural strategic studies are seen as the implementation of engineering studies (strengthening building stocks within innovative technological developments, infrastructure improvements, environmental regulations, etc.) to provide resistance and flexibility in physical structures or systems to reduce or prevent the possible effects of hazards. Hazards and risks must be identified when conducting structural works. Traditional methods such as field monitoring, physics-based models, expert research, and multi-criteria decision-making are applied to identify hazards and risk factors. Today, artificial intelligence techniques can quickly analyze large volumes of data to assess hazard risks promptly. There are extensive studies in which different methods of artificial intelligence are applied to develop susceptibility maps for different types of hazards (Sun et al., 2020). Non-structural strategic studies do not include engineering studies, such as legislative regulations, strengthening of non-governmental organizations, training, awareness-raising activities, and adopting awareness and behavioral patterns to reduce risks and their effects (Sun et al., 2012). Education and awareness-raising activities are considered the most critical social risk reduction strategies in modern integrated disaster management. In addition to this definition, land use, climate conditions and natural energy resources, migration movements, natural and cultural assets, industrial infrastructure, and demographic structure are considered subcategories of non-structural strategic studies (Bozkurt and Çiçekdağı, 2022).

Disaster Risk Financing and Insurance

Disasters destroy human populations, physical structures, economic assets, and fragile environments and restrict sustainable development (Thirawat et al., 2017). Uninsured disasters can devastate human welfare and economic growth, especially in underdeveloped countries (De Janvry, 2015). Insurance is one of the critical elements of disaster risk management, as it is financially sustainable and can influence behavioral changes toward disaster prevention

(Thirawat et al., 2017). Governments can use appropriate and financially efficient strategies to finance disaster losses. These;

- reserve fund (or ex-ante budget allocation);
- line of contingent credit;
- emergency ex-post budget reallocation;
- ex-post sovereign borrowing;
- insurance (or any other risk transfer instrument).

A reserve fund is a fund that the government sets aside at the beginning of its fiscal year and distributes only in case of a disaster. Line of contingent credit: These are pre-arranged loans that can be withdrawn in case of disaster. These are available from the World Bank, the IMF, development banks, multilateral financial institutions, national governments, and commercial lenders. Emergency ex-post budget reallocation: It is the rapid reallocation of budgets by the government after a disaster. For example, post-sovereign borrowing allows the state to finance expenditures by issuing additional loans after a disaster (Clarke et al., 2017). Insurance is a contractual transaction that guarantees financial protection against potentially significant losses in exchange for a premium. If the insured experiences a loss, the insurer pays a pre-agreed amount (Warner et al., 2009).

The Recovery and Rebuilding Process

The rescue and reconstruction process in disasters are studies carried out within the scope of the intervention and recovery phases of integrated disaster management, with the aim of limiting injuries, losses of life, and structural and environmental damage that may occur before, during, and immediately after the disaster and to meet the basic livelihood needs of the affected people. These studies begin with the understanding that a hazardous event is approaching and may continue for a long time, depending on the magnitude of the disaster. Rescue and recovery efforts are generally thought to begin immediately after the disaster occurs. Coordination and actual intervention can be initiated as risk reduction studies for the danger signs of disasters before the disaster occurs. For example, before volcanic eruptions occur, the extent of the danger is determined according to volcanic movements, and the evacuation of the people living in the region is carried out within the scope of rescue efforts. This work is also carried out as a prevention work before storms and floods. Before Hurricane Katrina hit the USA in 2005, preventive measures began to be taken according to the speed of the hurricane. As part of preventive measures, evacuation plans have been implemented for those living in the states in the Gulf. Voluntary or mandatory evacuations have been implemented for 1.2 million people along the Gulf Coast (Renne and Mayorga, 2022).

The first aim of rescue efforts is to save the lives of those affected by the disaster or those likely to be affected by the risk of danger. This phase, which includes search and rescue, emergency medical aid, and evacuation, can last for days or weeks, depending on the type, severity, and impact of the disaster. Rescue efforts are first initiated amateurishly by people in the disaster area and then continue by well-equipped professional teams. Perry (2004) emphasizes that the capabilities of rescue efforts should be improved under the unpredictable conditions of a disaster, and response personnel should be quick in the decision-making process (Perry, 2004, p. 72). Katoch (2006) stated that operational standards should be established for disasters, which is against Perry's emphasis on flexibility for rescue efforts (Katoch, 2006). Rescue operations consist of three stages: activation, implementation, and logistics. Activation phase: It is the collection of data as soon as possible after the emergence of danger signs (e.g., volcanic activity) or after a disaster occurs (e.g., earthquake) and converting them into information that will facilitate studies. Application phase: These operational studies follow the information received during the activation phase and field scans. The logistics phase covers the efforts to meet all the basic needs of both the disaster victims and the response teams in the region (Zibulewsky, 2001: 147).

Reconstruction process: The long-term implementation of actions and programs is needed to help individuals and communities return to normal after disasters. Studies carried out during this period are defined as assisting victims and their families, repairing institutions to sustain economic growth and security, rebuilding destroyed properties, and restructuring government operations and services (AFAD, 2014: 94; FEMA, 1996). The reconstruction aims to cover many main issues, such as communication, transportation, water, electricity, long-term shelter, education, and socialization efforts to return life to normal in the disaster area. Search and rescue efforts replace efforts to prepare people for social life (Coppola, 2015, p. 405).

Collaboration and Governance

Coordination and cooperation are vital in disasters and emergencies to save as many lives as possible and minimize disaster impacts (Bisri, 2016). After disasters, disaster victims can be at risk for illness, injury, and even death. For this reason, local, regional, national, or international cooperation may be needed immediately after disasters, depending on the magnitude of the disaster (Toyoda, T. 2016). Before disasters, cooperation and solidarity between countries can contribute to proactive disaster management. Furthermore, since countries' disaster risk potentials vary, they experience different types of disasters. Exchanging opinions and sharing experiences of experts in different types of disasters can enable the construction of more successful models that can be used in managing disaster risks. Both national and international cooperation reduce the effects of disasters and contribute to economic sustainability. Additionally, disasters may not be limited to one country. Thus, a disaster can affect more than one country. Therefore, international cooperation and communication for disaster risk reduction and mitigation can be considered a mandatory component of disaster risk management (Hagen et al., 2013).

Assessment and Learning

Disasters are extreme events that can shatter lives, devastate communities, and undo years of economic development in hours or even minutes. The frequency of disasters is increasing worldwide, and more people are at risk. Disaster assessments and reports must be conducted with transparency to reduce losses and damages in future events. Improving the quality of disaster assessments and strengthening accountability ensures social resilience (Buttenheim, 2010). After disasters occur, the disaster is scaled according to the potential size of damages and losses. With this scaling, impact evaluation analyses are carried out. Although crisis management studies are generally evaluated against the destructive power of disasters, capacity-increasing risk management studies provided by a resilient physical and social structure must be evaluated.

The conduct of disaster assessments is an internationally specialized field of study. Comprehensive studies show the necessity of creating an evaluation structure under three subheadings in disaster evaluations. First, it creates a coordination framework with three stages: before, during, and after an emergency or disaster. The previous or post-disaster phases will influence the future or pre-disaster phases. The second subheading is the creation of evaluation guides by expanding the three stages mentioned under the first heading. This guide creates successful examples by determining resilient areas and structures under the headings of danger, risk, and damage. Under the third subheading, the evaluation of coordination, governance, security, risk management, and capacity development is carried out within the scope of strengthening resilience (Wong et al., 2017).

Discussion and Conclusion

Today, there are significant technological developments that offer opportunities to reduce disasters. Innovative technological developments add a new dimension to disaster management by visualizing disaster risks with the ability to predict and analyze disasters (Orimoloye et al., 2021). For example, disaster risk maps have been one of the most essential tools for visualizing and reporting disasters in recent years. Regional maps that define places that may be negatively affected by natural disasters are called natural disaster risk maps (such as the Turkey Earthquake Risk Map) (Jia et al., 2016).

Geographic information systems (GIS) play a supporting role in disaster decision-making, especially in cases where data is scarce, and also provide the opportunity to ensure the participation of multiple stakeholders in disasters (Brandt et al., 2020). It is also a significant technological development in collecting critical information, such as monitoring losses and damages, search and rescue activities, and delivering humanitarian aid materials. Today, many geographic information systems are used effectively in disaster risk management. For example, EM-DAT, ReliefWeb, Desinventar Sendai, Hayat Eve Sığar (HES) Application (Dinçer, 2023).

Within the disaster risk management framework, many different technological developments are effectively used in the mitigation, preparation, response, and recovery stages, and their use is increasing rapidly. Other technologies used in disaster risk management include social media, mobile applications, drones and uncrewed aerial vehicles, solar panels, floating shelters, marine hospitals, portable telecommunications facilities, robotic systems, information and communication technologies, remote sensing systems, exoskeleton use, and data usage systems (Şahinöz, 2023).

Disclaimer

None.

References

AFAD, (2014). Açıklamalı Afet Yönetimi Terimleri Sözlüğü. Afet ve Acil Durum Yönetimi Başkanlığı: Ankara.

Alcántara-Ayala, I. (2002). Geomorphology, natural hazards, vulnerability and prevention of natural disasters in developing countries. *Geomorphology, 47*(2-4), 107-124. https://doi.org/10.1016/S0169-555X(02)00083-1.

Bisri, M. B. F. (2016). Comparative study on inter-organizational cooperation in disaster situations and impact on humanitarian aid operations. *Journal of International Humanitarian Action, 1*(1), 8.

Bozkurt, Ö., Çiçekdaği H. İ. (2022). İl Afet Risk Azaltma Planları (İRAP) Sonrası Yapılacak Risk Azaltma Yatırımlarında Best-Worst Metodu (BWM) ile Kriter Önceliklendirme. *Afet ve Risk Dergisi, 5*(1), 109-121.

Brandt, K., Graham, L., Hawthorne, T., Jeanty, J., Burkholder, B., Munisteri, C., & Visaggi, C. (2020). Integrating sketch mapping and hot spot analysis to enhance capacity for community-level flood and disaster risk management. *The Geographical Journal, 186*(2), 198-212. https://doi.org/10.1111/geoj.12330.

Buttenheim, A. (2010). Impact evaluation in the post-disaster setting: a case study of the 2005 Pakistan earthquake. *Journal of Development Effectiveness,* 2(2), 197-227.

Clarke, D. J., Mahul, O., Poulter, R., & Teh, T. L. (2017). Evaluating sovereign disaster risk finance strategies: a framework. *The Geneva Papers on Risk and Insurance-Issues and Practice, 42*, 565-584.

Coppola, D. P. (2015). The management of disasters. In D. P. Coppola (Ed.), Introduction to international disaster management (pp. 1-39). Butterworth-Heinemann. https://doi.org/10.1016/B978-0-12-801477-6.00001-0.

CRED (2016). Annual Disaster Statistical Review 2015.

De Janvry, A. (2015). Quantifying through ex-post assessments the micro-level impacts of sovereign disaster risk financing and insurance programs. World Bank Policy Research Working Paper, (7356).

Dinçer, S. (2023). Afetlerde coğrafi bilgi sistemleri, Afet Yönetimi ve Teknoloji, İçinde (176-188), İstanbul Tıp Kitabevi, İstanbul.

EM-DAT, (2024), https://public.emdat.be/data, Accessed: 15.04.2024.

FEMA, (1996). Guide for All Hazards Emergency Operations Planning (SLG, 1996, p. 101), p. 1–3) Retrieved from https://www.fema.gov/pdf/plan/slg101.pdf.

Galloway, G. E. (2014). Emergency preparedness and response: Lessons learned from 9/11. In *The geographical dimensions of terrorism* (pp. 27-34). Routledge.

Granberg, T. A. (2013). Preparedness measures for emergency and disaster response. In *Humanitarian and Relief Logistics: Research Issues, Case Studies, and Future Trends* (pp. 59-75). New York, NY: Springer New York.

Gülsoy, A. & Uyan, Y. (2023), Afetlere genel bakış, Afet Yönetimi ve Teknoloji, İçinde (1-19), İstanbul Tıp Kitabevi, İstanbul.

Hagen, J. C., Chhetri, M. B. P., & Steiner, N. (2013, May). Promoting cooperation and coordination in international disaster management. In *2013 International Conference on Collaboration Technologies and Systems (CTS)* (pp. 268-272). IEEE.

Heijmans, A. (2015). From vulnerability to empowerment. In *Disaster Prevention* (pp. 14–29). Routledge.

Herstein, J. J., Schwedhelm, M. M., Vasa, A., Biddinger, P. D., & Hewlett, A. L. (2021). Emergency preparedness: What is the future? *Antimicrobial Stewardship & Healthcare Epidemiology, 1*(1), e29.

Hsu, E. B., Li, Y., Bayram, J. D., Levinson, D., Yang, S., & Monahan, C. (2013). State of virtual reality-based disaster preparedness and response training. *PLoS Currents, 5.*

Jacquet, G. A., Obi, C. C., Chang, M. P., & Bayram, J. D. (2014). Availability and diversity of training programs for responders to international disasters and complex humanitarian emergencies. *PLoS Currents, 6.*

Jia, H., Pan, D., Wang, J. A., & Zhang, W. C. (2016). Risk mapping of integrated natural disasters in China. *Natural Hazards, 80*, 2023-2035.

Katoch, A. (2006). The responders' cauldron: The uniqueness of international disaster response. *Journal of International Affairs, 59*(2), 153–172. Retrieved from http://jia.sipa.columbia.edu/.

Mohamed Shaluf, I. (2007). An overview on disasters. Disaster Prevention and Management: *An International Journal, 16*(5), 687-703.

Mojtahedi, M., & Oo, B. L. (2017). Critical attributes for proactive engagement of stakeholders in disaster risk management. *International Journal of Disaster Risk Reduction, 21*, 35-43.

Mora, S. (2009). Disasters are not natural: risk management, a tool for development. Geological Society, London, *Engineering Geology Special Publications, 22*(1), 101-112. https://doi.org/10.1144/EGSP22.7.

Morgan, M. G., Henrion, M., & Small, M. (1990). *Uncertainty: A guide to dealing with uncertainty in quantitative risk and policy analysis*. Cambridge University Press.

Orimoloye, I. R., Ekundayo, T. C., Ololade, O. O., & Belle, J. A. (2021). Systematic mapping of disaster risk management research and the role of innovative technology. *Environmental Science and Pollution Research, 28*, 4289-4306.

Papathoma-Köhle, M., Promper, C., & Glade, T. (2016). A common methodology for risk assessment and mapping of climate change-related hazards—implications for climate change adaptation policies. *Climate, 4*(1), 8.

Perry, R. W. (2004). Disaster exercise outcomes for professional emergency personnel and citizen volunteers. *Journal of Contingencies & Crisis Management, 12*(2), 64-75. DOI: 10.1111/j.0966-0879.2004.00436.x.

Renne, J. L., & Mayorga, E. (2022). What has America learned Since Hurricane Katrina? Evaluating evacuation plans for carless and vulnerable populations in 50 large cities across the United States. *International journal of disaster risk reduction, 80*, 103226.

Rosselló, J., Becken, S. ve Santana-Gallego, M. (2020). The Effects of Natural Disasters on International Tourism: A Global Analysis. *Tourism Management, 79*, 104080. https://doi.org/10.1016/j.tourman.2020.104080.

Shen, G., Zhou, L., Xue, X., & Zhou, Y. (2023). The risk impacts of global natural and technological disasters. *Socio-Economic Planning Sciences, 88*, 101653.

Shi, P., (2019). Hazards, disasters, and risks. *Disaster Risk Science,* 1-48. Springer, Singapore.

Sun, D., Zhang, D., & Cheng, X. (2012). Framework of national non-structural measures for flash flood disaster prevention in China. *Water, 4*(1), 272-282.

Sun, R., Gao, G., Gong, Z., & Wu, J. (2020). A review of risk analysis methods for natural disasters. *Natural Hazards, 100*(2), 571-593.

Sun, W., Bocchini, P., & Davison, B. D. (2020). Applications of artificial intelligence for disaster management. *Natural Hazards, 103*(3), 2631-2689.

Şahinöz, S. (2023), Afet yönetimi ve teknoloji,1. Baskı, İstanbul Tıp Kitabevi, İstanbul.

Thirawat, N., Udompol, S., & Ponjan, P. (2017). Disaster risk reduction and international catastrophe risk insurance facility. *Mitigation and Adaptation Strategies for Global Change, 22*, 1021-1039.

Toyoda, T. (2016). The framework of international cooperation for disaster management and Japan's contribution. In *Asian Law in Disasters* (pp. 315-328). Routledge.

UNISDR (2009), The United Nations International Strategy for Disaster Reduction, Geneva, Switzerland, May 2009.

Warner, K., Ranger, N., Surminski, S., Arnold, M., Linnerooth-Bayer, J., Michel-Kerjan, E.,Kovacs P. & Herweijer, C. (2009). Adaptation to climate change: Linking disaster risk reduction and insurance. United Nations International Strategy for Disaster Reduction, Geneva.

Wong, D. F., Spencer, C., Boyd, L., Burkle, F. M., & Archer, F. (2017). Disaster metrics: a comprehensive framework for disaster evaluation typologies. *Prehospital and Disaster Medicine, 32*(5), 501–514.

Yu, M., Yang, C., & Li, Y. (2018). Big data in natural disaster management: a review. *Geosciences, 8*(5), 165. https://doi.org/10.3390/geosciences8050165.

Zibulewsky, J. (2001). Defining Disaster: The Emergency Department Perspective, Baylor University Medical Center *Proceedings, 14* (2), 144–149, doi: 10.1080/08998280.2001.11927751.

Chapter 9

Technologically Supported Educational Arrangements in Developing Disaster Awareness in Early Childhood Education

Tuğba Kontaş Azaklı[1,*] PhD
and Gülşah Günşen[2], PhD
[1]Department of Preschool Education, Education Faculty, Ordu University, Ordu, Türkiye
[2]Department of Preschool Education, Education Faculty, Trakya University, Edirne, Türkiye

Abstract

In recent years, the frequency of natural disasters has increased, affecting all people, especially the most vulnerable groups. Among these vulnerable and defenseless groups, preschool children are particularly affected by natural disasters, including death and injuries, in a lasting way. Natural disasters have short- and long-term effects on preschool children. Educators and caregivers play an important role in reducing the impact of natural disasters on preschool children. These individuals can provide children with information about natural disasters, teach them coping skills, and support them. Additionally, it is important to develop disaster preparedness and response plans in schools. These plans should be designed to ensure the safety of children and help maintain their emotional well-being. Digital technology, which is becoming increasingly widespread today, can be used to protect children from the effects of disasters. It can be used for "information and education about natural disasters" before disasters and for "coping skills training," "support and communication," "assisting with normalization," and

---

* Corresponding Author's Email: tugbakontas@odu.edu.tr.

In: Disaster Management in the Digital Age and Emerging Technologies
Editors: Ayşe Asiltürk and Ayşegül Sarıoğlu Kemer
ISBN: 979-8-89530-213-2

"continuation of remote learning processes" during possible disaster situations to support children's well-being.

Keywords: early childhood education, disaster risk reduction, digital technology

Introduction

In recent years, the observed increase in the number and severity of natural disasters has been causing serious concern. Since 2000, there has been a seven-fold increase compared to the 1980s, with an average of 560 disasters occurring annually. This situation has resulted in over $2.5 trillion in economic damage and the loss of 1.2 million lives. Natural disasters significantly weaken the recovery capacity of communities by causing major damage to infrastructure and food security. Especially vulnerable groups are disproportionately affected by these disasters (UNDRR, 2023).

The research results clearly show that women and children are the most affected by natural disasters (Save the Children, 2023; Winangsih and Kurniati, 2019). This situation indicates that these groups are more vulnerable to the negative effects of disasters and need more support. Preschool children, who are among the vulnerable and defenseless groups, are significantly affected by natural disasters, primarily in terms of death and injuries.

Children are faced with drought, hurricanes, earthquakes, landslides, forest fires, and other natural disasters (Dyregrov, Yule, and Olff, 2018). Natural disasters have short- and long-term effects on preschool children. Because children are disproportionately affected by such events, understanding the impact of disasters on child development is very important in terms of providing the necessary support (Corrarino, 2008).

Natural disasters negatively affect the well-being of children and leave lasting marks on them. The extent of the negative impact of natural disasters on children is influenced by factors such as the level of destruction, proximity to the disaster, subjective response during the event, and loss of loved ones (Furr et al., 2010). It is stated that disasters can cause stunting in children and affect their nutritional status (Gaire et al., 2016), symptoms such as unwillingness to go to school, fear, anxiety, lack of self-confidence, regression in age, sleep terror can be seen after disasters (Limoncu and Atmaca, 2018), children may have post-traumatic stress disorder (PTSD), behavioral and emotional problems and functional impairments after a disaster (North et al.,

2018), negative life events, including stress caused by natural disasters, can make children more vulnerable to mental health problems such as depression, anxiety, and behavioral problems (Kingston et al., 2019), and it is stated that disasters can leave long-term traumatic effects on children; affecting their mental, physical, and social health (Kinoshita and Woolley, 2015).

The early childhood period is critical for initiating disaster preparedness education. Research emphasizes that children, who are future adults, can be the pioneers of actions to reduce disaster risk in the societies they live in (Inten et al., 2021; Midtbust, Dyregrovand, and Djup, 2018). However, it is thought that disaster education is not only related to reducing immediate risks but also to minimizing long-term psychological effects such as trauma and anxiety that can affect children's overall development (Widyaningrum, Nurhadiyanti, and Widana, 2020).

The concept of Disaster Risk Reduction (DRR), which gained momentum with the International Decade for Natural Disaster Reduction (IDNDR) in the 1990s by the United Nations, aims to raise awareness and promote international cooperation in managing disaster risks. DRR is a multidimensional field that requires cooperation among various sectors, including education, science, technology, and community participation, to effectively reduce disaster risks and increase resilience at the local, national, and global levels. The DRR is critically important for saving lives, accelerating recovery, reducing vulnerability, and managing disaster risks for children and women (UNICEF, 2013). The Sendai Framework for Disaster Risk Reduction 2015-2030, adopted by the UN, serves as a global plan to significantly reduce disaster risk and losses in the coming decades. The Sendai Framework for Disaster Risk Reduction (SFDRR) was developed to renew the global commitment to increasing people's resilience, health, and well-being and was adopted at the World Conference on Disaster Risk Reduction in 2015 after various negotiations and meetings (Aitsi-Selmi et al., 2015).

Ensuring that children are prepared for disasters is a critical aspect of Disaster Risk Reduction (DRR). Therefore, it is emphasized that the education sector plays a significant role in the effectiveness of DRR (Seddighi et al., 2021). In different regions of the world, disaster coping methods have been developed for early childhood. Research has shown that Disaster Risk Reduction (DRR) programs play a significant role in providing earthquake education to children from an early age, enabling them to acquire the necessary knowledge to cope with earthquakes (Izadkhah and Heshmati, 2007). Furthermore, it has been stated that investments in DRR programs will minimize the vulnerability of children and women by strengthening resilience against disasters (UNICEF, 2013). It has been revealed that the participation of small children in disaster preparedness programs can significantly increase their resilience and intervention capabilities against natural disasters, thereby potentially saving lives by developing a basic understanding and intervention strategies in disaster scenarios (Hakim, Inten, and Mulyani, 2020). Moreover, it has been proven that they can significantly increase their understanding and preparation for coping with disasters and also learn appropriate responses that could potentially save their lives (Kurniah, Agustriana, and Kusumah, 2023). It has been determined that the disaster risk reduction program (DRR) prepared for children in early childhood contributes to the creation of a suitable and safe learning environment for children and that there is a significant increase in the knowledge of children participating in the program compared to non-participants, including safe behaviors and basic intervention strategies during a disaster (Proulx and Aboud, 2019).

Research emphasizes the integration of disaster education into standard curriculums for early childhood education (Widyaningrum, Nurhadiyanti, and Widana, 2020), stating that there is a significant need for early childhood educators, especially in disaster-prone areas, to develop their information literacy skills (Oktaria, Putra, Windah and Haerudini, 2023).

Research on children's disaster education indicates a remarkable gap in converting knowledge into real preparedness actions, even if children's awareness and knowledge levels increase on this subject, and it is unclear whether these effects are continuous and long-term. It is stated that more research is needed to understand how changes in knowledge and attitudes can translate into behavior changes in real-world disaster contexts (Midtbust, Dyregrovand, and Djup, 2018; Ronan et al., 2015).

Materials and training approaches used in disaster risk reduction training processes play a significant role in preparing teachers and students for disasters. Research shows that active learning methods and participatory approaches are effective in training (Gökmenoğlu et al., 2021). One of the methods that teachers can benefit from in the training process is the use of digital technologies. Research shows that educators initially use traditional print media and some digital media to gather information about disasters, but they need more comprehensive training on digital literacy to expand their ability to effectively use online resources (Oktaria, Putra, Windah, and Haerudini, 2023). It is stated that the use of video games instead of traditional methods in Disaster Risk Reduction Education could be effective, and it is expressed in the literature that this is insufficient (Midtbust, Dyregrovand, and Djup, 2018). In the studies conducted, it is seen that digital games on the subject of disaster are mostly two-dimensional, the curriculum is not adhered to, physical simulations close to reality are used limitedly, and the interaction dimension is not used much (Çoban, 2017).

Use of Digital Technologies in Disaster Education

The effective use of digital technology in the process of developing disaster awareness in preschool children can raise children's awareness of disasters. In this context, digital technologies can be used for education and information before disasters, for communication and support during disasters, and to support normalization after disasters. The use of digital technologies in disaster education is shown in Table 1.

Table 1. The use of digital technologies in disaster education

Pre-Disaster Education and Information	Digital Games and Applications	Touch-screen Games: Touch-screen technology is ideal for games and applications where children can interact with simple touch and slide movements. These games can teach basic information about what to do in disaster situations in a fun way. At this point, children can develop digital games and applications that will allow them to learn basic information about disasters in a fun way. These games can include topics such as finding a fire exit and what to do during a tremor. For example: Wordwall application, My Earthquake Plan, Ready for Anything, Disaster Hero, AFAD Child, etc. available on Playstore and Appstore.
	Educational Videos and Animations	Informative animations and videos about disasters and necessary precautions can be prepared and presented to children. These contents should be designed by experts in a way that will attract children's attention, and a simple, understandable language should be used. Audiobooks and Stories: Audiobooks and stories are effective tools for explaining disaster awareness topics to children who cannot read or write. Children can learn by listening and interacting with voice commands.

Table 1. (Continued)

	Virtual Reality Experiences (VR)	Realistic Simulations: VR provides children with the opportunity to experience disaster scenarios in a safe virtual environment. For example, finding shelter safely during an earthquake or finding escape routes from home during a fire. These simulations help children develop practical skills that can be applied in real-life in disaster situations. Educational Games: Educational games designed in a VR environment allow children to have fun while learning about disaster awareness topics. Scientific Discoveries: VR allows children to go on virtual field trips, for example, exploring the inside of a volcano or the center of a storm. These kinds of experiences provide the opportunity to discover the mechanisms of natural disasters and explore scientific concepts.
	Augmented Reality Experiences (AR)	Interactive Learning Materials: AR, by adding digital information to the physical world, allows children to explore and learn about the world around them. For example, using a tablet or smartphone, children can scan the QR code found on a book page and view 3D images or animations related to the disaster described on that page. This helps to make the learned information tangible and better understood. Informative Stories and Images: Books and images enriched with AR technology provide an interactive way for children to understand important information about disaster preparedness. For example, when children scan book pages with their tablets or smartphones, they can view the inside of a volcano or the layers of the Earth in three dimensions, which will help them understand a volcanic eruption.
	Experience with Artificial Intelligence (AI)	Personalized Learning Experiences: Generative AI can provide educational content tailored to children's learning styles and needs. This allows for the development of materials suitable to children's interests and learning speeds in disaster education. Question-Answer Systems and Chatbots: Chatbots powered by Generative AI can answer children's questions about disasters in real-time. This allows children to receive immediate feedback on topics they are curious about, making the learning process more interactive. Behavior Analysis and Safety Training: Generative AI technology can analyze children's behavior during disaster simulations and provide specific recommendations for better preparation. This analysis provides important insights into how children can act more effectively in disaster situations.

		Real-time feedback: Artificial intelligence can provide real-time feedback on how children respond to disaster scenarios, helping them correct their mistakes.
Communication and Support During a Disaster	Emergency Applications	Mobile applications and programs on smartphones and wearable technologies specially designed for emergencies can help families stay in touch with their children. Safe meeting points can be determined through the application, and emergency information can be shared. For example: AFAD Emergency Call, I'm Safe, tracking applications, etc.
	Digital Communication Tools	Digital communication tools such as SMS, social media platforms, and email can be used to inform during and after a disaster. Especially during a disaster, quick information can be provided about safe places and what needs to be done. For example, many people trapped under the rubble in the earthquake that occurred in Türkiye on February 6, 2023, were rescued using digital communication technologies.
Helping to Normalization	Psychological Support Applications	After a disaster, digital platforms and applications can be activated to reduce the trauma experienced by children. These applications can be designed to provide professional psychological support and help children express their feelings.
	Online Education Platforms	In cases where physical school environments become unusable in disaster situations, children's education can continue over online platforms. These platforms can be used to present educational content before, during, and after the disaster.
	Digitalization of Educational Materials	Digital versions of textbooks and other educational materials provide easy access during distance education processes.

* Prepared by the authors.

The use of digital technologies increases children's disaster awareness while simultaneously making the learning process more enjoyable and interactive. These technologies allow disaster education to be personalized according to the children's ages and learning styles, thereby contributing to raising more prepared and conscious individuals against disasters. The integration of digital technologies into the education process allows children to explore, question, and understand abstract scientific concepts during the pre-disaster process, while also helping to develop their critical thinking and problem-solving skills during and after a disaster (Bulut, 2020; Maulida, et al., 2023; Mitsuhara et al., 2017; Solfiah, et al., 2020; Türkoğlu, 2023).

Technology-Supported Disaster Training for Families

Digital technologies can be used in various ways for the survival, health tracking, care, and education of preschool-aged children in the post-disaster process. Some of these could be as follows (Bulut, 2020; Merdeka & Alamsyah, 2023; Setiadi, & Frederika, 2022).

Reaching Families and Finding Lost Children

Artificial Intelligence-Supported Recognition Systems

Artificial intelligence-supported face recognition and voice recognition systems can help find missing children and family members more quickly. These systems can make matches by analyzing photographs and voice recordings of missing individuals in post-disaster recovery efforts and temporary shelters.

Augmented Reality Applications

AR applications can help people in disaster areas by providing guiding signs to find each other or reach safe gathering points. Additionally, it can enable the visual presentation of critical information, such as the last-seen locations of lost children.

Tracking and Care of Children's Health

Smart Health Monitoring

Smart bracelets and wearable devices can be used to monitor children's health status. These devices can track children's body temperature, heart rate, and other vital signs and can send alerts to parents or healthcare providers if necessary. This is especially crucial for children with special needs.

Virtual Reality Training

VR technology can provide practical training for parents and caregivers on topics such as post-disaster childcare, first aid, and psychological support. This helps parents acquire the necessary skills to better care for their children after a disaster.

It is believed that the effective use of these technologies can greatly help children and families overcome the challenges they face in the post-disaster process.

A digital game proposal that can be developed to instill earthquake awareness in preschool children and that they can learn by playing at school or with their parents is presented in Table 2.

Table 2. Suggestion for a digital game that can be developed with technological support to raise awareness of disasters in preschool children

Game Name	Little Earth Scientists
The Concept of the Game:	An interactive mobile game that allows children aged 4-6 to discover how earthquakes occur through Augmented Reality (AR) and Artificial Intelligence (AI) technologies could be developed. The game could provide an educational and fun experience that teaches children the causes of earthquakes and ways to stay safe during natural disasters.
AR Visualizations:	Children can create a virtual model of tectonic plates on the floor of their homes or classrooms using their tablets or smartphones. AR technology can show the movements of tectonic plates and how they lead to earthquakes with real-time 3D visualizations. In addition, scientific videos and animations that support the learning process can be offered within the game to be developed. These videos can help children to concretize by explaining the scientific reasons for earthquakes, the structure of the earth, and how tectonic plates move in simple terms.
AI Supported Interaction:	Artificial intelligence can answer children's questions within the game and guide them to the correct information. AI personalizes the learning process for children and can make recommendations based on each child's interests and learning pace. Providing Interactive Game Support: In addition to learning about how earthquakes occur, children can move tectonic plates themselves, create earthquake simulations, and see the results of these simulations. These games can help children better understand the concept of earthquakes. Safety Training Module: The game can include a section that teaches ways to stay safe during an earthquake. Children can have the opportunity to learn the correct actions to take during an earthquake and find safe places in a virtual house.
Pedagogical Approach:	The game adopts an approach that triggers the curiosity of children and provides them with practical scientific knowledge. "Little Earth Scientists" can include elements such as storytelling, task completion, and reward winning to enhance children's learning experiences. The game can allow children to learn at their own pace and provide support for the development of scientific thinking skills.

* Prepared by the authors.

Discussion and Conclusion

Implementing disaster awareness and education programs for preschool children who cannot read and write requires the use of visual and auditory materials, as well as interactive and sensory learning techniques. Using digital technologies effectively for children in this age group can attract their attention, support the learning process, and provide support for the concretization of scientific events that remain quite abstract to them. Furthermore, it is very important not to drift away from real-life experiences with technology for children in this period. At this point, children should be supported to learn by experiencing disaster situations through role-play games. The preparation of educational toys and activity kits related to disaster preparation allows children to physically interact and learn, in addition to technology. Furthermore, for children who cannot read and write and have special needs, sign language and universal symbols can be used to explain basic concepts related to disasters. For example, distinct symbols and signs can be used for concepts such as "exit" or "safe area."

Each of these methods can be integrated with technology as part of programs aimed at increasing the awareness of disasters in preschool children.

Disclaimer

None.

References

Aitsi-Selmi, A., Egawa, S., Sasaki, H., Wannous, C., & Murray, V. (2015). The sendai framework for disaster risk reduction: renewing the global commitment to people's resilience, health, and well-being. *International Journal of Disaster Risk Science, 6*, 164-176. 10.1007/s13753-015-0050-9.

Bulut, A. (2020). Raising awareness of disaster and giving disaster education to children in preschool education period. *Acta Educationis Generalis, 10*(2), 162-179.https://doi.org/10.2478/atd-2020-0016.

Corrarino, J. E. (2008). Disaster-related mental health needs of women and children. *McN: The American Journal of Maternal/Child Nursing, 33*(4), 242-248. https://doi.org/10.1097/01.nmc.0000326079.26870.e3.

Çoban, M. (2017). *Effects of Earthquake Education via Three-Dimensional Earthquake Game on Primary School Students' Academic Achievement, Motivation, and Views"* PhD diss., University of Atatürk.

Dyregrov, A., Yule, W., & Olff, M. (2018). Children and natural disasters. *European Journal of Psychotraumatology, 9* (sup2). https://doi.org/10.1080/20008198.2018.1500823.

Furr, J. M., Comer, J. S., Edmunds, J., & Kendall, P. C. (2010). Disasters and youth: a meta-analytic examination of posttraumatic stress." *Journal of Consulting and Clinical Psychology, 78* (6), 765-780. https://doi.org/10.1037/a0021482.

Gaire, S., Delbiso, T. D., Pandey, S. K., & Guha-Sapir, D. (2016). Impact of disasters on child stunting in Nepal. *Risk Management and Healthcare Policy, 113.* https://doi.org/10.2147/rmhp.s101124.

Gökmenoğlu, T., Sonmez, E. D., & Yavuz, İ. (2021). A cascade teacher training model in disaster education: A case of Turkey. *International Journal of Curriculum and Instructional Studies, 11*(1), 23-42. https://doi.org/10.31704/ijocis.2021.002.

Hakim, A., Inten, D. N., & Mulyani, D. (2020). The Literation of Disaster Mitigation for Early Childhood. *In 2nd Social and Humaniora Research Symposium (SoRes 2019)* (pp. 287-290). Atlantis Press.10.2991/assehr.k.200225.060.

Inten, D. N., Mulyani, D., Khambali, K., Tiwi, D., & Lichandra, F. (2022). Disaster Mitigation Learning Using Katumbiri Model in Early Childhood Education. *In 4th Social and Humanities Research Symposium (SoRes 2021)* (pp. 320-325). Atlantis Press.10.2991/assehr.k.220407.065.

Izadkhah, Y. O., & Heshmati, V. (2007). Applicable Methods in Teaching Earthquakes to Preschool Children. *In Proceedings of fifth international conference on seismology and earthquake engineering*, Iran. https://www.unisdr.org/files/10882_10882SEE5KINDERGARTEN1.pdf.

Kurniah, N., Agustriana, N., & Kusumah, R. G. T. (2023). Applied Science for Early Childhood: Disaster Mitigation in Bengkulu City Kindergarten. *In Mathematics and Science Education International Seminar 2021 (MASEIS 2021)* (pp. 225-232). Atlantis Press.10.2991/978-2-38476-012-1_30.

Kingston, D., Mughal, M. K., Arshad, M., Kovalchuk, I., Metz, G. A. S., Wynne-Edwards, K. E., … & Arnold, P. (2019). Prediction and understanding of resilience in albertan families: longitudinal study of disaster responses (purls) – protocol. *Frontiers in Psychiatry, 10.* https://doi.org/10.3389/fpsyt.2019.00729.

Kinoshita, I. and Woolley, H. (2015). Children's play environment after a disaster: the great east Japan earthquake. *Children, 2*(1), 39-62. https://doi.org/10.3390/children2010039.

Limoncu, S., & Atmaca, A. B. (2018). Child-Centered Disaster Management. *Megaron, 13*(1). 10.5505/megaron.2017.49369.

Maulida, A., Hanif, H., Kamal, M., & Oktari, R. S. (2023). Roblox-Based Tsunami Survival Game: A Tool to Stimulate Early Childhood Disaster Preparedness Skills. *In E3S Web of Conferences (Vol. 447, p. 02003).* EDP Sciences.

Merdeka, P. H., & Alamsyah, D. (2023). Enhancing sustainability in coastal areas: ımproving early childhood education and integrating disaster mitigation in spatial planning. *Indonesian Journal of Earth and Atmospheric Sciences, 1*(1), 20-29.https://journals.eduped.org/index.php/ijeas/article/view/555.

Midtbust, L. G. H., Dyregrov, A., & Djup, H. W. (2018). Communicating with children and adolescents about the risk of natural disasters. *European Journal of Psychotraumatology, 9* (sup2). https://doi.org/10.1080/20008198.2018.1429771.

Mitsuhara, H., Iguchi, K., & Shishibori, M. (2017). Using digital game, augmented reality, and head mounted displays for ımmediate-action commander training. *International Journal of Emerging Technologies in Learning (iJET), 12* (02), 101–117. https://doi.org/10.3991/ijet.v12i02.6303.

North, C. S., Mendoza, S., Simic, Z., & Pfefferbaum, B. (2018). Parent-reported behavioral and emotional responses of children to disaster and parental psychopathology. *Journal of Loss and Trauma, 23*(4), 303-316. https://doi.org/ 10.1080/15325024.2018.1443710.

Oktaria, R., Putra, P., Windah, A., & Haerudin, N. (2023). Needs Analysis of Disaster Mitigation Learning Design Based on Information Literacy in Efforts to Increase Disaster Self-awareness of Early Childhood." *In 3rd Universitas Lampung International Conference on Social Sciences (ULICoSS 2022)* (pp. 816-826). Atlantis Press.10.2991/978-2-38476-046-6_78.

Proulx, K., & Aboud, F. (2019). Disaster Risk Reduction in Early Childhood Education: Effects on Preschool Quality and Child Outcomes. *International Journal of Educational Development, 66,* 1-7. https://doi.org/10.1016/j.ijedudev.2019.01.007.

Ronan, K. R., Alisic, E., Towers, B. et al. (2015). Disaster preparedness for children and families: a critical review. *Curr Psychiatry Rep, 17,* 58. https://doi.org/ 10.1007/s11920-015-0589-6.

Save The Children. (2023). *Japon- One Year On: Save The Children's Response And Recovery Program.* https://resourcecentre.savethechildren.net/pdf/5747.pdf/.

Seddighi, H., Yousefzadeh, S., López, M. L., Sajjadi, H., Vameghi, M., Rafiey, H., … & Khankeh, H. (2021). Disaster risk reduction in ıranian primary and secondary school textbooks: a content analysis. *Disaster Medicine and Public Health Preparedness, 16*(4), 1503-1511. https://doi.org/10.1017/dmp.2021.161.

Setiadi, R., & Frederika, R. (2022). Family financial planning for disaster preparedness: a case study of North Semarang, Indonesia. *International Journal of Disaster Risk Reduction, 82*, 103332. https://doi.org/10.1016/j.ijdrr.2022.103332.

Solfiah, Y. S., Risma, D., & Kurnia, R. (2020). Early childhood disaster management media through picture story books. *Jurnal Pendidikan Usia Dini, 14*(1), 141-155.https://doi.org/10.21009/141.10.

Türkoğlu, B. (2023). Preschool teachers' opinions about disaster education in the preschool period. *Sakarya University Journal of Education, 13* (4) (Special Issue-Disaster Education and Education in Disaster Regions), 759-784.https://doi.org/10.19126/suje.1384145.

UNICEF. (2013). *EU-UNICEF Child Rights Toolkit.* https://www.unicef.org/media/press-centre.

United Nations Office for Disaster Risk Reduction. (2023). *Global Assessment Report On Disaster Risk Reduction: Special Report 2023, Mapping Resilience For The*

Sustainable Development Goals. UN.https://www.undrr.org/gar/gar2023-special-report.

Widyaningrum, N., Nurhadiyanti, A. P., & Widana, I. D. K. K. (2020). The Importance of Disaster Knowledge in Early Childhood for the Future Investment." *In Proceeding International Conference on Science and Engineering* (Vol. 3, pp. 605-609). https://doi.org/10.14421/icse.v3.573.

Winangsih, I., & Kurniati, E. (2020). Disaster Mitigation in Early Childhood Education." *In International Conference on Early Childhood Education and Parenting 2019 (ECEP 2019)* (pp. 296-301). Atlantis Press.0.2991/assehr.k.200808.058 oy.

Chapter 10

Civil Defense and Security in Disasters

Selin Bitirim Okmeydan*
Department of Public Relations and Publicity, Faculty of Communication, Ege University, Bornova, İzmir, Türkiye

Abstract

Civil defense and security in disasters are an extremely important and priority area for countries. Disaster management aims to protect societies and states from the destructive effects of disasters, to overcome disasters with minimal damage, and to be prepared for disasters before they occur. Investments in civil defense and security in disaster management include protecting society; mitigating damages; timely emergency response; effective coordination; and the sustainability of state affairs before, during, and after disasters. To ensure effective civil defense and security in disasters, new information technologies are crucial in disaster preparedness, mitigating damages, improving response processes, and coordinating rapid recovery after disasters. Based on the literature review, this chapter focuses on current examples of digital transformation in civil defense and security applications in disasters and its benefits in disaster management. In the future, disaster management is expected to consist of more integrated practices in the field of civil defense and security. Therefore, digital technologies, active participation, global cooperation, competitive strategies, innovation, and sustainability are expected to play vital roles in the field of civil defense and security in disaster management.

Keywords: civil defense, disaster, disaster management, information technologies, security

* Corresponding Author's Email: selin.bitirim.okmeydan@ege.edu.tr.

In: Disaster Management in the Digital Age and Emerging Technologies
Editors: Ayşe Asiltürk and Ayşegül Sarıoğlu Kemer
ISBN: 979-8-89530-213-2

Introduction

Civil defense and security in disasters have a direct impact on the existence of countries. Accordingly, almost every country in the world makes various investments to improve civil defense and security as part of their efforts to prepare for disasters. Natural or man-made disasters can threaten the lives of all living things, especially human beings, and have a profound impact on the environment and the economy. As a result, the independence of countries could be at serious risk. According to Neerunjun (2023), taking civil defense and security measures based on new technologies and determining proactive strategies before disasters occur play an important role for people and countries. Although disaster management aims to protect societies and states from the devastation of disasters before they occur, it also aims to be prepared for the foreseeable consequences and to reduce damage in cases where disasters cannot be prevented. Accordingly, disaster management consists of three stages, pre-disaster, during disaster, and post-disaster.

All three phases of disaster management are closely related to new information technologies that support civil defense and security measures. Current practices based on digital transformation that stand out in the field of civil defense and security in disasters can be grouped under eight headings (Chang, 2017; Damaševicius et al., 2023; Due, 2023; Ray et al., 2017; Llovido and Palaoag, 2020; Sakurai and Murayama, 2019; Tomaszewski, 2021; Yu et al., 2018), Modeling and Early-and-Advanced Warning Systems, Mobile Applications and Interactive Platforms, Geographic Information, and Satellite Systems, Emergency Management Systems and Decision Support Systems, Unmanned Aerial Vehicles and Remote Sensing, Big Data and Artificial Intelligence Analysis, Robotic-Autonomous Technologies and Rapid Prototyping, Social Media and Virtual Education Systems. This classification reveals the areas of new technologies used in civil defense and security and highlights the benefits of digitalization in disaster management processes.

The Scope of Civil Defense and Security in Disasters

Disasters have negative impacts on people, nature, and society. For this reason, the concept of disaster management has been developed to prevent disasters before they occur and to mitigate their negative effects with rapid interventions in cases where prevention is not possible. Disaster management is defined as the coordinated effective actions, decisions, and policies aimed at reducing the effects of disasters before, during, and after the disaster. These three phases are part of integrated disaster management, which consists of risk mitigation, preparedness, response, and recovery. This approach sheds light on a new risk management-centered strategy (Değirmenci and Yılmaz, 2023).

Disaster management, which has become increasingly important globally, follows a standardized process for all countries with global cooperation supported by international civil defense and security strategies. Civil defense and security encompass a range of unarmed measures, policies, and initiatives aimed at minimizing the loss of life and property during natural disasters in peacetime and addressing enemy attacks during wartime (Jordá, 2011; Nasar et al., 2023).

The scope of civil defense and security in disaster management encompasses the improvements necessary for the state's official and private facilities and organizations to continue their duties, the adoption and support of defense efforts by large masses of people, the mobilization of society by directing the civilian population toward correct behaviors, and various practices to maintain the morale and motivation of the public (Nkombi and Wentink, 2022). However, currently, the scope of civil defense and security in disasters is gradually expanding due to developments in information technologies resulting from digitalization.

Digital Foundations of Disaster Management: New Technology-Based Applications in Civil Defense and Security

In disaster management, new information technologies have an important share in preparing society for disasters and increasing disaster resilience. With the increasing importance of disaster management globally, the 2015-2030 Sendai Framework for Disaster Risk Reduction, adopted at the United Nations World Conference in 2015, emphasizes the importance of integrating information technologies into the disaster management process and supports risk management that leverages new technologies (Kelman, 2015). There is an aim to invest in information technologies to add resilience to civil defense and security in disasters and to predict and reduce risks (Aydın and Genç, 2023).

The contemporary paradigm in disaster management, focusing on civil defense and security, encourages the use of information technologies at every stage of disaster management for a sustainable society. Furthermore, the global artificial intelligence market has grown significantly in recent years, especially with the increasing adoption of AI-based solutions in civil defense and security (Yugandhara, 2022). In the age of digital transformation that we live in, artificial intelligence is mentioned in many different fields such as information technologies, health, entertainment, disaster management, and the defense industry. Additionally, artificial intelligence, which has become the common ground of different disciplines, facilitates human life in every field (Okmeydan, 2022).

The industrial transformation shaping artificial intelligence is a socio-technological phenomenon, referred to as Industry 5.0. This new stage in industrial transformation presents a humanized vision of technology and defines the future. Industry 5.0 is advancing under the leadership of artificial intelligence, emphasizing human-centricity, anticipating risks, and taking proactive action, and highlighting a resilient digital ecosystem against unexpected situations and conditions (Barata and Keyser, 2023). Therefore, emerging artificial intelligence-based technologies under the Industry 5.0 axis have great potential to build resilient and sustainable communities against disasters.

However, weakened or damaged digital systems in disasters leave governments open and vulnerable to cybersecurity threats. In such cases, artificial intelligence applications developed specifically for cyber-attacks make it possible to secure bank accounts, hospital records, medical data, personal information on computers and mobile devices, as well as information on public and commercial organizations during disasters (Adıgüzel, 2022). Furthermore, investments in digital technologies and new applications improve and strengthen every stage of the disaster management process.

Civil Defense and Security-Oriented Digital Transformations in Disaster Management: Current Examples

The current understanding of disaster management emphasizes information technologies based on artificial intelligence that support civil defense and security policies. Developing day by day due to global competition, technologies like artificial intelligence, cloud computing, deep learning, machine learning, sensors, remote sensing, the Internet of Things, big data, virtual and augmented reality, robotics, collaborative robots (co-bots), cyber security, and network technology are being developed. These technologies contribute to web-based digital technologies and bring new dimensions to disaster management (Acar and Çağdaş, 2019; Memiş and Babaoğlu, 2020). Civil defense and security practices, combined with innovative information technologies in disaster management, have become more effective and functional within the digital ecosystem.

Information technologies developed under the leadership of artificial intelligence aim to address inadequacies and deficiencies resulting from the fragility of human resources and operationalize disaster management practices. The digital technologies used in disaster management enhance the effectiveness of disaster response by implementing new applications in the field of civil defense and security. These technologies, which support civil defense and security in disaster management, can be classified as follows based on their current usage areas and purposes (Chang, 2017; Damaševicius et al., 2023; Due, 2023; Ray et al., 2017; Llovido and Palaoag, 2020; Sakurai and Murayama, 2019; Tomaszewski, 2021; Yu et al., 2018).

Modeling and Early-and-Advanced Warning Systems

Modeling based on real-time data analytics is used in disaster management, particularly in areas such as earthquakes, floods, forest fires, volcanic eruptions, landslides, and climate change-related weather events, where early and advanced warning systems are crucial. These applications support the risk prevention and mitigation, response and coordination, and recovery and support phases of disaster management (Ray et al., 2017). Using modeling based on data collected and analyzed with sensors and the Internet of Things, disaster risks can be identified in advance, potential damages predicted, necessary precautions taken promptly, resources allocated during and after a disaster, and rapid responses provided. Many countries employ early and advanced warning systems, sensors, and modeling to detect tsunamis, seismic activity, climate data, and hurricanes early on (STM ThinkTech, 2018). Disaster modeling is conducted using data collected from sensors via the Internet of Things, allowing a focus on the pre-disaster period, which involves particularly the pre-detection of risks and preparation against possible hazards (Damaševicius et al., 2023).

The alarm system used by the Federal Emergency Management Agency (FEMA) in the US, known as the Integrated Public Alert & Warning System (IPAWS), serves early warning purposes. This system promptly transmits life-saving information, including weather updates, climate change alerts, and storm tracking data, to mobile phones, televisions, radios, and NOAA (National Oceanic Atmospheric Administration) Weather Radios, effectively informing and warning the public about imminent disasters (Sadiq et al., 2023). Early and advanced warning systems, along with modeling, play a crucial role in safeguarding lives and properties by providing timely alerts to individuals regarding potential disasters. These systems form the cornerstone of civil defense and security by emphasizing pre-disaster preparedness.

Mobile Applications and Interactive Platforms

User-friendly disaster tracking applications on mobile phones with simple interfaces offer low-cost and rapid solutions that support civil defense and security in disaster management. Applications such as the Centers for Disease Control & Prevention (CDC), My Hurricane Tracker, the National Oceanic Atmospheric Administration (NOAA), and QuakeFeed, available for download from the Apple App Store, quickly deliver emergency warnings and real-time data to authorities while providing essential information on disaster preparedness (STM ThinkTech, 2018). In Japan, the satellite-based J-Alert system operated by the national government sends instant alerts to mobile phones nationwide for earthquakes, tsunamis, ballistic missile attacks, acts of terrorism, and other emergencies. This practice alerts the public to imminent dangers and enables appropriate measures, including evacuation, to be taken based on the nature of the disaster (Sakaki and Lukner, 2013).

In recent years, mobile applications designed as disaster training simulation games have gained popularity in disaster management. Tokyo authorities, for instance, have developed a mobile device-based training game called "Nige-Tore" (Personal Tsunami Evacuation Drill) to enhance disaster preparedness among citizens (Yamori and Sugiyama, 2020). Similarly, the e-LAHOK (Legazpi Alisto Handang Opinyon Kaligtasan) Disaster Risk Reduction and Management Platform was created for the Philippines, a country prone to disasters due to its geographical location. This platform encourages civil defense and participatory governance by gathering disaster risk reduction strategies and incident reports from the local population (Llovido and Palaoag, 2020). These interactive platforms, which facilitate active and online learning through information technologies, promote participation and support civil defense and security practices in disaster management.

Geographic Information and Satellite Systems

Geographic information systems (GIS) often rely on satellite data processing output. GIS processes and updates geographical data from various sources, including satellite systems, to create visual output such as versatile maps, tables, and graphics (Arca, 2012). The use of current data processed by GIS enables the creation of maps, land condition assessment, environmental change monitoring, real-time risk identification, natural disaster prediction, and risk monitoring. GIS supports analysis, planning, timely data management, situational awareness, field operations, and forecasting for floods, droughts, landslides, and man-made disasters such as terrorism and road/traffic accidents (Tomaszewski, 2021).

In the USA, NASA monitors, tracks, and manages disasters around the world using satellite systems. NASA processes daily data and images from the MODIS (Moderate Resolution Imaging Spectroradiometer) satellite and integrates them into GIS. GIS is widely used to monitor natural disasters and assess damage on a global scale. Similarly, the Japan Meteorological Agency (JMA) processes data from the next-generation satellite system Himawari to update GIS data and assess disaster risks (Zhao et al., 2020). These systems are developed to prepare for natural and man-made disasters and respond quickly to emergencies, thereby supporting disaster management. They play a crucial role in helping local and national governments strengthen civil defense and ensure public safety.

Emergency Management Systems and Decision Support Systems

Emergency management systems (EMS) and decision support systems (DSS) collect, process, and analyze large amounts of data generated before, during, and after disasters. They play an important role in disaster preparedness, mitigation, response, and recovery processes by providing accurate information to managers for strategic planning. Emergency calls sent by FEMA (Federal Emergency Management Agency) and EMI (Emergency Management Institute) in the USA to disaster management personnel through integrated digital platforms or applications that provide resource management, personnel monitoring, and disaster coordination are included in the EMS (Jordá, 2011).

California has developed an emergency system called FIRESCOPE specifically designed to combat forest fires and respond to various disasters. FIRESCOPE aims to establish a rapid emergency response system not only for wildfires but also for all types of natural and man-made disasters (Chang, 2017). By integrating artificial intelligence technologies into data analysis, more efficient and effective decisions can be made, contributing to improved search and rescue operations and overall disaster management (Nasar et al., 2023). Sahana, developed by programmers in Sri Lanka after the 2004 Indian Ocean earthquake and tsunami, operates as an information-sharing system for humanitarian aid during disasters and has been used in evacuation operations. Using free and open-source software, the Sahana information system demonstrated increased effectiveness during the 2010 Haiti earthquake and subsequent disasters (Sakurai and Murayama, 2019). The availability of situational information before, during, and after disasters supports civil defense and security measures through the use of information technologies.

Unmanned Aerial Vehicles (UAVs) and Remote Sensing

UAVs are used to detect damage in disaster areas and support rescue operations. UAVs and remote sensing technologies accelerate reconstruction processes by providing analysis, mapping, and imaging in the post-disaster damage assessment. With these technologies, it is possible to conduct activities at points that had to be stopped or were inaccessible during the disaster (Vermiglio et al., 2021). Remote sensing and monitoring technologies, such as UAVs and sensor network systems, also help reduce disaster risk. A sensor network system installed on a large bridge and road infrastructure can report on the state of the fracture and the durability of the structure (Sakurai and Murayama, 2019). Remote sensing technologies improve the ability to perform real-time situational analysis and thus become effective in preparing for and preventing disasters. With data from remote sensing systems, it is possible to detect three-dimensional terrain models, along with topographic and geological maps, in real-time. This enables the observation, mapping, and updating of all these features (Arca, 2012).

Autonomous technologies, such as UAVs, support rapid reconnaissance and rescue operations in disaster areas. These systems collect data from dangerous disaster areas that are unsafe for humans to enter and deliver aid to affected areas. In Türkiye, Bayraktar AKINCI and TB2 UAVs flew more than 2000 hours of missions after the February 6, 2023, Kahramanmaraş earthquakes. In these flights, which provided uninterrupted instant data and images, the security of disaster areas was maintained by conducting rapid mapping operations for damage assessment, security, search and rescue, and coordination support (STM ThinkTech, 2023). Such new technologies play an important role in managing the preparedness, mitigation, response, and recovery phases of the disaster management cycle.

Big Data and Artificial Intelligence Analysis

These technologies, including big data, deep learning, machine learning, and artificial intelligence, are utilized in every stage of natural disasters such as earthquakes, floods, tsunamis, fires, gas explosions, and hurricanes. Big data plays a significant role, especially in risk detection, the creation of early and advanced warning systems, and the analysis of communication before, during, and after disasters (Yu et al., 2018). Alibaba, based in China, has integrated big data and artificial intelligence into the 'City Brain Project', using cloud infrastructure and data analytics in China. Initially developed in 2016 to address traffic issues in Hangzhou, one of the largest cities in China, the app has also enhanced emergency management within the city. The system infrastructure, updated within the scope of the City Brain Project, sends instant automatic alerts to smartphones and emphasizes smart disaster management (Caprotti and Liu, 2022).

Big data and artificial intelligence analytics handle complex real-time datasets from various sources such as cloud computing, the Internet of Things, sensors, telecommunication networks, and social media. This enables better and more accurate decision-making before, during, and after a disaster (Yu et al., 2018). They also provide significant support for post-disaster logistics operations. Adıgüzel (2022) noted that logistics activities were disrupted due to road damage caused by the Wenchuan earthquake in China in 2008. The activation of the Artificial Emergency-Logistics-Planning System was highlighted, using artificial intelligence technology to manage emergencies and expedite aid and rescue efforts.

Robotic-Autonomous Technologies and Rapid Prototyping

Robots and robotic technology are used for post-disaster emergency response and search and rescue missions. An example is EMILY (Emergency Integrated Lifesaving Lanyard), a robotic system designed by the US company Hydronalix. EMILY can safely navigate through dangerous or restricted areas to assist emergency responders and search and rescue dogs in reaching disaster victims. It can transport victims to safe areas and facilitate the evacuation process (STM ThinkTech, 2023). In the aftermath of the 2011 earthquake and tsunami in Japan, which caused a leak at the Fukushima Daiichi nuclear power plant, specialized robots were used to operate in radioactive environments. These included the double-armed heavy machine-type robot ASTACO-SoRa, the shape-shifting robot PMORPH, and the submersible robot Gengo-ROV/Tri-diver. These flexible and autonomous robotic technologies can operate in high-radiation areas to scan and inspect nuclear leak sites, even underwater, where living creatures cannot enter (Okada et al., 2020). Similarly, Robot Spot, developed by Boston Dynamics, features various sensors and 360-degree video cameras, along with advanced artificial intelligence algorithms. This robot-dog (Robodog) technology autonomously coordinates micro-movements and is designed for research, and search and rescue purposes (Due, 2023).

3D printing technology fulfills the need for rapid prototyping in disaster situations. This technology, commonly used in industrial design, manufacturing, and engineering, can rapidly produce essential materials for post-disaster recovery, construct temporary shelters for disaster victims, manufacture fundamental necessities, such as clothing, and urgently needed medical supplies in disaster areas, repair damaged infrastructure, and reconstruct residential areas (Espíndola and Beltagui, 2018). Due to its fast and flexible production capability, 3D printing has the potential to functionally assist disaster victims and expedite the recovery of disaster areas through rapid prototyping.

Social Media and Virtual Education Systems

Social media manages the crisis and supports social interaction with fast and two-way communication after natural disasters (Demirel and Durgeç, 2022). Social media platforms such as Twitter (X), Flickr, and YouTube are increasingly used in disaster management by both civil society organizations and government agencies. It is possible to determine the feelings and reactions of the public toward the disaster through social media, to ensure multi-directional communication and information flow through online platforms, and to make disaster response, reaction, and recovery efforts efficient through crisis crowdsourcing (Yu et al., 2018). Hurricane Sandy, recorded in the Atlantic Ocean in 2012, resulted in hundreds of deaths, thousands of people left homeless, and nearly 50 billion dollars in damage. Millions of tweets during this hurricane, during prolonged power outages and transportation restrictions, provided information about the current situation. Through these tweets, it was possible to reach disaster victims and coordinate aid (STM ThinkTech, 2018).

Furthermore, virtual reality (VR) can serve as a visual training system to enhance civil defense and security before disasters. The VR system demonstrates how individuals should respond in a disaster and how to safely evacuate from hazardous environments. Training in emergency procedures through VR provides individuals with a simulated experience in disaster management (Sakurai and Murayama, 2019). Various reality technologies are encompassed within the broader category of extended reality (XR). XR technology offers innovative solutions for disaster education within communities, providing a realistic training tool for experiencing real disasters. Promoting awareness and educating the public about disasters are crucial aspects of effective disaster management (Khanal et al., 2022). This technology not only informs communities about disaster risks but also contributes to building a disaster-resilient society, supporting civil defense and security prior to disasters.

The Future of Civil Defense and Security in Disaster Management: Emerging Trends, Insights, and Sustainability

The new information technologies in disaster management have a direct impact on the sustainability of disaster management. At the same time, sustainable disaster management is also the key to creating a disaster-resilient society. New trends and insights that bring civil defense and security to the forefront in disaster management can be listed as follows (Acar and Çağdaş, 2019; Jordá, 2011; Memiş and Babaoğlu, 2020; Neerunjun, 2023; Walton et al., 2021).

Multi-Threat Approach and Sustainability

Future disaster management is expected to adopt a more holistic approach to multiple threats rather than focusing on a single threat. Different types of disasters, such as the climate crisis, pandemics, and cyber-attacks, threaten the physical and biological health of individuals and lead to devastating impacts on society (Walton et al., 2021). However, sustainable disaster management based on a multi-threat approach has the effect of increasing the resilience of societies (Kusumastuti et al., 2014). These interventions aim to reduce the impacts of natural and man-made disasters, bringing sustainability to civil defense and security in disaster management.

Community-Based Preparedness and Disaster Resilience

In the future, focusing on community-based disaster management, raising awareness within communities about disasters, and strengthening social solidarity will enable more effective responses to disasters (Neerunjun, 2023). This approach increases society's capacity to deal with disasters by promoting community participation in disaster management (Nkombi and Wentink, 2022). Strengthening society's resilience to disaster management makes it more sensitive to civil defense and ensures a safer life.

Local- and National-Level Integration

The relationship between disasters and development must consider local and national integration. There is a growing tendency to promote the participation of local actors in future disaster management (Memiş and Babaoğlu, 2020).

Ensuring local and national integration in disaster management is crucial to prepare for, respond to, and recover from disasters (Jordá, 2011). This integration also plays a significant role in supporting civil defense and security policies.

Communication, Technology, and Information Management

In the future, information systems and information-sharing platforms are expected to become more integrated and user-friendly (Acar and Çağdaş, 2019). There is also an expectation of expanded access to accurate information and technology-based disaster education, impacting civil defense and security in disaster management (Khanal et al., 2022). At this point, there is a trend for artificial intelligence-supported applications to strengthen integration and social communication in all stages of disaster management.

Discussion and Conclusion

In this chapter, the role of new information technologies in the civil defense and security process is discussed in the context of disasters. Disasters pose a significant threat to humanity, and evolving national and international civil defense and security practices worldwide are being shaped by technology (Neerunjun, 2023). The integration of information technologies into every stage of the disaster management process, from preparation to mitigation, response, and recovery, aims to protect against the devastating consequences of disasters.

In disasters, new information technologies are used to prevent their occurrence, reduce their effects when prevention is not possible, provide fast and effective responses for civil defense teams, and strengthen security measures in disaster management (Tomaszewski, 2021). Research shows that effective civil defense and security measures greatly reduce the impact of disasters on people (Sakurai and Murayama, 2019). As the civil defense and security dimensions in disaster management evolve with new technologies, artificial intelligence-supported applications stand out. At this point, especially with artificial intelligence-supported data analysis, the potential effects of disasters can be predicted in advance, allowing civil defense teams to prepare accordingly (Vermiglio et al., 2021).

Shedding light on how digital transformation is taking place in civil defense and security in disaster management helps societies and governments better prepare for disasters. Additionally, digital transformation leads to more effective interventions before, during, and improves recovery processes after disasters. The development driven by new technologies and digitalization enhances the overall effectiveness of disaster management. Consequently, steps can be taken with the help of emerging technologies to strengthen civil defense and security in disaster management, develop effective strategies in this field, enhance preparedness before disasters, mitigate damage during disasters, expedite recovery from disaster damage, and respond more effectively to future disasters.

Disclaimer

None.

References

Acar, M., and Çağdaş, D. Y. (2019). Endüstri devriminin ışığında "Afet 4.0." E. S. B. Meydanoğlu (Ed.), in *Dijital Dönüşüm Trendleri* (pp. 231-257). İstanbul, Filiz Kitabevi.

Adıgüzel, S. (2022). Afet durumlarında yapay zekâ teknolojisi ile lojistik yönetimi örnekleri. *Akademik İzdüşüm, 7*(1), 47-70.

Arca, D. (2012). Afet yönetiminde coğrafi bilgi sistemi ve uzaktan algılama. *Karaelmas Fen ve Mühendislik Dergisi, 2*(2), 53-61.

Aydın, G., and Genç, F. N. (2023). İklim değişikliğine bağlı afetler ve sürdürülebilir kalkınma. *Çankırı Karatekin Üniversitesi İİBF Dergisi*, 13(2), 563-590.

Babaoğlu, C. (2017). Afet yönetimi politikalarına sivil toplum örgütlerinin katılımı sorunsalı. Ö. Önder and M. Yaman (Eds.), in *Afet Yönetimi* (pp. 163-170). Bursa, Ekin Yayınevi.

Barata, J., and Kayser, I. (2023). Industry 5.0. *Procedia Computer Science, 219*, 778-788.

Caprotti, F., and Liu, D. (2022). Platform urbanism and the Chinese smart city, The co-production and territorialisation of Hangzhou City Brain. *Geo Journal, 87*,1559-1573.

Chang, R. (2017). A literature review and analysis of the incident command system. *International Journal of Emergency Management, 13*(1), 50-67.

Damaševicius, R., Bacanin, N., and Misra, S. (2023). From sensors to safety, Internet of emergency services (IoES) for emergency response and disaster management. *Journal of Sensor and Actuator Networks, 12*(41), 1-45.

Değirmenci, F., and Yılmaz, D. V. (2023). Afet öncesinde, sırasında ve sonrasında kadın sağlığı hemşiresinin rol ve sorumlulukları. N. Ulutaşdemir and M. K. Aykut (Eds.), in *Afetlerde Kadın ve Çocuk Sağlığı* (pp. 5-20). Ankara, İKSAD.

Demirel, S. D., and Durgeç, P. (2022). Dijital medyada deprem haberlerini infografikler üzerinden okumak, Anadolu Ajansı ve TRT Haber örneği. *Journal of Communication, Sociology and History Studies, 2*(2), 27-44.

Due, B. L. (2023). A walk in the park with robodog, navigating around pedestrians using a Spot Robot as a "Guide Dog." *Space and Culture, 1,* 1-17.

Espíndola, O. R., and A. Beltagui. (2018). "Can 3D printing address operations challenges in disaster management?," in *Proceedings Book of 25th Euroma Conference,* pp. 1-10.

Jordá, A. (2011). Emergency and civil protection, Curriculum design for 30 hours of ınstructional time. *Journal of Homeland Security and Emergency Management, 8*(2),1-15.

Kelman, I. (2015). Climate change and The Sendai Framework for disaster risk reduction. *International Journal of Disaster Risk Science, 6*(2), 117-127.

Khanal S., Medasetti, U. S., Mashal, M., Savage, B., and Khadka, R. (2022). Virtual and augmented reality in the disaster management technology, A literature review of the past 11 years. *Frontiers in Virtual Reality,* (3), 1-22.

Kusumastuti, R. D., Husodo, Z. A., Suardi, L., and Danarsari, D. N. (2014). Developing a resilience index towards natural disasters in Indonesia. *International Journal of Disaster Risk Reduction,*10, 327-340.

Llovido, J. L., and Palaoag, T. D. (2020). e-LAHOK, An e-participatory platform for disaster risk reduction and management. *Materials Science and Engineering,* 803, 1-8.

Memiş, L. and Babaoğlu, C. (2020). Afet yönetimi ve teknoloji. M. Yaman and E. Çakır (Eds.), in *Farklı Boyutlarıyla Afet Yönetimi* (pp. 163-178). Ankara, Nobel.

Nasar, W., Torres, D. R., Gundersen, O. E., and Karlsen, A. T. (2023). The use of decision support in search and rescue, A systematic literature review. *International Journal of Geo-Information, 12*(182), 1-31.

Neerunjun, H. H. (2023). Constructing a composite disaster resilience index towards natural disasters in Mauritius. *IJEGEO, 10*(3), 1-14.

Nkombi, Z., and Wentink, G. J. (2022). The role of public participation in disaster risk reduction initiatives, The case of Katlehong Township. *Journal of Disaster Risk Studies,14*(1), 1-12.

Okada, S., Hirano, K., Kobayashi, R., and Kometani, Y. (2020). Development and application of robotics for decommissioning of Fukushima Daiichi Nuclear Power Plant. *Hitachi Review, 69*(4), 148-155.

Okmeydan, S. B. (2022). Yapay zekâdan öte evrene (metaverse) yeni teknolojilerin halkla ilişkilere yansımaları. K. Ağın (Ed.), in *Sosyal, Beşerî ve İdari Bilimler Alanında Uluslararası Araştırmalar-XVIII* (pp. 811-838). Konya, Eğitim Yayınevi.

Ray, P. P., Mukherjee, M., and Shu, A. L. (2017). Internet of Things for disaster management, State-of-the-art and prospects. *IEEE Access, 5,* 18818-18835.

Sadiq, A. A., Okhai, R., Tyler, J., and Entress, R. (2023). Public alert and warning system literature review in the USA, Identifying research gaps and lessons for practice. *Natural Hazards, 117,* 1711-1744.

Sakaki, A., and Lukner, K. (2013). Japan's crisis management amid growing complexity, In search of new approaches. *Japanese Journal of Political Science, 14*(2), 155-176.

Sakurai, M., and Murayama, Y. (2019). Information technologies and disaster management-Benefits and issues. *Progress in Disaster Science, 2,* 1-4.
STM ThinkTech (2018). *Afet yönetiminde teknoloji.* Ankara, Teknolojik Düşünce Merkezi.
STM ThinkTech (2023). *Kahramanmaraş merkezli depremler odağında doğal afetlerde teknoloji kullanımı.* Ankara, Teknolojik Düşünce Merkezi.
Tomaszewski, B. (2021). *Geographic information systems (GIS) for disaster management.* UK, Routledge.
Vermiglio, C., Noto, G., Bolivar, M. P. R., and Zarone, V. (2021). Disaster management and emerging technologies, A performance-based perspective. *MEDAR,* 1-26.
Walton, A. A., Marr, J., Cahillane, M. J., and Bush, K. (2021). Building community resilience to disasters. *Sustainability, 13,* 1-31.
Yamori, K., and Sugiyama, T. (2020). Development and social implementation of smartphone app Nige-Tore for improving tsunami evacuation drills. *International Journal of Disaster Risk Science, 11*(6), 751-761.
Yu, M., Yang, C., and Li, Y. (2018). Big data in natural disaster management, A review. *Geosciences, 8*(5), 1-26.
Yugandhara, R. Y. (2022). *Commercial AI market trends and growth.* India, SNS Insider.
Zhao, L., Zhao, C., Wang, Y., Wang, Y., Yang, Y. (2020). Evaluation of cloud microphysical properties derived from MODIS and Himawari-8 using in situ aircraft measurements over the Southern Ocean. *Earth and Space Science, 7,* 1-18.

Chapter 11

Social Work in Disasters: Working with Vulnerable Groups and the Use of Technology

Betül Çolak*, **PhD**
and Dilek Özsoy, PhD
Department of Social Work, Faculty of Economics, Administrative and Social Sciences, İstanbul Nişantaşı University, İstanbul, Türkiye

Abstract

Disasters represent significant life occurrences capable of exerting substantial negative impacts on individuals' lives. Mastering effective disaster management is crucial to minimizing the adverse impacts of disasters on individuals and nations, meeting needs, and implementing preventive measures. Social work interventions play an essential role in carrying significant responsibilities and duties within this process. Disaster-affected individuals need support at various stages, including search and rescue, meeting basic needs, and rebuilding their lives, with social workers actively involved. Disasters create chaos and confusion, particularly impacting vulnerable groups with special needs, including those with chronic illnesses, disabilities, the elderly, and children requiring care. This book chapter will encompass insights and suggestions regarding the duties and roles of social work, post-disaster support for special needs groups, and the use of technology in these efforts.

---

* Corresponding Author's Email: betul.colak@nisantasi.edu.tr.

In: Disaster Management in the Digital Age and Emerging Technologies
Editors: Ayşe Asiltürk and Ayşegül Sarıoğlu Kemer
ISBN: 979-8-89530-213-2

Keywords: social work in disasters, social workers' roles in disasters, vulnerable groups in disasters

Introduction

Disaster generally refers to the aftermath of natural, technological, or human-induced events causing physical, economic, social, and environmental losses, disrupting normal life and activities in communities, and surpassing the community's ability to cope with its own resources. For an event to be labeled a disaster, it requires losses in human lives, exceeding local resources, and disrupting human activities (Ergünay, 2009).

Disaster management is one of the key concepts of disaster policies in the processes of preventing disasters and minimizing damages. In the disaster management process, the pre-disaster includes prevention, risk reduction, and preparedness during a disaster, personal behavior and protection issues, and post-disaster, intervention, and reconstruction stages are included (Tercan, 2018).

Türkiye frequently faces natural disasters due to its geographical, tectonic, and climatic structure. Among these disasters, earthquakes are the most important. Türkiye ranks third in the world in terms of human casualties and eighth in terms of the number of people affected by earthquakes (AFAD, 2023). In the last 99 years before the August 17, 1999 earthquake, 146 damaging earthquakes were recorded. It is estimated that 65,882 lives were lost, 128,000 injured, and 540,000 buildings were destroyed or damaged in these earthquakes (TBMM, Parliamentary Research Commission Report). Between 1900 and 2023, 269 earthquakes have caused loss of life and economic damage. The top three earthquakes in terms of casualties and economic damage are the 2023 Kahramanmaraş earthquake, the 1939 Erzincan earthquake, and the 1999 Marmara earthquake. According to the "Türkiye Earthquake Recovery and Reconstruction Assessment" (TERRA), the impact of the Kahramanmaraş earthquake is equivalent to USD 103.6 billion, or 9% of Türkiye's estimated GDP for 2023 (Strategy and Budget Office, 2023).

Social Work

The International Federation of Social Workers (IFSW), updated in 2014, defines social work as "a practice-based profession and academic discipline that promotes social change and development, social cohesion, empowerment, and liberation of people." In this definition, the principles of social justice, human rights, collective responsibility, and respect for diversity are at the center of social work. The definition includes all those affected by disasters at the national or regional level regarding life challenges. The inherent vulnerability and need resulting from the structure of disasters prominently elevates the discipline of social work. In this regard, social work plays a vital role in mitigating the adverse impacts of disasters through pre-, during-, and post-disaster social work interventions within disaster management.

Türkiye's seismic activity highlights earthquake-related disasters' significance in social service practices. In spite of the calculability of the material damage among the negative consequences of the physical destruction of the earthquake, it is not possible to quantify the destruction experienced on a humanitarian level.

The 1992 Erzincan earthquake was a turning point in terms of pioneering many innovations in disaster management practice. The 1992 earthquake is notable for pioneering the investigation of social and psychological issues alongside the physical destruction caused by the event. In this context, the social and psychological impacts of disasters and earthquakes were first studied under a scientific program. The integration of scientific methodologies in disaster management, coupled with university partnerships with authorized institutions, signifies a crucial development highlighting the significance of scientific research (Şengün and Güleryüz, 2016).

In disaster management, the importance of the social work profession is evident in the need for competent, prepared individuals with organizational skills and interdisciplinary experiences before, during, and after disasters.

Disaster Management in Türkiye: A Social Work Perspective

Türkiye frequently experiences natural or human-induced disasters, resulting in significant human and material losses. The most important reason is that most of the urban areas and vulnerable structures are built near earthquake fault zones or in other disaster-prone areas. Türkiye's high disaster damages also stem from reactive policies addressing post-disaster wounds rather than proactive disaster policies fostering disaster-prepared lifestyles. (Tercan, 2018).

The experiences following disasters have significantly influenced the development of social policies related to disasters in Türkiye. Various disciplines have contributed to formulating regulations to address post-earthquake challenges, approached from a multidimensional perspective. Post-earthquake discussions initially focus on rescue operations and subsequently shift to meeting the basic needs of affected individuals, including food, clothing, and shelter. Following this process, debris removal works are carried out. Efforts to meet the basic needs of earthquake survivors coincide with the investigation into factors contributing to the disaster's occurrence. Discussions held within this process focus on issues such as problems encountered in rescue operations, difficulties in meeting the basic needs of earthquake survivors and providing shelter, as well as factors believed to have contributed to the loss of life and property destruction caused by the earthquake. This process involves different disciplines due to the breadth of the subject. Within the scope of disaster management, multidimensional studies and arrangements are made in terms of settlement, economic, social, and psychological dimensions.

Until the end of the 1990s in Türkiye, primarily reactive policies, which focused on post-disaster efforts, dominated the field of disaster management. However, following the 1999 Marmara earthquake, proactive policies emphasizing prevention efforts gained prominence. Disaster management in Türkiye changed from crisis management to risk management (Tercan, 2018). In 2009, with Law No. 5902, the "Disaster and Emergency Management Presidency" (AFAD) was established under the Prime Ministry, and the authorities and responsibilities of the institutions working on disasters were gathered under a single roof (AFAD, 2023). Comprehensive execution and management of all processes related to disasters are provided by the government. AFAD, as the responsible public institution for disasters in Türkiye, is tasked with preparing, training, and organizing interventions, establishing logistic centers, and forming intervention teams (Özçelik, 2020).

Social Work in Disaster Management

In disaster-affected areas, survivors are vulnerable, and the period following search and rescue operations is critical to their survival. Preparations made before disasters play a crucial role in quickly mitigating and minimizing experienced problems. Social work interventions during and after the disaster will be effective in providing support to individuals and communities directly and indirectly affected by the disaster. Today, the consensus is unquestionable regarding the creation of a comprehensive strategy encompassing pre-, during, and post-disaster services. From a social work perspective, determining the involvement of disaster policies at different stages and identifying effective interventions to mitigate social and economic losses from disasters are critical considerations. The social work approach requires preparation before disasters and coordinated work of adequate teams in terms of training and equipment. Following emergency medical assistance in disasters, the discipline that immediately intervenes and facilitates the smoothest possible process for disaster survivors is social work (Özçelik, 2020).

Disaster-related activities typically unfold across five stages: risk mitigation, preparation, intervention, recovery, and reconstruction. These stages are interrelated and sequential, where the activities of the preceding stages influence those of subsequent stages. The first two stages pertain to pre-disaster preparations, while the subsequent three stages involve actions during and after the disaster. (Ergünay, 2009).

The primary aim of recovery phase activities following disasters is to address the essential needs of affected individuals at a minimum level. These include economic and social requirements, long-term temporary housing, communication, transportation, water, electricity, sewerage, education, and psycho-social support. This process is envisaged by some researchers to continue until the needs of those affected by the disaster are met at least at the pre-disaster level or even at a higher level (Ergünay, 2009).

In all five stages identified concerning disasters, social work entails certain duties and responsibilities. An assessment of these stages can be viewed through the conceptualization of professional roles within the generalist approach to social work. In these processes, macro-level roles play a predominant role in damage mitigation, pre-disaster preparation, and reconstruction. Planning and community engagement are essential for achieving societal awareness and institutional preparedness against disasters. In the stages of rescue, first aid, and recovery, the roles of social workers such as establishing connections, linking, advocacy, mediation, empowerment, and education become increasingly significant (Tuncay, 2004).

Since disasters affect disaster survivors in the narrowest scope, the family environment of the survivors at the medium level, and the inhabitants of the region in the broad sense, social work activities are structured according to this situation. The breadth of the impact area makes it necessary for social workers to engage in multidimensional planning and process management. Social work interventions should encompass dynamic, multi-faceted approaches that integrate service provision in the field with comprehensive planning for present and future conditions (Özçelik, 2020).

As stated in the definition of disaster, disasters are the subject of social policies due to the losses they cause and the negative consequences they create. In this respect, the concept of vulnerability after disasters becomes crucial in planning the recovery process. Vulnerability is typically defined as the incapacity of individuals or social groups to anticipate the impacts of hazards, mitigate potential damages, withstand their consequences, and restore normalcy to life. (Ergünay, 2009). The concept of vulnerability in disasters can be divided into four groups, such as physical, social, economic, and environmental vulnerability. Physical vulnerability, a quantifiable concept, varies based on human-made structures and individuals' physical capacities, allowing for the estimation of damage and loss. Previous disaster analyses have shown that the elderly, children, disabled individuals, and women are disproportionately affected by disasters due to the varying levels of physical vulnerability among social groups. Social vulnerability, a new definition of vulnerability, reflects society's lack of resilience in disasters, shaped by factors like age, gender, awareness, knowledge, education, and living conditions. Social vulnerability is extremely challenging to measure and quantify. In disaster terminology, social vulnerability denotes the preparedness of individuals, groups, institutions, or societies to mitigate disaster risks, respond rapidly and effectively, and recover from their impacts. In economic vulnerability, a direct correlation has been established between poverty and increased physical and social vulnerability, resulting in greater susceptibility to disasters among low-income groups (Ergünay, 2009).

Nowadays, psychosocial services have become a standard procedure as part of the social work provided after disasters. The psychosocial support provided after disasters aims to alleviate and improve mild and moderate mental problems after the disaster and to prevent the formation of major problems that will affect a larger part of society. Post-disaster psychosocial support is an important area for the stabilization of the society affected by the disaster. Psychosocial support will contribute to the intervention of future concerns of individuals due to the adversities caused by the disaster and to the reconstruction of society (Aydın, 2012).

Considering all social policies related to the disasters in Türkiye, it is seen that the predominant focus is mainly aimed at compensating for economic and physical damages. The Ministry of Family and Social Services has a fundamental role in the improvement of social vulnerability. The Department of Psychosocial Support in Migration, Disasters, and Emergencies is responsible for carrying out all activities related to psychosocial support services within the scope of the Türkiye Disaster Response Plan.

While the principles of the activities to be carried out before, during, and after the disaster are laid down, institutional arrangements and operational principles in terms of social services will be effective in reducing the economic and social impact of disasters through sequentially conducted activities.

The Role of Social Workers in Disaster Management

Social workers are equipped with essential roles and functions throughout the stages of disaster management, acquired through both their educational training and fieldwork experiences. The generalist social work approach requires social workers to perform the functions of counseling with the role of counselor, resource management with the role of resource manager, and education with the role of educator at various stages of disaster management. For instance, addressing post-earthquake basic needs falls under resource management, while tackling educational issues of children and youth is within the realm of counseling functions. Existing regulations such as TAMP and TARAP delineate the duties of social workers in the context of these roles and functions.

Counseling, as a professional endeavor, aims to rectify disruptions in social functionality by restoring individuals to a state of equilibrium. In disaster contexts, counseling functions entail both micro-level interventions, like post-disaster psychosocial support, and macro-level efforts, such as formulating programs, plans, and policies, and contributing to legal frameworks for disaster management systems. Within the counselor role, social workers have the roles of enabler, facilitator, and planner (Miley et al., 1998). In the role of enabler, social workers provide psychosocial support to help disaster victims cope. As facilitators and planners, they collaborate within organizational structures to address system weaknesses and ensure the smooth delivery of services. In disasters, effectively storing and sorting humanitarian aid for distribution to those in need demands adept planning skills in a challenging environment.

In the resource management function, social workers facilitate clients' access to existing resources, help them obtain new resources, and promote the development of unavailable resources (Duyan, 2003). According to Miley et al. (1998), in resource management, social workers act as brokers and advocates at the micro level, convenors and mediators at the mezzo level, and activists at the macro level. In disaster contexts, resource management plays a crucial role in addressing the physical, psychological, and social needs of survivors, while upholding principles of equality and social justice.

The educational function within social work entails fostering an empowering exchange of information between the social worker and the client, as highlighted by Duyan (2003). Miley et al. (1998) highlight the diverse roles of social workers, serving as educators at the micro level, trainers at the mezzo level, and engaging in outreach efforts at the macro level. The educational function of social workers spans various activities, including direct client education, dissemination of information in the community and professional contexts, and participation in research. Social workers conduct macro-level training and micro-level informational activities to enhance knowledge and awareness about disasters in the pre-disaster stage. They also conduct post-disaster research on disasters and share the results with society, all within the scope of their roles.

The defined roles of social workers in disaster management underscore their vital contributions throughout all phases of the disaster continuum-micro, meso, and macro levels-aimed at effectively addressing the multifaceted biopsychosocial needs of disaster survivors.

The Duties of Social Workers Across Phases of Disaster Management

According to the stages of disaster management, social workers have certain duties before, during, and after the disaster. These duties according to disaster phases are as follows:

Pre-Disaster Duties

Educating the Community to Be Prepared for Disasters

These training courses can take place within various institutions (schools, community centers, youth centers, etc.), or social workers can provide information during home visits.

Actively Participating in Disaster Planning, Programs, and Legal Frameworks

In preparation for potential disasters, TAMP details the step-by-step operational procedures for social workers' areas of responsibility, outlining relevant institutions, individuals, and resources at regional and local levels, and determining their active involvement areas. Following the post-disaster response to the 2023 Kahramanmaraş earthquake, it is evident that there is a pressing need to enhance the efficacy of plans and the coordination among various actors and teams to function cohesively in the chaotic environment typical of earthquake zones. Social workers' participation in new planning initiatives, sharing their field experiences, and contributing insights are crucial.

Supporting and Strengthening NGOs

Non-governmental organizations also play important roles in disasters. In large-scale disasters, NGOs support various aspects from search and rescue to social reorganization. As per the "Guidelines for Accreditation of Psychosocial Support in Disasters and Emergencies" issued by the Ministry of Family and Social Services, civil society organizations meeting the Ministry's criteria are accredited and incorporated into potential disaster plans. The accreditation criteria for NGOs and their employees include participation in disasters within the last five years, commitment to fielding at least one team of twenty to a maximum of five teams of one hundred volunteer members or employees for psychosocial support, completion of psychosocial support training programs conducted by the Ministry, graduation from relevant university departments (psychology, social work, child development, psychological counseling and guidance, sociology) or having a medical specialty in psychiatry, possessing a minimum of two years of professional experience, and demonstrating proficiency in individual and group counseling skills. In this context, the Social Workers' Association, known for its efficacy and active engagement in Türkiye, has obtained accreditation. It initiated field operations within the first week of the Kahramanmaraş earthquakes and continues its efforts. Likewise, AFAD is currently accrediting NGOs capable of participating in search and rescue missions. Resource management and training are key priorities for social workers in pre-disaster efforts, encompassing training initiatives, coordination with resource entities, and active engagement in disaster management processes (Artan and Özkan, 2020).

Duties During and Post-Disaster

In 2014, the Türkiye Disaster Response Plan (TAMP) was prepared to ensure effective response to disasters in line with the experiences gained from previous disasters. TAMP outlines the duties of working groups and coordination units for disaster intervention, ensuring the establishment of intervention planning principles before, during, and after disasters. In this context, TAMP includes ministries, institutions and organizations, the private sector, NGOs, and real people that will be involved in the intervention to potential disasters and emergencies of all types and scales (Strategy and Budget Office , 2023).

The Türkiye Disaster Risk Reduction Plan 2022-2030 (TARAP), which covers the years 2022-2030, was prepared under the coordination of AFAD. This plan delineates the roles and responsibilities of public institutions, local administrations, the private sector, NGOs, and universities involved in risk reduction activities for all types and scales of disasters in Türkiye. The strategic priorities of TARAP have been determined in accordance with the internationally accepted strategic priorities of the Sendai Framework[1] (2015-2030). The TAMP will be able to function effectively with all responsible and relevant stakeholders taking ownership, relevant institutions and organizations working in coordination with each other, and the actions and targets set are realized on time and in accordance with the general principles of the plan (Strategy and Budget Office , 2023).

The Ministry of Family and Social Services has been designated as the main solution unit in the field of social work in the National Disaster Response Organization of TAMP, in the Disaster Psychosocial Support Group within the "Pre-Recovery Working Groups" and in the Disaster In-Kind Donation Warehouse Management and Distribution Group within the "Logistics and Maintenance Service." In this context, the Ministry is tasked with preparing a psychosocial analysis report after a disaster. Additionally, the Ministry is a

---

[1] The Sendai Framework for Disaster Risk Reduction 2015–2030 was adopted at the Third United Nations World Conference on Disaster Risk Reduction, held from 14 to 18 March 2015 in Sendai, Miyagi, Japan. The Sendai Framework aims to reduce disaster risk and significantly reduce losses to the lives, livelihoods, health and economic, physical, social, cultural and environmental assets of individuals, businesses, communities and countries. The Sendai Framework (2015-2030) is the first major agreement of the post 2015 development agenda, calling on member states to take concrete steps to protect development gains from disaster risks. The Sendai Framework is coordinated with other 2030 Agenda agreements, namely the Paris Agreement on Climate Change, the Addis Ababa Development Finance Action Agenda, the New Urban Agenda and the Sustainable Development Goals (UNDRR, 2023).

support solution partner in Disaster Evacuation and Relocation Planning, Disaster Sheltering, and Disaster Nutrition Groups (TAMP, 2014).

In TARAP, the Ministry of Family and Social Services is designated as the responsible organization for analyzing the post-disaster needs of special needs groups and determining social and cultural needs. The Ministry of Family and Social Services is the relevant organization in various activities, raising the awareness of the community regarding the reduction of disaster risks, raising awareness of reducing disaster risks for special needs groups planning risk reduction and preparedness activities for effective intervention, determining the risks related to mass migration, and ensuring the adaptation of people arriving with mass migration into the community (TARAP, 2022).

According to TAMP, the activities to be conducted in the Disaster Psychosocial Support Group, which is one of the tasks assigned to the Ministry as the primary solution partner, are specified in eight articles under the title of Psychosocial Intervention Tools. Psychosocial Intervention Tools in Disasters (Guide to Psychosocial Support Services in Disasters, 2008) are (1) Needs and Resource Assessment, (2) Psychological First Aid, (3) Referral and Guidance, (4) Establishing Information Center, (5) Community Mobilization, (6) Social Projects, (7) Educations, and (8) Support for Field Personnel.

Needs and Resource Assessment

It is the first psychosocial intervention after the disaster occurs. This encompasses the formulation and execution of a response strategy addressing the impact assessment of the disaster on affected populations, the determination of their needs, and the utilization of exisiting resources and institutions to fulfill those needs. In this context, activities such as meeting the basic needs of the people affected by the disaster, taking special needs groups (children, elderly, and disabled people) under protection and care, assisting the families of special needs groups, relocating them to existing sound institutions in the region or, if not possible, transferring them to relevant public institutions outside the region, storing, and regularly distributing aid are carried out.

Psychological First Aid

Psychological first aid starts at the same time as needs and resource assessment. Psychological First Aid is designed to foster a supportive environment wherein individuals affected by the disaster can freely articulate their emotions and experiences, facilitating their initial coping mechanisms and averting isolation during this critical period.

Referral and Guidance

This intervention involves directing individuals grappling with significant and severe psychological distress post-disaster to psychologists/psychiatrists or pertinent healthcare facilities within the locality for therapeutic intervention.

Establishing an Information Center

In the chaotic aftermath of a disaster, facilitating access to accurate information is imperative for both the professional personnel operating in the area and the general public. This serves not only to ensure efficient resource accessibility but also to mitigate the dissemination of erroneous information that can incite public unrest. These works are conducted with the objective of establishing information centers strategically positioned to offer accessible provisions of fundamental necessities, healthcare services, and the presence of skilled personnel, catering to the needs of the general populace.

Community Mobilization

The main objective of this work is to increase individuals' coping and self-help skills. This tool can be considered a biopsychosocial empowerment of disaster survivors throughout the whole process.

Social Projects

Social projects are programmatic efforts carried out through the use of financial resources and technical support to directly solve the problems of disaster victims, mostly with the initiative of civil society. For example, project calls and support provided by universities and international non-governmental organizations during the reconstruction process after the Kahramanmaraş earthquakes can be considered in this context. The community centers opened by the United Nations Development Program (UNDP) in Hatay, Adıyaman, and Kahramanmaraş are examples of social projects. These centers provide emergency care services to the elderly, the disabled, and other groups that have become vulnerable and disadvantaged due to the disaster.

Education

Training may be in the form of providing information to those affected by the disaster on different subjects such as ways of coping, post-disaster family relations, ensuring hygiene in the new living environment, as well as providing information to the professional staff who will work in the region on working conditions, works to be done, and transfer of planning.

Support for Field Personnel

Psychosocial interventions targeting professionals from diverse sectors engaged in post-disaster operations include personalized counseling services, therapeutic group sessions facilitating mutual sharing of experiences, identification, and mitigation of adverse impacts on staff well-being.

The Use of Technology and Digital Platforms Within the Scope of Social Work Duties in Disasters

The imperative of uninterrupted communication and navigation systems to facilitate post-disaster response, as evidenced by the 2023 Kahramanmaras earthquakes, underscores the criticality of ensuring operational continuity to mitigate potential disruptions. The interruption of communication during disasters obstructs crucial search and rescue efforts and vital psychosocial support activities, especially in the immediate aftermath. This breakdown in communication increases the challenges faced by disaster victims, increasing the risk of loss of life and exacerbating biopsychosocial adversities. Hence, it is imperative to prioritize the establishment of resilient communication and navigation infrastructures capable of operating effectively in diverse disaster scenarios during the preparatory phases preceding earthquakes. Additionally, efforts should focus on exploring alternative technological methods that can function independently of Internet connectivity or closed circuits. Seamless communication among social worker teams, access to resources, and coordination with local authorities are essential for uninterrupted operations.

Another important issue is the records kept by social workers for their assessment of the needs of disaster victims, and access to stock information of relief supplies in warehouses or arriving in the region. Initially, social workers should have access to inventory details to prevent aid wastage and ensure targeted delivery to those in need. Hence, upon aid arrival, social workers engaged in storage and sorting can update inventory. This information can then be enhanced through a digital system managed by social workers conducting needs assessments, facilitating expedited matching of resources with disaster victims. The loss of family members of vulnerable groups in the aftermath of disasters makes their situation more difficult both in terms of loss of life and not knowing where they are. Examples include children rescued from rubble and taken to hospitals without family present, as well as instances of individuals with ill intentions falsely claiming kinship to abscond with children. To mitigate such adversities, comprehensive records of vulnerable groups (children, elderly, disabled) encountered during search and rescue operations should be meticulously maintained and made accessible to social workers. Upon completion of search and rescue operations, social workers are responsible for monitoring the victim's situation and providing necessary support as part of psychosocial assistance. To facilitate this, an integrated digital system accessible to both search and rescue and psychosocial support teams should be implemented in preparatory efforts. This approach can contribute to preventing the disappearance of children and the elderly.

Discussion and Conclusion

Social workers use technology in their duties through social networks. This proves pivotal in addressing the needs of disaster victims, both within the affected area and those relocated to other provinces in the aftermath of the disaster. The establishment of sharing and information groups on various social media platforms enables quick results to be achieved. However, what is important here is the possibility that the public who wants to help may be misled about topics that fall within the field of social work, and some ethical problems can arise. It is paramount to avoid causing secondary trauma to vulnerable disaster victims in the aftermath of the event. Therefore, it is crucial to inform the public about such sensitive situations during preparation activities. The platforms established by psychosocial support specialists, including social workers, psychologists, and psychiatrists, prove highly effective. For instance, in scenarios such as providing access to an adolescent specialist for post-traumatic stress disorder experienced by a youth relocated from the disaster area to another province following an earthquake, prompt referrals to the appropriate expert or institution can be facilitated efficiently.

Another notable concern is the limitation of long-term, structured psychosocial support available to disaster victims post-disaster, primarily due to the scarcity of experts in the affected area following a major event. Likewise, in the aftermath of the Kahramanmaraş earthquakes, disaster victims from both the affected area and other provinces expressed their interest in receiving online psychosocial support, which social workers facilitated. Hence, as part of pre-disaster planning, it is prudent to undertake developmental initiatives and establish a technological infrastructure to facilitate the delivery of online therapies.

Disclaimer

None.

References

AFAD. (2023). AFAD ve Tarihçesi. https://www.afad.gov.tr/afad-hakkinda.

Afet ve Acil Durumlarda Psikososyal Destek Akreditasyon Kılavuzu, Aile ve Sosyal Hizmetler Bakanlığı. https://www.aile.gov.tr/uploads/athgm/uploads/pages/goc-afet-ve-acil-durumlarda psikososyal-destek/psi-kososyal-destek-nedi-r.pdf (Accessed: 04.03.2024).

Afetlerde Psikososyal Destek Hizmetleri Rehberi (2008). Aile ve Sosyal Hizmetler Bakanlığı, Göç Afet ve Acil Durumlarda Psikososyal Destek Hizmetleri Daire Başkanlığı. https://www.aile.gov.tr/uploads/athgm/uploads/pages/goc-afet-ve-acil-durumlarda-psikososyal-destek/psi-kososyal-destek-nedi-r.pdf.

Aktaş, A. (2003). Kriz Durumlarında Sosyal Hizmet Müdahalesi. *Kriz Dergisi*, 11(3), 37-44.

Artan, T., Özkan, A. O. (2020). Afetlerde Sosyal Hizmet. *Journal of ADEM*, 1(1), 47-54.

Aydın, D. (2012). Afet Sonrasi Psikososyal Destek Uygulamalari. İnsani ve Sosyal Araştırmalar Merkezi, 1-9.

Duyan, V. (2003). Sosyal Hizmetin İşlev ve Rolleri. *Toplum ve Sosyal Hizmet, 14*(2), 1-22.

Ergünay, O. (2009). *Afet Yönetimi: Genel İlkeler, Tanımlar, Kavramlar.* Ankara.

International Federation of Social Workers. (2024). https://www.ifsw.org/what-is-social-work/global-definition-of-social-work/.

Miley, K. K., O'Melia, M. & DuBois, B. (1998). *Generalist Social Work Practice: An Empowering Approach.* Allyn and Bacon.

Özçelik, E. (2020). Afetlerde Sosyal Hizmetler. *Afet ve Risk Dergisi, 3*(1), 46-55.

Strategy and Budget Office. (2023). *Türkiye Earthquakes Recovery And Reconstruction Assessment.* Ankara.

Şengün, H., Güleryüz, D. (2016). Türkiye'de Afet Yönetiminde Bir Dönüm Noktası: 1992 Erzincan Depremi. *Uluslararası Erzincan Sempozyumu, 2*, 857-867, Erzincan.

Türkiye Afet Müdahale Planı (TAMP). (2014). T. C. İçişleri Bakanlığı, Afet ve Acil Durum Yönetimi Başkanlığı(AFAD).https://www.afad.gov.tr/kurumlar/afad.gov.tr/e_Kutuphane/Planlar/TAMP.pdf.

Türkiye Afet Risk Azaltma Planı TARAP. (2022). İçişleri Bakanlığı Afet ve Acil Durum Yönetimi Başkanlığı.https://www.afad.gov.tr/kurumlar/afad.gov.tr/e_Kutuphane/Planlar/28032022-TARAP-kitap_V6.pdf.

TBMM, Meclis Araştırma Komisyonu Raporu. (Accessed: 03.04.2024).

Tercan, B. (2018). Türkiye'de afet politikaları ve kentsel dönüşüm. *Abant Kültürel Araştırmalar Dergisi, 3*(5), 63-74.

Tuncay, T. (2004). *Afetlerde Sosyal Hizmet: 1999 Yılı Marmara ve Bolu-Düzce Depremleri Sonrasında Gerçekleştirilen Sosyal Hizmet Uygulamaları.* Ankara: Hacettepe Üniversitesi Yayınları.

Yıldırım, Ş. (2016). Sosyal Hizmette Krize Müdahale Yaklaşımının Kullanımı Krize Müdahale. *CBU-SBED, 3*(4):481-486.

Chapter 12

Education Management in Disasters

Sefa Mızrak*, PhD
and Eyyüp Yildiz, PhD
Emergency Aid and Disaster Management Department, Faculty of Health Sciences, Gümüşhane University, Gümüşhane, Türkiye

Abstract

The effects of disasters on the community increase with increasing population and urbanization, the destruction of the environment, and climate change. This situation increases the need for disaster education at every stage of integrated disaster management. Effectively planned disaster education for preparation, mitigation, response, and recovery periods will ensure that the community is more resilient to disasters. The aim of this study is to investigate how the use and prevalence of new developing technologies can be used more effectively to increase the disaster education capacity of the community. Within the scope of the study, the use of technologies such as virtual reality, social media, simulations, and mobile games in disaster education will be explained. The public administration policies that need to be implemented to increase the use of these tools will be explained. In particular, suggestions will be presented to increase the sustainability and safety of the community through the use of innovative technology in disaster education. It is expected that this study will give ideas to scientists and decision-makers about the use and importance of technology in disaster education.

* Corresponding Author's Email: sefamizrak1@gmail.com.

In: Disaster Management in the Digital Age and Emerging Technologies
Editors: Ayşe Asiltürk and Ayşegül Sarıoğlu Kemer
ISBN: 979-8-89530-213-2

Keywords: disaster education, mobile applications, social media, virtual reality

Introduction

One of the strategies available to avoid disasters, and perhaps the most important one, is disaster education. According to Petal and Izadkhah (2008), the view that "building a resilient society depends on the success of disaster education" is shared by many. This view is quite appropriate considering the importance of education, specifically disaster education for people.

There are also some obstacles that disaster education must overcome. Perhaps the most important of these is how to ensure people's participation in disaster education. Although those who were directly or indirectly exposed to the disaster were found to be more inclined to preparedness behavior than those who never experienced it (Mileti et al., 2014; Becker et al., 2017), in general, people are not willing to participate in disaster education even if they live in areas of high disaster risk (Kitagawa, 2019).

The importance of adopting behavior is emphasized in disaster education. In fact, acquiring behavior in preparation for disasters is seen as beyond acquiring knowledge. Petal (2009, p.287) emphasized that there is a need for disaster education that equips people to produce their own solutions in dealing with disasters, beyond knowing the disaster risks of the place they live in. Such an understanding of disaster education implicitly suggests a "lifelong learning" process in which formal, informal, or non-formal learning types are not separated but all together (Takeuchi et al., 2011). Disaster education researchers have also stated that it would be more appropriate to apply all of these three different learning types together in disaster education, rather than emphasizing their advantages or disadvantages (Nielsen & Lidstone, 1998; Bhandari, 2014).

Although there is no fully agreed-upon definition, education has been defined as the transfer of knowledge and values of a society (Marshall, 2006; Nakosteen et al., 1999). The term meaning of education is related to knowledge, experience, learning, and teaching (Doharey et al., 2023). It is a process that continues from the beginning to the end of an individual's life. Moreover, education is necessary not only for the development of the individual but also for social well-being.

It is widely accepted that education is a "deliberate, systematic, and sustained" effort (Chazan, 2022: p.15). Based on this, disaster education can be defined as systematic and ongoing training related to every phase of a disaster for the safety of people.

Disaster education is an emerging topic within educational studies and research (Kitagawa, 2021). However, the devastating disasters that have occurred in recent years have led to increased interest in disaster education. The contributions of disaster education, especially in the context of disaster risk reduction, have been frequently emphasized in the United Nations Conferences on Disaster Reduction (IDNDR, 1994; United Nations, 2005). At the 2005 World Conference on Disaster Risk Reduction, the use of information, innovation, and education for building societal safety and resilience at all levels was proposed. Furthermore, it has been emphasized that disaster risk reduction knowledge should be integrated into appropriate sections of school curricula, children and youth should be informed through other formal and informal channels, and disaster risk reduction should be supported as an integral part of the United Nations Decade of Education for Sustainable Development (United Nations, 2005, p.10).

The Importance of Disaster Education

Disaster education focuses on traditional knowledge, local resources, practices, skills, and supporting structures for local preparedness and mitigation, while also supporting the formation of social networks (Takeuchi et al., 2011). Social networks make serious contributions to the dissemination of disaster education. In this way, it ensures the transfer of all kinds of knowledge, skills, and experience, starting from the individual to the society. Beyond these benefits of disaster education, Bhandari (2014) stated that disaster education is a necessity in building a resilient, knowledgeable, and confident society that is determined to cope with disasters using all possible ways.

Fu and Zhang (2024, p.2) explained the importance of disaster education in combating disasters as follows:

- Education serves as a foundation on which societies build their resilience.
- Education catalyzes community (resilient) building efforts.
- Education instills a sense of preparedness.
- Education fosters a culture of proactive intervention.

The importance of disaster education has been emphasized in many national and international activities, especially the International Decade for Disaster Reduction (IDNDRR), the Hyogo Framework for Action (HFA), and the Sendai Framework for Disaster Risk Reduction (2015-2030) (IDNDR, 1994; United Nations, 2005; United Nations Office for Disaster Risk Reduction Sendai Framework for Disaster Risk Reduction 2015– 2030, 2015). Since the first of these activities, public education activities have been carried out in many countries; Disaster education materials are developed and implemented in schools for children and the general public (Shaw, Takeuchi, and Shiwaku, 2011). Promoting public awareness and education concepts for disasters are also among the results of these activities (Shaw, Takeuchi, Ru Gwee, et al., 2011). These efforts demonstrate that the importance of disaster education in reducing disaster risks is recognized both nationally and internationally (Nielsen and Lidstone, 1998). Disaster education also has an instrumental role in the implementation of other disaster management strategies. Kitagawa (2021) stated that education (disaster education) is a tool to fulfill the four priorities specified in the Sendai Framework for DRR 2015-2030 document.

The need for a holistic approach to disaster education is emphasized in the "education and training" section of the Hyogo Framework for Action (2005-2015) (United Nations, 2005, pp.9-10).

Disaster education researchers also emphasize the importance of a holistic approach to disaster education. According to Bhandari (2014), considering the differences in society such as rich-poor, young-old, healthy-disabled, etc., instead of a standard and static disaster education, a dynamic disaster education that incorporates all learning strategies together should be adopted. Thus, disaster education can effectively address the needs of a wide range of individuals or communities with diverse backgrounds.

Types of Disaster Education

There are essentially three types of disaster education, each addressed as subheadings in this section.

Formal Disaster Education

The "formal" type of education is mostly institutionalized, purposeful, and systematically provided by public institutions or recognized private organizations. Upon completion of formal education, the mentioned organizations provide learners (participants) with a certificate (UNESCO Institute for Statistics, 2012). Formal education is structured education (Colardyn and Bjornavold, 2004).

It has been stated that formal disaster education is quite successful in creating disaster risk perception and awareness (Ronan, 2001; Shiwaku et al., 2007). However, the basic dynamics of education are not limited to creating awareness or perception. The essence of education is the desired behavioral change in the targeted subject. As an example of this situation, in a study conducted on a student group where 70% of the participants received disaster training, it was found that even though 80% had an understanding of risk-related issues, only 30-40% actually implemented preparedness measures (Shaw et al., 2004). Other studies have also reported that disaster education (informal) obtained from society and family contributes more directly to good disaster preparedness (Shaw, Takeuchi, Ru Gwee, et al., 2011; Shiwaku et al., 2007). However, this does not make formal disaster training unnecessary. Indeed, schools serve not only as places where knowledge is acquired by students but also as venues for its dissemination throughout society (Takeuchi et al., 2011). School-age children who acquire knowledge, in some way, spread it to their families and from there to the entire community.

Formal education is a system with its institutions, instructors, learners, and their families. Within this system, schools can play important roles in community teaching and building partnerships among diverse groups (Shaw and Kobayashi, 2001). It also plays important roles in creating awareness among schools, students, teachers, and parents (Takeuchi et al., 2011).

Shaw and Kobayashi (2001) stated that schools play a crucial role not only during emergencies but also before and after disasters by providing community education and building partnerships among various groups. Additionally, researchers emphasizing the concept of "school safety" have noted that it involves dimensions of physical reinforcement, education, and the socialization of disaster efforts (p.9). The physical reinforcement of schools will not only benefit students in protecting them from potential disasters but will also serve as a safe gathering area or shelter in the aftermath of a disaster. The involvement of students, teachers, and the community in disaster education, as well as the dissemination of these efforts throughout society, will contribute to raising awareness and building capacity.

Petal (2008) stated that elective disaster risk reduction courses could be integrated into existing school curricula as formal disaster education. Several approaches have been suggested for integrating these courses, including (a) curricular integration, (b) extra-curricular integration, (c) curriculum infusion, (d) the integration or infusion of disaster risk reduction into a wide range of subjects, (e) stand-alone courses, (f) curriculum resource materials (Petal and Izadkhah, 2008, p.n.p.).

An important issue in incorporating disaster education into the formal education system is what will be taught in classes. Several suggestions have been made on this subject: (I) disaster mechanism and causes, (II) disaster management cycle, experiences, problems, lessons learned, and measures taken at each stage of the disaster management cycle, (III) roles of government, communities, individuals and other institutions in disaster management, and (IV) nature and social environment (Shiwaku, 2007 as cited in Shaw et al., 2010, pp. 55-56).

Another method to implement disaster education in schools is through clubs. Koichi and Glen (2011) stated that science clubs have very positive effects, especially on young people. Based on this, new clubs can be established in higher education institutions, or activities related to disaster science can be conducted in existing clubs.

In a case study on school disaster education, three recommendations were made for formal disaster education: (I) innovative concepts and school disaster education programs, (II) the possibility of integrating disaster education into the school curriculum, and (III) integrating school disaster education between in-school and out-of-school contexts (Koichi and Glen, 2011, p.72).

Informal Disaster Education

Informal learning, simply put, is learning that occurs outside of school (Rogoff et al., 2016), has no specific structure, and happens randomly. Because of that, learners cannot obtain an official document (Dib, 1988).

In informal disaster education, fun teaching methods can be used to teach skills, competence, and critical information to children of all ages (Petal and Izadkhah, 2008). Informal education methods can also be used in adult disaster education.

It has been stated that informal disaster education, as an activity carried out outside normal school time, has positive results in solving disaster problems in the context of new and innovative strategies (United Nations Office for Disaster Risk Reduction, 2006). Because traditional education is largely based on classroom activities. However, research shows that disaster education is more successful when it is experiential and action-based (Takeuchi et al., 2011).

Another benefit that informal disaster education can provide is that it allows participants to discover their own solutions and strengths through their own experiences. For example, Shaw et al., 2011) stated that disaster education should enable people to generate power and solutions on their own.

Petal and Izadkhah (2008, p.n.p.) listed various ways that can be used in informal disaster education as follows:

- Distributing written materials.
- Creating creative educational materials.
- Organizing cultural and performing arts.
- Participating in after-school safety clubs, scouting badges, and project activities.
- Doing projects that bring students into contact with the local community and government.
- Organizing races and giving prizes.
- Involving parents and local communities.
- Conducting disaster drills.

Non-Formal Disaster Education

Although non-formal education includes elements of formal and informal learning, it differs from them (Dib, 1988). The fact that it takes place within a specific plan and program, and that it takes place purposefully and deliberately for the learner, under the guidance of an educator with an institutional structure (Colardyn and Bjornavold, 2004) makes it similar to formal education. However, it does not have a long-lasting gradual structure and is short-term, which distinguishes it from formal education. Moreover, at the end of non-formal education, the student is often given a certificate similar to formal education, although not at the level of a diploma. The defining feature of non-formal education is that it is an addition, complement, and/or alternative to formal education in people's lifelong learning processes (UNESCO Institute for Statistics, 2012, p.11).

In addition to the formal education system, providing disaster education in settings such as seminars, workshops, etc., under the guidance of an instructor exemplifies non-formal disaster education. It has greater potential in terms of reaching a larger audience compared to the formal education system. Because disaster education can be provided to people of almost every age, gender, and region in this way.

Advantages of Technological Developments in Disaster Education

Technological tools such as mobile games, social media, Wiki, Blogs, mobile devices, animations, simulations, e-books, virtual reality, augmented reality, and robotics have begun to be used increasingly in education with the development of technology (Lai and Bower, 2019). The use of technology in education saves time, money, and energy and provides a collaborative, flexible, interactive, accessible, innovative, and up-to-date learning environment (Haleem et al., 2022). During the COVID-19 epidemic, the distance education method was applied for security and health reasons, and this epidemic showed that technological developments in education need to be increased (Saltürk and Güngör, 2020).

Distance and online education were used after the earthquakes in Türkiye in 2023 (Telli Yamamoto and Altun, 2023). Computer-based flood protection games increased students' desire to learn about floods (Tsai et al., 2015). Shyr et al. (2021) found that disaster education provided with technology action puzzles was more effective than disaster education provided with traditional classroom teaching. Rahiem et al. (2020) state that combining digital technologies and folklore in early childhood disaster education will make disaster education more interesting and fascinating. In an experimental study, Şahan and Dinç (2021) investigated the effect of the simulation teaching method and the traditional teaching (direct instruction, demonstration, and making) method on students' disaster preparedness. As a result, after the training, they revealed that the students who received earthquake training with the simulation method were more prepared than the students who received training with the traditional method.

The mobile applications that thousands of people have downloaded to their mobile phones or computers are designed for education about disasters and disaster management (Sukhwani and Shaw, 2020). For example, a mobile application was developed to support tsunami evacuation training (Leelawat et al., 2018). Kirci et al. (2023) developed a mobile application called DepApp to inform people in the event of an earthquake. An Android-based mobile educational game has improved the disaster awareness and disaster response skills of elementary students, household owners, police officers, firefighters, and IT experts (Catedrilla et al., 2021). The emergency mobile application developed by Kilinç et al. (2023) aims to inform people about disaster types, past devastating disasters, disaster preparation, early warning systems, damage assessment, insurance transactions, first aid, and gathering areas.

Virtual reality is an effective training method that can be used for emergency response and preparation, thanks to its immersive, engaging, and safe methods of learning (D'Amico et al., 2023). Disaster training provided with virtual reality eliminates the effects of real-life hazards (Sudiarno and Wahyuni, 2020). Feng et al. (2021) revealed that virtual reality-supported education was more effective than traditional education methods for educating children and increased their knowledge and self-efficacy about earthquakes. It motivated people in Australia to assess their emotional coping capacity, make plans, and increase their preparedness when dealing with fire (Molan et al., 2023). D'Amico et al. (2023) found that virtual reality applications increased flood-related safety knowledge and self-efficacy. Disaster education provided with virtual reality can increase people's concerns about floods and increase their behavior in taking protective measures (Mol et al., 2022). Virtual reality can bring together people from different geographies and organizations by providing a multi-site model for training in complex situations such as hurricanes and earthquakes (Hsu et al., 2013).

Social media is used for communication, monitoring, disaster education, and awareness activities in disaster management (Kankanamge et al., 2020). Social media is used to receive and provide information about disaster preparedness and early warning before disasters, to disseminate disaster response information during disasters, and to inform society about the impact and process after disasters (Luna and Pennock, 2018; Houston et al., 2014). Hidayat (2023) found that young people's access to content about disaster risks on social media increased their sharing of disaster information on social media. After the earthquakes that occurred in Türkiye in 2023, people obtained information by following the posts of experts on social media and asking questions to experts on social media (Yıldırım and Kurban, 2023). Hall et al. (2017) found that social media was an important source of information about the tsunami for students.

Discussion and Conclusion

The increasing number and severity of disasters indicate that it is critical that people participate in disaster education. Comprehensive disaster education should be updated according to the needs of all individuals in the community and environmental hazards. Taking advantage of technological innovations will make education more effective and efficient. The video and visual opportunities offered by technological tools will make it easier to understand the risks of disasters. Additionally, thanks to technological tools, disaster information will reach more people in a short time. For this reason, decision-makers responsible for disaster management should ensure the integration of technological developments into disaster education to increase the resilience of society through disaster education. Strategies should be developed for the public to learn, use, and follow technological developments in disaster education.

Disclaimer

None.

References

Becker, J. S., Paton, D., Johnston, D. M., Ronan, K. R., & McClure, J. (2017). The role of prior experience in informing and motivating earthquake preparedness. *International Journal of Disaster Risk Reduction, 22* (July 2016), 179–193. https://doi.org/10.1016/j.ijdrr.2017.03.006.

Bhandari, R. K. (2014). Disaster Education: The Silver Bullet. In *Disaster Education and Management* (pp. 1–27). Springer India. https://doi.org/10.1007/978-81-322-1566-0_1.

Catedrilla, G. M. B., Lerios, J. L., Sapin, S. B., Lanuang, M. C., & Buama, C. A. C. (2021). *An android-based mobile educational game for disaster preparedness : an input to risk reduction management*. *22*(2), 936–943. https://doi.org/10.11591/ijeecs.v22.i2.pp936-943.

Chazan, B. (2022). What Is "Education"? In B. Chazan (Ed.), Principles and Pedagogies in Jewish Education (pp. 13–21). Springer International Publishing. https://doi.org/10.1007/978-3-030-83925-3_3.

Colardyn, D., & Bjornavold, J. (2004). Validation of Formal, Non-Formal and Informal Learning: Policy and practices in EU Member States 1. *European Journal of Education, 39* (1), 69–89. https://doi.org/10.1111/j.0141-8211.2004.00167.x.

D'Amico, A., Bernardini, G., Lovreglio, R., & Quagliarini, E. (2023). A non-immersive virtual reality serious game application for flood safety training. *International Journal of Disaster Risk Reduction*, *96*(February), 103940. https://doi.org/10.1016/j.ijdrr.2023.103940.

Dib, C. Z. (1988). Formal, non-formal and informal education: Concepts/applicability. AIP Conference Proceedings, 173, 300–315. https://doi.org/10.1063/1.37526.

Doharey, R. K., Verma, A., Verma, K., & Yadav, V. (2023). Education: Meaning, definition & Types. In *Agriculture Extension Education* (pp. 1–6). S. R. Scientific Publication.

Feng, Z., González, V. A., Mutch, C., Amor, R., & Cabrera-Guerrero, G. (2021). Instructional mechanisms in immersive virtual reality serious games: Earthquake emergency training for children. *Journal of Computer Assisted Learning*, *37*(2), 542–556. https://doi.org/10.1111/jcal.12507.

Haleem, A., Javaid, M., Qadri, M. A., & Suman, R. (2022). Understanding the role of digital technologies in education: A review. *Sustainable Operations and Computers*, *3* (February), 275–285. https://doi.org/10.1016/j.susoc.2022.05.004.

Hall, S., Pettersson, J., Meservy, W., Harris, R., Agustinawati, D., Olson, J., & McFarlane, A. (2017). Awareness of tsunami natural warning signs and intended evacuation behaviors in Java, Indonesia. *Natural Hazards*, *89*(1), 473–496. https://doi.org/10.1007/s11069-017-2975-3.

Hidayat, Z. (2023). Environmental disaster education and communication in Indonesia: A survey among generation Z near Mount Krakatoa. *Kasetsart Journal of Social Sciences*, *44* (1), 231–242. https://doi.org/10.34044/j.kjss.2023.44.1.26.

Houston, J. B., Hawthorne, J., Perreault, M. F., Park, E. H., Goldstein Hode, M., Halliwell, M. R., Turner McGowen, S. E., Davis, R., Vaid, S., McElderry, J. A., & Griffith, S. A. (2014). Social media and disasters: a functional framework for social media use in disaster planning, response, and research. *Disasters*, *39* (1), 1–22. https://doi.org/10.1111/disa.12092.

Hsu, E. B., Li, Y., Bayram, J. D., Levinson, D., Yang, S., & Monahan, C. (2013). State of Virtual Reality Based Disaster Preparedness and Response Training. *PLoS Currents*, *APR 2013*, 1–6. https://doi.org/10.1371/currents.dis.1ea2b2e71237d5337fa53982a38b2aff.

IDNDR. (1994). Yokohama Strategy and Plan of Action for a Safer World: Guidelines for natural disaster prevention, preparedness and mitigation, World Conference on Natural Disaster Reduction [Conference Output]. https://www.preventionweb.net/files/8241_doc6841contenido1.pdf.

Kankanamge, N., Yigitcanlar, T., Goonetilleke, A., & Kamruzzaman, M. (2020). Determining disaster severity through social media analysis: Testing the methodology with South East Queensland Flood tweets. *International Journal of Disaster Risk Reduction*, *42*, 101360. https://doi.org/10.1016/j.ijdrr.2019.101360.

Kilinc, E., Altindal, S., & Kocakalay, S. (2023). *Emergency Application for Natural Disasters : AfetAp Doğal Afetler için Acil Durum U ygulaması : AfetAp Emergency Application for Natural Disasters: AfetAp. 5* (December), 301–313. https://doi.org/10.46464/tdad.1351700.

Kirci, P., Arslan, D., & Dincer, S. F. (2023). *A Communication , Management and Tracking Mobile Application for Enhancing Earthquake Preparedness and Situational Awareness in the Event of an Earthquake*.

Kitagawa, K. (2019). Exploring 'everyday-life preparedness': Three case studies from Japan. *International Journal of Disaster Risk Reduction, 34* (August), 265–274. https://doi.org/10.1016/j.ijdrr.2018.11.025.

Kitagawa, K. (2021). Conceptualising 'Disaster Education'. *Education Sciences, 11* (5), 233. https://doi.org/10.3390/educsci11050233.

Koichi, S., & Glen, F. (2011). Roles of school in disaster education. In Shaw, Rajib, Shiwaku, Koichi, & Takeuchi, Yukiko (Eds.), Community, Environment and Disaster Risk Management Volume 7 *Disaster Education* (Vol. 1, pp. 137–153). Emerald Group Publishing Limited.

Lai, J. W. M., & Bower, M. (2019). How is the use of technology in education evaluated? A systematic review. *Computers and Education, 133* (May 2018), 27–42. https://doi.org/10.1016/j.compedu.2019.01.010.

Leelawat, N., Suppasri, A., Latcharote, P., Abe, Y., & Sugiyasu, K. (2018). International Journal of Disaster Risk Reduction Tsunami evacuation experiment using a mobile application : A design science approach. *International Journal of Disaster Risk Reduction, 29* (February 2017), 63–72. https://doi.org/10.1016/j.ijdrr.2017.06.014.

Luna, S., & Pennock, M. J. (2018). Social media applications and emergency management: A literature review and research agenda. *International Journal of Disaster Risk Reduction, 28* (January), 565–577. https://doi.org/10.1016/j.ijdrr.2018.01.006.

Marshall, J. D. (2006). The Meaning of the Concept of Education: Searching for the Lost Arc. *Journal of Thought, 41* (3), 33–37. http://www.jstor.org/stable/42589880.

Mileti, D. S., Brien, P. W. O., Mileti, D. S., & Brien, P. W. O. (2014). Warnings During Disaster: Normalizing Communicated Risk*. 39(1), 40–57. https://doi.org/10.2307/3096912.

Mol, J. M., Wouter Botzen, W. J., & Blasch, J. E. (2022). After the virtual flood: Risk perceptions and flood preparedness after virtual reality risk communication. *Judgment and Decision Making, 17* (1), 189–214. https://doi.org/10.1017/s1930297500009074.

Molan, S., Weber, D., & Kor, M. (2023). Understanding the intention to stay and defend during a bushfire: An application of virtual reality to improve awareness of predictors associated with behavioural response. *International Journal of Disaster Risk Reduction, 84* (November 2022), 103444. https://doi.org/10.1016/j.ijdrr.2022.103444.

Nakosteen, M. K., Lawson, R. F., & Browning, R. (1999). Education. Online Encyclopaedia. https://www.britannica.com/topic/education/The-Byzantine-Empire.

Nielsen, S., & Lidstone, J. (1998). Public education and disaster management: Is there any guiding theory? *The Australian Journal of Emergency Management, 13* (3), 14–19.

Petal, M. (2009). Education in disaster risk reduction. In R. Shaw & R. R. Krishnamurthy (Eds.), *Disaster Management Global Challenges and Local Solutions* (pp. 285–320). Universities Press (India) Private Limited.

Petal, M., & Izadkhah, Y. O. (2008). Concept note: Formal and informal education for disaster risk reduction. *Proceedings of the International Conference on School Safety,* Islamabad, Pakistan, 1416.

Rahiem, M. D. H., Rahim, H., Subchi, I., Zulkifli, Sururin, & Khalimi. (2020). Digital Folklore in Early Childhood Disaster Education. *2020 8th International Conference on Cyber and IT Service Management, CITSM 2020.* https://doi.org/10.1109/CITSM50537.2020.9268864.

Rogoff, B., Callanan, M., Gutiérrez, K. D., & Erickson, F. (2016). The Organization of Informal Learning. *Review of Research in Education, 40* (1), 356–401. https://doi.org/10.3102/0091732X16680994.

Ronan, K. R. (2001). School children's risk perception and preparedness: A hazard education survey. *The Australasian Journal of Disaster and Trauma Studies.*

Şahan, C., & Dİnç, A. (2021). *Simülasyon Öğretim Yönteminin Ortaokul Öğrencilerinin Afetlere Karşı Hazırlık Durumlarına Etkisi** The Effect of Simulation Teaching Method on Middle School Students' Preparedness for Disasters. 5*(1), 21–36. https://doi.org/10.32569/resilience.807289.

Saltürk, A., & Güngör, C. (2020). COVID-19 Döneminde Öğrencilerde Topluluk Hissi ve Çevrimiçi Uzaktan Eğitimin İncelenmesi. *İnsan ve Toplum Bilimleri Araştırmaları Dergisi, 9* (5), 4204–4221. https://doi.org/0000-0003-4143-8917.

Shaw, R., & Kobayashi, M. (2001). Role of schools in creating earthquake-safer environment. OECD Workshop, Thessaloniki, 2001.

Shaw, R., Shiwaku Hirohide Kobayashi, K., & Kobayashi, M. (2004). Linking experience, education, perception and earthquake preparedness. Disaster Prevention and Management: *An International Journal, 13* (1), 39–49. https://doi.org/10.1108/09653560410521689.

Shaw, R., Takeuchi, Y., & Shiwaku, K. (2011). Chapter 8 "Tsunagaru. In *Disaster Education* (pp. 153–162). Emerald Group Publishing Limited.

Shaw, R., Takeuchi, Y., Ru Gwee, Q., & Shiwaku, K. (2011). Chapter 1 Disaster education: An introduction. In *Disaster education* (pp. 1–22). Emerald Group Publishing Limited.

Shiwaku, K. (2007). Towards innovation in school disaster education: Case research in Kathmandu, Nepal.

Shiwaku, K., Shaw, R., Chandra Kandel, R., Narayan Shrestha, S., & Mani Dixit, A. (2007). Future perspective of school disaster education in Nepal. Disaster Prevention and Management: *An International Journal, 16* (4), 576–587. https://doi.org/10.1108/09653560710817057.

Shyr, W. J., Chen, I. M., Lee, J. C., & Su, T. J. (2021). Applying interactive teaching experience and technology action puzzles in disaster prevention education. *Sustainability (Switzerland), 13* (9). https://doi.org/10.3390/su13094788.

Sudiarno, A., & Wahyuni P, A. D. (2020). Analysis of Human Factors and Workloads in Earthquake Disaster Evacuation Simulations Using Virtual Reality Technology. *IOP Conference Series: Materials Science and Engineering, 1003* (1). https://doi.org/10.1088/1757-899X/1003/1/012082.

Sukhwani, V., & Shaw, R. (2020). Operationalizing crowdsourcing through mobile applications for disaster management in India. *Progress in Disaster Science, 5*, 100052. https://doi.org/10.1016/j.pdisas.2019.100052.

Takeuchi, Y., Mulyasari, F., & Shaw, R. (2011). Roles of Family and Community in Disaster Education (Chapter 4). In R. Shaw, Y. Takeuchi, & K. Shiwaku (Eds.), Community, Environment and Disaster Risk Management Volume 7 *Disaster education* (pp. 77–94). Emerald Group Publishing Limited.

Telli Yamamoto, G., & Altun, D. (2023). Türkiye'de Deprem Sonrası Çevrimiçi Öğrenmenin Vazgeçilmezliği. *Üniversite Araştırmaları Dergisi, 6* (2), 125–136. https://doi.org/10.32329/uad.1268747.

Tsai, M. H., Chang, Y. L., Kao, C., & Kang, S. C. (2015). The effectiveness of a flood protection computer game for disaster education. *Visualization in Engineering*, *3* (1). https://doi.org/10.1186/s40327-015-0021-7.

UNESCO Institute for Statistics. (2012). International standard classification of education: ISCED 2011. In *Comparative Social Research* (Vol. 30). UNESCO Institute for Statistics Montreal.

United nations office for disaster risk reduction Sendai framework for disaster risk reduction 2015– 2030. (2015).

United Nations Office for Disaster Risk Reduction. (2006). Disaster risk reduction begins at school: 2006-2007 World Disaster Reduction campaign. United Nations.

United Nations. (2005). Hyogo Framework for Action 2005-2015: Building the Resilience of Nations and Communities to Disasters. In World Conference on Disaster Reduction. UN.

Yıldırım, S., & Kurban, E. A. (2023). Deprem bilincinin oluşmasında coğrafi konulara ait sosyal medya paylaşımlarının incelenmesi. *Türk Coğrafya Dergisi, 83* (83), 155–162. https://doi.org/10.17211/tcd.1302712.

Chapter 13

Breeding and Training Search and Rescue Dogs for Disaster Response

Onur Özsever*, MA, Vet
Department of Medical Parasitology, Faculty of Medicine, Van Yüzüncü Yıl University, Tuşba, Van, Türkiye

Abstract

Efficient use of time is crucial for the successful extraction of people trapped under debris, buried in avalanches and landslides, or stranded in water during search and rescue operations. The equipment used in search and rescue operations must be functional. Equipment should include concrete breaking tools, motorized handsaws, portable toilets, food, water, shelter, power sources, medical supplies, seismic listening devices capable of detecting faint sounds in rubble piles, digital cameras, and highly trained dogs. When evaluating search and rescue teams established by trainers and handlers, the performance of the dog, the performance and knowledge of the handler, and the performance of the dog and the handler as a team are taken into account. The performance of dogs is the result of training that can last up to two years. Most of the dogs used come from shepherds or hunting breeds. They must be physically, mentally, and anatomically healthy. Dogs are tested in both wilderness and urban environments. In Türkiye, the initial goal should be to reach the presence of at least three search and rescue-cadaver and blood dogs in each district, totaling up to three thousand dogs. Then, we need to quickly increase this number of trained animals to ten thousand. In the earthquake centered on Kahramanmaraş, thirteen million four hundred twenty-one thousand people were affected in ten provinces, with

* Corresponding Author's Email: oozseveren@gmail.com.

In: Disaster Management in the Digital Age and Emerging Technologies
Editors: Ayşe Asiltürk and Ayşegül Sarıoğlu Kemer
ISBN: 979-8-89530-213-2

thousands killed or injured. Naturally, in cities like Istanbul where earthquakes are expected, more people are at risk. Considering that procurement and training take at least two years, it is crucial to establish search and rescue teams without delay.

Keywords: disaster, dog, rescue, search

Introduction

Disaster is defined as an event that disrupts life, cannot be managed with local resources, leads to physical damage, and causes economic and social losses. It can be natural or human-induced, involving cases of missing persons, and the time elapsed between their disappearance and their safe retrieval is termed as "search," while the time taken to transport them to a safe place is termed as "rescue" (Uymaz, 2023). Uymaz (Uymaz, 2023) categorizes the fundamental elements of search and rescue under six main headings:

1. Physical Search: After the search and rescue team identifies and marks the incident site, a general scan is conducted horizontally and vertically in circular and parallel directions within the debris area. This is conducted by observation, calling out, and listening to team members. To reach the victim in the debris area and establish communication with them, except for the search personnel on duty, others should be removed from the environment to ensure silence.
2. Search Using Dogs: Trained dogs play a role in the location of disaster victims and in the rescue of those whose locations have been identified. The dogs search the debris and rubble area to identify the locations of living, deceased, or unconscious victims, and they enter areas inaccessible or unsafe to the rescue team members. The areas indicated by the dogs are marked, listened to using listening devices, and efforts are made to obtain clearer information.
3. Search Operations for Casualties in Emergencies or Disasters: Dogs respond similarly to both recently deceased and living disaster victims. Especially in areas where fires may occur in the debris, the paws of the dogs should be protected. Concrete layers, elevator shafts, and wind presence can confuse the dogs' sense of smell and direction, leading to deviations in locating the victim. Therefore,

seismic/acoustic listening devices should be used simultaneously to visually and audibly detect individual signals.

4. Visual Search: Devices that capture images and establish audio communication are used to locate the position of the victim under the debris and subsequently determine their position under the debris. The teams carrying these devices approach the victim step by step and conduct the most appropriate extraction operations.
5. Victim Rescue Operations: Rescue operations are divided into four stages. The first stage involves rescuing those who are not trapped under the debris, the second stage involves rescuing those who are visible under the debris, the third stage involves rescuing non-survivors, and the final stage involves rescuing those whose locations are unknown under the debris. Victims who are not trapped under the rubble and can be rescued quickly are provided the necessary first-aid support and immediately transported to healthcare centers. Typically, these individuals are rescued by bystanders or neighbors who can provide information about the disaster situation and assist in the operation.
6. Recovery of Non-Survivors: Bodies that can be easily retrieved from the upper parts of the debris are immediately removed from the scene. The locations of bodies requiring operations for extraction from the debris are marked, and they are retrieved by rescue teams (Uymaz, 2023).

Dog Training Sessions

Their training encompasses various specialties such as search and rescue, cadaver search, bomb detection, mine detection, law enforcement operations, narcotic detection, fire detection, tobacco detection, tracking, and food detection. These training sessions take place at the Search and Rescue Training Center.

a. Search and Rescue: Searching and rescuing individuals trapped under debris in disasters such as earthquakes, landslides, and avalanches, as well as conducting search and rescue operations for illegal human trafficking.

b. Cadaver Search: Finding the bodies of missing individuals in forensic cases, as well as searching for organic remains of bodies in incidents such as collapses and disasters.
c. Bomb Detection: Searching for explosive materials inside buildings, various modes of transportation, luggage, etc., with the purpose of detecting explosive substances.
d. Mine Detection: Using the olfactory sensitivity of dogs to detect mines and explosive materials buried underground.
e. Public Security: Patrol and surveillance services, prison security, intervention in social incidents, assistance to law enforcement forces, apprehension of resisting or fleeing individuals, neutralization, protection of critical facilities and areas for security purposes.
f. Operation: Facilitating the work of units in special operational activities, neutralizing suspects.
g. Drug Detection: Searching for evidence of individuals who use, transport, or trade narcotics or psychotropic substances in their belongings, vehicles, homes, and premises.
h. Fire Detection: Identifying the location of combustible, burned, or unburnt materials in cases of suspected arson in houses, warehouses, vehicles, buildings, or forest fires.
i. Tobacco Detection: Locating illegally transported, smuggled, or produced tobacco and tobacco products.
j. Tracking: Following the tracks of criminals and their accomplices, locate perpetrators of theft, escape, sabotage, smuggling, illegal immigration, human trafficking, and other public security incidents by tracing scent trails.
k. Food Supplies Detection: Uncovering hidden shelters and hideouts in counter-terrorism operations, identifying food and supplies hidden by terrorists in the field. A dog unit typically consists of one dog and one dog handler, either a specialized gendarmerie or a specialist corporal. Dog breeds trained for this purpose include Belgian Malinois, German Shepherd, Golden Retriever, German Pointer, and Labrador breeds (Jakem., 2023). Cadaver dogs have been deployed in Africa to search for the bodies of victims of tribal massacres, in Yugoslavia to locate mass graves, and in the Turkish Republic of Northern Cyprus to find missing individuals. When loved ones go missing, especially in mountainous or forested areas, their families often seek to recover their bodies intact. Following search and rescue operations after disasters such as floods, avalanches, and earthquakes, the deployment

of cadaver dogs becomes essential. These trained dogs can detect bodies even after they have decomposed several meters below the ground, sometimes up to 1-2 years after death. Due to their exceptional abilities, dogs continue to be used to locate mass graves, even years after burial (Eker, 2023).

Search and Rescue Dogs

Search and rescue dogs are being trained. In Dog Training Centers, an intensive two-year training activity is conducted for candidate dogs, which includes basic skills such as learning simple commands such as eating and using the toilet, basic obedience training, and subsequently advanced obedience where complex commands are applied. Search and rescue activities cannot be performed alone; a search-focused dog can accomplish a task that would require 15-20 people in a much shorter time on its own. Dogs are the greatest helpers in search and rescue operations during disasters (Sakarya, 2023). In Search and Rescue, there are seventeen live search dogs, five tracking dogs, and one cadaver search dog. At the Samsun Provincial Disaster and Emergency Management Directorate Dog Training Center, three live search dogs, one tracking dog, and one cadaver search dog are on duty. Belgian Shepherd dogs, which are ambitious, stubborn, courageous, and have strong psychological and physical levels, work under the supervision of trainers in the center to find missing persons in all kinds of incidents and disasters. As part of the team, three dog trainers and one veterinarian conduct intensive work with search and rescue dogs at the center. The dog named Ovit is a tracking dog, Kito, Çiçi, and Mali are live search dogs, and Abrek serves as a cadaver search dog within the team (Koçer, 2021). Abrek is ambitious, stubborn, fearless, and enjoys playing games. His training involves working with animal blood and bone fragments. Abrek can find corpses and body parts up to ten years old under all conditions. These conditions include factors such as the use of chemicals to erase blood, the use of chemicals, and the burning of the corpse, none of which hinder Abrek from finding it. This situation is also an indication that there is no better technology than the sense of smell of dogs in today's world (Stanevich, 2023).

Dogs in Disasters

Dogs are not only man's best friend but also our most valuable asset in disasters such as floods, hurricanes, and forest fires. The empathy and affection they exude during disasters provide incredible assistance to humans. Every search and rescue dog handler decides how their dogs will find people based on tracking, trailing, and air scenting. Air scenting means that dogs do not need a scent article to start searching. They are taught to scan the air for any human scent. Tracking dogs are scent discriminators, meaning they will search for and follow a scent when given a scent article. Dogs have the ability to find bodies or body parts hidden in car trunks, trash cans, or oil barrels, or otherwise concealed (Handwerk, 2021). Training a certified disaster dog requires hundreds of kilometers, a significant amount of money, and continuous training. Disaster dogs should have sensitive noses and be well-socialized, obedient, and agile, such as Golden and Labrador Retrievers, Border Collies, and their mixes. Five thousand dollars are spent annually on travel, equipment, and dog care. Inexperienced dogs are at high risk of injury and can lose their limbs. Through continuous training on rubble piles, dogs' pads harden (Akyürek, 2023). Dogs used for urban search and rescue purposes use their noses to find live victims trapped in cracks and voids in rubble piles resulting from earthquakes, hurricanes, or explosions. Rescuing people trapped under the rubble increases their chances of survival. Therefore, disaster dogs can focus on their search without distraction, even in the presence of other search dogs, humans, and cadavers, while navigating through piles of rubble (Akyürek, 2023). For dogs searching for survivors in the rubble in the middle of the night, the environment is dangerous and equally dark. Rescue dogs can be of different breeds and can be trained for various behaviors. Dogs search for human scent in a specific area and bark to alert the rescue team if they detect signs of human breath. Trailing dogs can follow the path of a missing person, as like traditional trailing dogs. Some can even work with boats to find human remains underwater. Dogs are trained to see their search activities as a game supported by appropriate rewards. This game is often challenging. A boat accident where someone has drowned is also a game for the dog. When human remains are found, family members often want to bond with the search dog, pet it, thank it, and express how good a dog it is to find their loved one (Atasoy, 2017).

The Importance of Dogs in Earthquakes

Despite the earthquakes that occurred in Gölcük and Düzce in 1999 in Türkiye, the presence of trained search and rescue dogs was questioned in the 2023 earthquake centered in Kahramanmaraş. Officials from the Search and Rescue Association, veterinarians, and academic experts in the field have stated that the most accurate and reliable data is provided with the assistance of trained search and rescue dogs. During the earthquake centered on Kahramanmaraş, teams from Türkiye, along with search and rescue teams from all around the world, including the United States, Mexico, Greece, Germany, Switzerland, Spain, and Japan, worked with nearly a hundred dogs they brought along (Erdem, 2020). Atasoy (Welsh, 2023) highlighted the value of dogs in his book "Adli Koku" (Forensic Scent). Dogs undergo rigorous training to distinguish between the scent of living and non-living beings and to locate individuals under debris. The breeds most commonly chosen for training due to their olfactory abilities include German Shepherd, Belgian Malinois, Labrador Retriever, English Springer Spaniel, and Greyhound (Beaubien & Balaban, 2023).

Search and Rescue Teams and Equipment

Rescue dogs undergo training for approximately 8-10 months. Following search and rescue training, dogs can discern between a live body and a deceased body, as well as between a live body and fresh clothing. Dogs have the ability to detect and differentiate odors from various sources, such as clothing, diapers, and scents associated with missing individuals (Layton&Gleim, 2023). Cadaver dogs are specifically trained to detect only corpses. Known as "cadaver dogs" or "HR" (human remains) dogs, most of these dogs are trained using food as a reward rather than toys. Real human body parts obtained from a facility studying the decomposition process of deceased individuals, as well as human placentas, are used in the training of HR dogs (Layton&Gleim, 2023). Trainers use thick and durable chew toys made from old fire hoses to reward search dogs for their successful work. When a live-find dog detects a scent, they approach as closely as possible and bark continuously. At this point, the handlers mark the location and direct rescue workers to start digging, while rewarding the dog by playing with them. Toys are used as rewards instead of food for logistical reasons (Hebard, 1993).

The equipment of search and rescue teams is also crucial. In the earthquakes centered on Kahramanmaraş, the US search and rescue team brought seismic listening devices capable of detecting faint sounds from within rubble piles, expandable digital cameras to explore openings, and highly trained dogs. Additionally, the team brought concrete breaking equipment, motorized saws, portable toilets, food, water, shelter, power sources, and rescue and medical equipment (Wegmann & Heines, 1997).

The Protection of Search and Rescue Dogs

When evaluating the dog and handler team, the performance of the dog, the performance and knowledge of the handler, and the performance of the dog and handler as a team are considered. The performance of dogs is the result of training that can last up to two years. Most of the dogs used at work come from shepherds or hunting breeds. They must be physically, mentally, and anatomically healthy. Dogs are tested in both wilderness and urban environments. Test conditions include loud noises, crowds, and smoke from the air. The dogs have twenty minutes to find four victims. Missing a victim or giving a false alarm means failing the assessment test. When testing the dog and handler as a team, it is important to observe the dog's responsiveness to commands or hand signals (Syrotuck, 2000). There are potential dangers and conditions in emergency training or real missions that could cause a dog to become sick or injured. Possible injuries in these trained dogs include abrasions, tears, sprains or strains, burns, frostbite, heatstroke, bruises, crush injuries, snake bites, and exposure to hazardous substances. Handlers should be partially prepared to authorize euthanasia for their canine partners (Augsburger, 2020; Fox et al., 2008).

Discussion and Conclusion

In disasters, we need search and rescue dogs for their exceptional ability to locate missing individuals under various conditions, including collapsed buildings, avalanches, or other disaster sites. Their keen sense of smell and agility enable them to detect human scents, even in challenging environments where human rescuers may struggle to access them. The presence of search and rescue dogs significantly enhances the efficiency and effectiveness of rescue operations, as they can cover large areas quickly and accurately.

While Search and Rescue (SAR) operates with nearly forty search and rescue dogs, the German Search and Rescue Dog Association is organized with more than seventy teams, more than two thousand members, and more than twelve hundred dogs. Thirty thousand dogs were used in the First World War, and two hundred thousand dogs were used in the Second World War. The number of trained dogs needs to be maintained according to the incidents and disasters faced. In this context, considering that there are fire departments, and police/gendarmerie organizations in each district, at least one search and rescue-cadaver and blood dog should be considered necessary for each of these institutions.

The need to increase the number of search and rescue dogs arises from several factors. First, disasters are unpredictable and can occur anywhere at any time, necessitating a sufficient number of trained canine teams to respond promptly. Additionally, search and rescue dogs have proven to be invaluable in locating survivors buried under rubble or debris, potentially saving lives that might otherwise be lost. Moreover, as disasters become more frequent and severe due to various factors such as climate change, the demand for search and rescue capabilities, including trained dogs, continues to grow. Therefore, increasing the number of search and rescue dogs and ensuring their availability in disaster response teams is essential to improve emergency preparedness and enhance the probability of successful rescue operations.

Disclaimer

None.

References

Ak, E. (2020). Köpekleri Kanser ve Kovid-19 Ön Tanısı Koymak Için Kullanabiliriz. (Accesed: 09.02.2024). https://www.indyturk.com/node/168476/haber/%E2%80%9Ck%C3%B6pekleri-kanser-ve-kovid-19-%C3%B6n-tan%C4%B1s%C4%B1-koymak-i%C3%A7in-kullanabiliriz%E2%80%9D.

Akyürek, Z. D. (2023). Enkazdan Hayat Kurtaran Köpeklerin Sırrı! Hiçbir Teknoloji Önüne Geçemiyor. (Accesed: 09.02.2024). https://www.milliyet.com.tr/gundem/enkazdan-hayat-kurtaran-kopeklerin-sirri-hicbir-teknoloji-onune-gecemiyor-6900417.

Atasoy, F., Erdem, E. (2014). Köpek duyuları. *Lalahan Hay. Araşt. Enst. Dergisi, 54*(1), 33-8.

Atasoy, F. (2017). Köpek Duyuları ve Yerli Irklar. ss: 81-99. Ed.: Ünver Y, Hancı İ. H. Adli Koku. Seçkin Kitabevi, Ankara. ISBN. 9789750241833. p231.

Augsburger, M. (2020). Mental health of voluntary canine search and rescue workers in Germany – Mission-related stressors and need for future interventions. A pilot investigation. *International Journal of Disaster Risk Reduction, 44*: 101409.

Beaubien, J., Balaban, S. (2023). Sniffer dogs offer hope in waning rescue efforts in Turkey (Accesed: 18.02.2024). https://www.npr.org/sections/goatsandsoda/2023/02/13/1156499861/sniffer-dogs-offer-hope-in-waning-rescue-efforts-in-turkey.

Bulanda, S. (2014). Ready! The Training of the Search and Rescue Dog. *Companion House Books,* p: 264.

Case, L. P. (2005). The Dog: Its Behavior, Nutrition, and Health." Second edition, Blackwell Publishing, p : 496. ISBN: 0-8138-1254-2. 9600 Garsington Road, Oxford OX4 2DQ, UK.

Cengiz, S. (2022). Possible hazards and risks that search and rescue (SAR) dogs may face in CBRN incidents. *GÜSBD, 11*(1), 300-7.

Curry, E., Skogen, M., Roth, T. (2021). Evaluation of an odour detection dog for non-invasive pregnancy diagnosis in polar bears (ursus maritimus): considerations for training sniffer dogs for biomedical investigations in wildlife species. *Journal of Zoo and Aquarium Research, 9*(1), 1-7.

Dalziel, D., Uthman, B., Mcgorray, S., Reep, R. l. (2003). Seizure-alert dogs: a review and preliminary study. *Seizure, 12*(2), 115-20.

Eker, G. (2013). Kadavra/Ceset Arama Köpekleri. (Accessed: 18.02.2024). https://www.goktaneker.com/2013/08/15/kadavra-ceset-arama-kopekleri-ve-kullanimi.

Fox, P. R., Puschner, B., Ebel, J. G. (2008). Assessment of acute injuries, exposure to environmental toxins, and five-year health surveillance of New York police department working dogs following the september 11, 2001, World Trade Center terrorist attack. *Journal of the American Veterinary Medical Association, 233* (1), 48-59.

Handwerk, B. (2021). How Search-And-Rescue Dogs Find Survivors. (Accessed: 18.02.2024) https://www.nationalgeographic.com/animals/article/rescue-dogs-behavior-animals.

Hebard C. Use of search and rescue dogs. Symposium on Disaster Medicine. *J Am vet Med Assoc.* 1993, 203(7): 999-1.

Jakem. Köpek Birlik Eğitimleri. (Accessed: 11.03.2024). https://www.jandarma.gov.tr/jakem/kopek-birlik-egitimleri.

Koçer, Y. (2021). Arama Kurtarma Köpekleri Samsun AFAD'ın Göz Bebeği. https://samsunsehirgazetesi.com/gundem/arama-kurtarma-kopekleri-samsun-afadin-goz-bebegi-1388.

Layton, J., Gleim, S. (2023). How Search-and-Rescue Dogs Work. https://animals.howstuffworks.com/ animal-facts/sar-dog1.htm.

Lorenzo, N., Wan, T., Harper, R. J., Hsu Y., Chow M., Rose S., Furton K. G. (2003). Laboratory and field experiments used to identify canis lupus var. familiaris active odor signature chemicals from drugs, explosives, and humans. *Analytical and Bioanalytical Chemistry, 376* (8), 1212-24.

Mehus-Roe, K. (2002). Disaster Search and Rescue Dogs. https://www.petfinder.com/helping-pets/ animals-and-disaster-relief/disaster-search-rescue-dogs/.

O'Connor, M. B., O'connor, C., & Walsh, C. H. (2008). A dog's detection of low blood sugar: a case report. *Irish Journal of Medical Science, 177*, 155-157.

Pickel, D., Manucy, G. P., Walker, D. B., Hall, S. B., & Walker, J. C. (2004). Evidence for canine olfactory detection of melanoma. *Applied Animal Behaviour Science, 89*(1-2), 107-116.

Rue N. 2019. Dogs provide help, support in natural disasters. Dogsbestlife.com. 20 Nisan 2019.

Sakarya. (2023). Türkiye'nin En Hassas Burunlu Arama Kurtarma Köpekleri Sakarya'da Belirleniyor. https://www.yenisafak.com/hayat/turkiyenin-en-hassas-burunlu-arama-kurtarma-kopekleri-sakaryada-belirleniyor-3865016.

Sanders, C. R. (2000). The impact of guide dogs on the identity of people with visual impairments. *Anthrozoös, 13*(3), 131-139.

Sergei, A. K., Barnes, P. J. (2006). Exhaled Biomarkers. *Chest, 130* (24), 1541-1546.

Sharun, K., Jose, B., Tiwari, R., Natesan, S., & Dhama, K. (2021). Biodetection dogs for COVID-19: An alternative diagnostic screening strategy. *Public Health, 197*, e10.

Shivik, J. A. (2002). Odor-adsorptive clothing, environmental factors, and search-dog ability. *Wildlife Society Bulletin, 30* (3), 721-727.

Stanevich, R. (2023). The Basics About Search and Rescue Dogs. https://www.ncadogs.orgsar1.html.

Stepień, I., Stepień, L., & Lubińska, E. (1990). Function of dog's auditory cortex in tests involving auditory location cues and directional instrumental response. *Acta Neurobiologiae Experimentalis, 50*(1-2), 1-12.

Syrotuck, W. G. (2000). Scent and the Scenting Dog. ISBN: 097004942. Barkleigh Productions, Inc.

Szetei, V., Miklósi, Á., Topál, J., & Csányi, V. (2003). When dogs seem to lose their nose: an investigation on the use of visual and olfactory cues in communicative context between dog and owner. *Applied Animal Behaviour Science, 83*(2), 141-152.

Uymaz, S. (2023). Arama ve Kurtarma Teknikleri. https://acikders.ankara.edu.tr/course/view. php?id=8468.

Wegmann, A., Heines, W. (1997). Such und Hilf- Hunde retten Menschenleben. ISBN: 3924008477. Kynos Verlag.

Welsh, N. (2018). Search Dogs Comb Hills of Montecito for Survivors and Remains. https://www. independent.com/2018/01/14/search-dogs-comb-hills-montecito-survivors-and-remains/.

Chapter 14

Food Logistics in Disasters and Emergencies

Alpaslan Alpaslanoğlu*, PhD
Department of Logistics Management, Iğdır University, Iğdır, Türkiye

Abstract

While logistics is a necessary infrastructure for all fields and activities, it also has a high level of technical subjectivity. Therefore, the necessity of using specific logistics practices and methods, which have become subjective for almost every sector, is also valid for disasters and emergencies. Disasters and emergencies refer to unusual and extraordinary conditions, and under these conditions, they require unusual, routine, and non-general logistics work, transactions and practices, and even infrastructures. On the other hand, even under normal conditions, food logistics already has a unique and subjective logistics infrastructure such as cold chain, refrigeration, and ATP. But disaster situations also involve extraordinary urgency. JIT (Just in Time), one of the basic principles in business logistics, is replaced by the main principle of BOT (Before on Time) in disaster situations. Furthermore, the high probability of encountering various infrastructure and transportation problems and obstacles is another subjective situation. Therefore, in disaster situations, logistics services and activities should be performed both subjectively and urgently.

This study is about the work and operations of food logistics that must be conducted in disaster situations. The reasons for the tendency to face more global and local disasters, the special place of food products, and the subjectivity and difficulties of logistics in disaster situations reveal the importance of the study. Although it always remains valid that preparing for disasters in advance and creating disaster-resistant

* Corresponding Author's Email: aalpaslanoglu@gmail.com.

In: Disaster Management in the Digital Age and Emerging Technologies
Editors: Ayşe Asiltürk and Ayşegül Sarıoğlu Kemer
ISBN: 979-8-89530-213-2

structures and systems is a priority, it has been concluded that in the face of ever-increasing number, type, and severity of disaster situations, our existing and relatively insufficient disaster logistics systems and operations need to be updated and new approaches are needed. It has been revealed that there is a need, in particular, to increase the resilience and integration of transportation networks, to make improvements in the layout and efficiency of storage, to increase the safety of priority products and equipment, and to increase transportation speeds. It is thought that the changing and increasingly complex nature of disasters will require more academic knowledge, especially in the field of disaster logistics, and therefore the contributions of logistics researchers can provide significant benefits.

Keywords: disaster logistics, emergency logistics, logistics

Introduction

Throughout history, humanity has encountered countless calamities, disasters, and emergencies expressed in various situations. Many times, some concepts are used - often interchangeably - to express such events. Broadly speaking, there are two types of event levels: emergencies and disasters (Center for Disaster Philanthropy, 2024a).

Emergency: It is a situation that affects the lives and well-being of many people or a significant percentage of the population and requires significant multisectoral assistance. An emergency may be acute (such as a cholera epidemic) or slow-onset (such as a drought), and more than one emergency may occur simultaneously. Sudden emergencies (World Health Organization, 2024) can be grouped under two headings: Natural disasters (earthquakes, tsunamis, floods, landslides or avalanches, extreme temperatures, advancing drought, and wildfires, etc.) and human-made hazards (armed conflict, civil unrest, terrorism, transportation accidents, structural fires, and industrial explosions, etc.).

A disaster event can be defined as "any natural disaster, act of terrorism, or other man-made disaster that causes extraordinary levels of death or damage or disruption, seriously affecting the population (including mass evacuations), infrastructure, environment, economy, national morale, or government functions in an area" (U.S. Department of Homeland Security, 2010). Disaster is a social event that occurs when a hazard strikes a vulnerable community in such a way that it exceeds or hinders the community's ability to cope and can cause serious harm to people's safety, health, welfare, property, or environment, which can be triggered by a naturally occurring event originating in the geophysical or biological environment, or by malicious or unintentional human action or error, including technological failures, accidents, and acts of terrorism (Public Safety Canada Search, 2024). Disaster is a serious disruption of the functioning of a community or society on any scale due to hazardous events interacting with conditions of exposure, vulnerability, and capacity, resulting in one or more of the human, material, economic, and environmental losses and impacts. Damage and loss can occur immediately, or the impact can last for a long time, often requiring years of recovery. The impact of the disaster can be immediate and local but is often widespread and may persist for a long period of time. A disaster may test or exceed the capacity of a community or society to cope using its own resources and, therefore, may require assistance from external sources, which may include local, statewide, national, or international responses. Types of disasters include technological crises (e.g., oil spills, explosions) or earthquakes, volcanoes, storms, floods, droughts, wildfires, epidemics, tsunamis, and mass movements (e.g., landslides). Disasters can occur suddenly (e.g., hurricanes, earthquakes) or have a slow onset (e.g., drought, desertification, epidemics) (Center for Disaster Philanthropy, 2024b).

Emergency situation: It can be defined as an actual or imminent event (Public Safety Canada Search, 2024) that requires immediate coordination of actions regarding persons or property to protect the health, safety, or welfare of people or to limit damage to property or the environment. Emergencies are generally smaller events for which a limited local role is sufficient, while major disasters are usually larger-scale events such as a powerful hurricane or earthquake (Center for Disaster Philanthropy, 2024a). EM-DAT classifies disaster types as natural and technological (human) caused. While natural disasters occur as hydrological and meteorological disasters, transportation sector-related disasters are by far the leader in technological disasters (EM-DAT, 2024a). According to the UN Office for Disaster Risk Reduction (UNDRR), disasters are classified into 5 categories (UNDRR, 2020).

Disasters affect millions of people every year. Between 1998 and 2017, there were approximately 7,250 disasters and more than 1.3 million people died. While floods and storms caused the most disasters, earthquakes caused the most deaths. During this period, nearly 750,000 people lost their lives in earthquakes (NASA, 2024). In 2023, the Emergency Events Database (EM-DAT) recorded a total of 399 disasters related to natural disasters. Due to these events, 86,473 deaths occurred, and 93.1 million people were affected.

Economic losses reached 202.7 billion USD. The earthquake in Türkiye and Syria in 2023 was the most devastating event of the year in terms of death and economic damage, with 56,683 deaths and damage worth 42.9 billion USD, affecting a total of 18 million people (EM-DAT, 2024b; USAID, 2023). Today, there are more than twice as many emergency cases as ten years ago (UNDRR, 2020), and as a result of these disasters, a total of 299.4 million people worldwide, the majority of them children, need humanitarian assistance to survive. This number increased to 168 million in 2020 and 235 million in 2021. Additionally, the nature of emergencies changes over time. For example, natural disasters caused by climate change occur faster, with longer lasting and more permanent results (World Vision International, 2024).

Disasters are often sudden and unpredictable events that cause serious damage, destruction, and human suffering (EM-DAT, 2024c) and can cause dramatic harm after their occurrence. Therefore, the needs for humanitarian assistance are also increasing as disasters create tremendous human suffering (UNDRR, 2023). Response time and aid quality, which are important factors in reducing social and economic losses that occur after disasters, are significantly affected by the degree of effectiveness of logistics processes (Schulz, 2009: 269). Indeed, the devastation wrought by Hurricane Katrina exceeded the capacity of national, regional, and local governments to respond and recover, and FEMA faced significant logistical challenges. Logistical deficiencies included shipments sent to the wrong destination, spoiled food that had to be thrown away, and wasted ice left to melt. The difficulty of intervening in such devastating events (U.S. Department of Homeland Security, 2010) and the serious and vital disruptions seen in the 2023 Türkiye earthquake (ITU, 2023; TMMOB, 2024) once again reveal the vital importance and priority of disaster management and disaster logistics and has also revealed the need for new approaches in disaster logistics.

Food Logistics in Disaster and Emergency Situations

Transportation, storage, supply, and distribution activities play key roles in logistics functions performed in disaster situations.

Disaster Logistics

Disaster logistics is the process of planning, preparing, implementing, and evaluating, in an effective, efficient, and coordinated structure, all functions that support the identification and evaluation of need and demand, purchasing, procurement, transportation, staging, storage, and distribution of needed disaster materials to appropriate end users. Logistics management functions include materials management, property management, facilities management, and transportation management (State of Michigan, 2012). Following a major disaster, the environment is often filled with uncertainty about the status of the damage, even after the initial damage assessment, and all parts of the supply chain operate under uncertainty. In some cases, demand often exceeds supply and capacity (food, water, transportation, storage space, etc.). For major incidents that require immediate search and rescue, involving basic survival needs such as triage and medical treatment, moving away from danger, and obtaining water and food, the urgency is very serious. Addressing these needs is a priority, while considerations of cost and resource efficiency are secondary until the next stage of the relief effort. Reviewing situational awareness is critical at every stage of assistance because priorities change with the transition from initial response to rescue (Transportation Research Board, 2013). In these conditions, disaster management can be handled in two stages. First-stage decisions cover disaster preparedness (food supply and stock level). The second stage is done after the disaster occurs and is represented by the food aid distribution plan (Yigezu and Sanders, 2012).

Disaster and emergency logistics management has emerged as a theme noticed worldwide due to disasters, whether artificial or natural, that can occur anywhere in the world and have huge consequences. Although such a disaster has gradually increased global humanitarian support in the field of aid delivery and rehabilitation, emergency logistics remains the key to efficient and effective mitigation of the immediate aftermath of the disaster. Although emergency logistics is of vital importance, it can involve many challenging problems (such as damaged infrastructures, supply problems, security, and speed) that cannot be solved as easily as business logistics (Sheu, 2007). Unlike business logistics, emergency logistics also differs in terms of the ultimate goal, dynamics, and variability of the problem. For example, in the face of sudden-onset disasters such as large-scale earthquakes, tsunamis, and meteorological disasters such as typhoons, practitioners or humanitarian organizations must cope with a highly dynamic and stochastic demand in terms of both volume and location. Typically, sub-members in business logistics may focus on cost and profit. In contrast, sub-members in emergency logistics focus on survival and immediate medical response. In emergency logistics, delivery times are very short and, therefore, a fast and efficient logistics and transportation network is needed to support the most important risk: human life (Kundu, Sheu, and Kuo, 2022).

Disaster logistics involves processes and systems for mobilizing people, resources, skills, and knowledge to provide humanitarian assistance to people in need. The main ones are evaluation, purchasing, transportation, tracking, customs clearance, local transportation, storage, and final distribution management (CARE International, 2024a). The effectiveness of disaster logistics is closely related to providing self-sustaining telecommunications, logistics, and operations support (FEMA, 2020) through personnel, equipment, and supplies that can be deployed immediately to support a large disaster field office and multiple field operations sites in the disaster area. The logistical deployment of resources to provide assistance to disaster victims and the proper planning of these activities are critical to reducing suffering. Furthermore, being able to cope in disaster situations and conduct successful operations largely depends on the cooperation of different organizations (Rodríguez-Espíndola, Albores, and Brewster, 2018). It is a must for this cooperation to carry out logistics operations effectively and in coordination. Logistics operation coordination can be structured between local logistics support centers, regional logistics support centers, national logistics support centers, and other stakeholders (Cabinet Office, 2009). However, after major disasters, the increasing support and assistance provided at the global level also requires a higher level of integration of logistics operations than ever before. Therefore, the integration and coordination of international official, semi-official, or private aid organizations such as the United Nations Humanitarian Response Depot (UNHRD) can also be included in logistics operations as international logistics support centers (World Food Programme, 2024a).

Disaster and Emergency Food Logistics

The term 'Food': It is defined as (1) items used as food or drink for humans or other animals, (2) chewing gum, and (3) items used for components of any such item (U.S. Government Publishing Office, 2023). Therefore, anything that is placed in the mouth for human consumption (food, drink, or other items that are swallowed or absorbed) is considered "Food." It refers to raw, cooked, or processed edibles, ice, beverages, or any content used or intended in whole or in part for human consumption or the use or sale of chewing gum (U.S. Government Publishing Office, 2023). An 'edible food product' is an item, beverage, or ingredient used or intended to be used, or offered for sale, in whole or in part, for human consumption (Arizona Secretary of State, 2023). In disaster situations, food in the amounts recommended by the World Health Organization as a minimum of 3 liters of water per day per person in need of vital water and 2 meals per day (per person) (Florida Division of Emergency Management, 2024) must be provided immediately and without interruption. Of course, it is imperative that both water (WHO, 2024) and food comply with the conditions of "food safety" (The World Bank, 2024) at the same time and precisely. Experiencing increasing levels of food safety problems on a global scale (UN, 2023) makes food security even more sensitive.

Food logistics: it includes i. purchasing, packaging, and/or preparing food, including food products, fresh foods, shelf-stable food products, and ready meals, ii. if conditions are so severe that food cannot be easily purchased, delivery of food, including hot and cold meals, if necessary, to distribution points and/or individuals, and iii. work and transactions regarding the rental of distribution and storage areas, vehicles, and necessary equipment (World Food Programme, 2024b).

In disasters and emergencies, the supply and delivery of food products are among the main priority tasks (National Library of Medicine, 2024). On the other hand, food and medical product supply chain capacity and challenges in disaster situations include challenges specific to the operations of these products, as well as requiring unique supply chain infrastructure and capacity (AusAID, 2007). In addition to their urgent priorities, foods can be considered an extremely important and sensitive area for disaster logistics due to their perishability, different lifespans, requiring a high level of reliability, a large number of varieties, and being the most important supporters of life. Therefore, one of the main goals in emergency situations is to ensure that basic food needs are met (World Food Programme, 2002). Unsafe foods can also cause serious problems such as disease spread, poisoning, and contamination, but the probability that these situations occur in other products - except medical products - is low. For all these reasons, food logistics operations in disaster situations involve differences and difficulties in many aspects and therefore require unique operational sensitivities.

Transport

During disasters, many relief organizations often face serious difficulties in transporting large quantities of various goods, such as food, clothing, medicines, medical supplies, machinery, and manpower, from various starting points to different destinations in disaster areas. Materials and relief personnel must be transported quickly and efficiently to optimize the likelihood of survival of affected persons while minimizing the expenses of such activities (National Library of Medicine, 2024). Transport activities in disaster situations (World Food Programme, 2002) generally consist of the following work and transactions:

- Making transport plans including transport routes, modes, travel time, capacity, planned throughput, notes (e.g., planned actions to reduce bottlenecks/increase efficiency),
- Highway transports; use and management of commercial, government, and other assistance fleets,
- Air/Rail/Water transportation,
- Fuel and maintenance operations for transport units.

Delivery of food products can be done in different ways: dropping from the sky, using lakes and rivers, or crossing nearly impassable terrain by truck, amphibious vehicles, or pack animals (World Food Programme, 2019). However, among transport modes, road transport appears to have a separate position (and at a relatively higher level). This is because, unlike the normal logistics and transportation flow, the ability of road transport to reach door-to-door is much more vital. In cases of disaster and emergency, the ability of those in need to meet their own needs almost disappears, the ability to use alternatives against infrastructure damage compared to other types of transportation, the existence of a wide variety of transportation vehicles (such as tractors, tracked vehicles, all-terrain vehicles, sleds) and the relatively low need for transfer put the road transportation in a special position. However, it is deemed necessary to use all transportation modes effectively and integrated to ensure the sustainability of the flow of necessary materials, increase their varieties and quantities, and meet the needs beyond local capacities.

Large-scale disasters can cause extensive damage to communications, power supplies, and transportation infrastructures, rendering them unusable for emergency relief operations. For example, disruption of transportation facilities, including ports, airports, roads, and bridges, can limit humanitarian access to disaster-stricken areas and impede communications. Therefore, these additional constraints need to be taken into account for emergency logistics operations (Yan and Shih, 2009). Disaster resilience of critical infrastructures is an absolute prerequisite. This is because natural disasters have the potential to cause serious damage to various components of critical infrastructure, disrupt essential services, and endanger public safety. These infrastructure systems, which form the backbone of modern society, span a wide range of sectors and include, transportation networks, which include ports, roads, bridges, railways, and airports, which are critical for the transportation of goods, emergency response teams, and the evacuation of affected populations during natural disasters. Damage to these networks can seriously hinder response efforts and slow recovery. Measures to increase the resilience of transport infrastructure, such as improved drainage systems, reinforced bridges, and pre-positioned emergency equipment, can help ensure that key roads remain accessible during and after disasters (EIS Council, 2024). The 2023 Türkiye earthquake revealed this situation, and infrastructures in all transportation modes (land, air, sea, and railways) suffered serious structural damage (Strategy and Budget Presidency, 2023; ITU, 2023). Naturally, in such situations where there is a harmful and dysfunctional transportation structure, logistic activities will also stop. For this reason, in disaster logistics operations, all transportation modes should be used urgently and in an integrated manner, at a level that can meet demands at their full capacity or even above capacity, and should be at least fully functional, ready, and resilient. Therefore, building the resilience of the emergency logistics network can not only increase its reliability in an environment of deep uncertainty but also reduce the restoration time caused by emergency damages (Jiang and Yuan, 2019). The World Food Program (WFP) Isaka Logistics Center (Tanzania) is an emergency food storage with specified features and was created to prevent losses when needed, maintain the nutritional safety and quality of food, and ensure rapid and efficient delivery to where it is needed. Isaka Logistics Center is connected to the seaport (Dar es Salaam port) by road and rail. Therefore, the sea, railway, and road are integrated very effectively. Moreover, direct loading and unloading can be performed on the train wagons located in front of the food storage with a direct railway line inside the Isaka logistics center, and thanks to the rails leading to the complex, it allows a fully loaded 20-wagon train carrying 800 tons of food products to be quickly unloaded right next to the storage structures. This setup is ideal for an extremely rapid response, providing a unique advantage in saving time and saving lives during emergency operations. The proximity of the storage area also minimizes the potential for food loss during unloading and transportation (World Food Programme, 2019).

Problems may also arise in ensuring the safety of the transportation of aid materials. For example, 38% of the USD 257 million worth of commodity shipments from FEMA to Puerto Rico were lost to visibility. There have been "significant disruptions" in goods shipments due to reasons such as FEMA's inability to track shipments and lack of oversight (Federation of American Scientists, 2022). Therefore, traceability of the supply chain, ensuring transportation safety, and efficiency of distribution are indispensable elements of food logistics and supply chain. GPS-based tracking technologies, minimizing distribution and transportation distances, and assigning logistics experts and operators can provide benefits.

Storage

The adequate and effective storage of food products is critical to efficient and fast logistics operations. The goods should be easily visible, arranged to ensure efficient operation and stored appropriately to prevent damage and expiration (Figure 3). In emergency situations, it is necessary to use all available storage areas correctly and efficiently (USAID, 2020). The storage should be any large indoor storage area available (ideally located in a secure/fenced area of 25,000 – 200,000 square feet) with adequate space to park and unload vehicles and other vehicles carrying disaster donations or goods. Examples of facilities that can be used as storage include empty commercial storage areas, storage facilities of public and non-governmental organizations, ammunition depots, fairground buildings, ice fields, highway/public works maintenance garages, airport hangars, and community recreation centers (State of Michigan, 2012). The issues to be considered in food logistics storage can be shown as the availability, capacity, and cost of storage facilities and transportation in the specified region, usual ways of bringing goods to the region, current problems, and expected changes (World Food Programme, 2002). Furthermore, storage must have flexible, practical, and simple operational processes, minimum product loss/damage, high traceability/labeling and security, self-sufficiency (energy, communication, etc.), cleanliness and hygiene, a variety of loading/unloading docks and stacking machines and be able to connect to as many disaster-resistant transportation modes and networks as possible.

The storage location (site selection) should generally be in open areas, well away from crowded places such as markets or hospitals, and preferably some distance from residences and food storage, especially in any area with high population density, and should be easily accessible for food delivery in all seasons. The sites should be surrounded by a fence and divided into separate sections. There should be emergency exits in areas for queuing, distribution, and food stocks. Water, shelter, sanitation facilities, and first-aid services should be available for staff as well as beneficiaries. There should be a well-drained and level ground above ground level where the food sacks will be unloaded and transported (World Food Programme, 2002). Sites should be as close to beneficiaries as possible, but multiplicity of sites should be avoided due to the costs and difficulties of monitoring and preventing people from presenting themselves on different sites and benefiting from them (World Food Programme, 2002). According to the United Nations approach regarding the number of sites, it is recommended to reduce transportation and purchasing costs by using on-site or nearby resources to minimize the total cost of food aid (World Food Programme, 2024b). Furthermore, to reduce congestion and transportation time, it would be extremely beneficial to place the 20 to 30 highest-circulating, fast-moving products in the storage area for easy and fast shipment as possible (World Food Programme, 2002).

Storage areas can be in the form of large storage or small stores, depending on the type of goods, quantity, and distance from the project site. Material and equipment stocks should be stored in safe storage areas (CARE International, 2024b), and the stocks and movements within the storage should be in a structure that can be carried out effectively and efficiently to provide the storage function as a strong logistics activity (World Food Programme, 2002).

Supply

After it is determined that a need item is necessary, elements can be used, such as 'Community resources' (resources provided by individuals, businesses, organizations, and community groups), 'Just-in-time'; (from the market and suppliers), 'Mutual aids' (agreements to share resources between organizations), 'Virtual stocks' (pre-arranged contracts for specific materials) and 'Physical stocks' (usually for bespoke products or products with long lead times or weak supply chains) (Cabinet Office, 2009). In terms of food logistics, different supply routes can be applied depending on the type of food. For example, physical stocks can be used for durable food products, just-in-time for ATP (Doğan, 2014: 388) foods, and mutual aid mechanisms for over-capacity demands. Natural disasters, especially floods, can compromise water supply systems, contaminating water and making it unsafe to drink. This scarcity of clean water can quickly turn into a humanitarian crisis, leading to the spread of waterborne diseases and worsening the emergency. A solid water supply and treatment infrastructure is essential to reduce these risks. Therefore, flood-resistant water facilities, backup treatment systems, and clean water stocks should also be available to meet the needs of affected communities (EIS Council, 2024).

Discussion and Conclusion

According to UN studies, it is predicted that medium and large-scale disasters will increase by 40% from 2015 to 2030 due to climate change alone (USA Facts, 2023; UN, 2022). According to the data, it is understood that the disaster risks of planet Earth are quite high and close to danger (EM-DAT, 2024d). Therefore, after pre-disaster preparations and resistance-building activities, the most important issue will be studies on the development, effectiveness, and efficiency of disaster and emergency logistics. It is because logistics, as in every field of civilization (Alpaslanoğlu, 2023), is a critical element of disaster preparedness and response. Disaster logistics also focuses on managing and coordinating resources to support government and private sector operations (State of California, 2024a). However, for strong all-round disaster logistics, close collaboration is required at local, regional, and national (State of California, 2024b), and international levels to ensure a highly integrated approach.

Food logistics is particularly important in disaster situations. In addition to taking priority among relief materials in disaster situations, food materials also constitute the majority. For example, food aid constituted 83% of the assets deployed to Texas in anticipation of Hurricane Gustav (Department of Homeland Security, 2009). Additionally, malnutrition reduces people's ability to recover from a crisis. It impairs cognitive functions, reduces immunity to disease, increases susceptibility to chronic diseases, limits income opportunities, and reduces the ability to interact with society. It weakens resilience and can increase dependence on ongoing support (Sphere Association, 2018: 160). Therefore, food support and availability can be seen as extremely important for disaster logistics. Therefore, ensuring a sufficient and sustainable flow of food products quickly and uninterruptedly in disaster situations can be considered the most important function of food logistics. For this reason, storage areas with high disaster resistance can be created as close as possible to disaster areas. These storage areas should have direct access to at least two different modes of transport and have loading/unloading ports based on rail and road transportation, and local, regional, and national - as much as possible, international - storages should be connected and integrated with each other.

In addition to existing disaster logistics storages, resilient disaster storages can be created in seaports, train stations, airports, and commercial logistics centers for priority hardware such as food products, medical products, search and rescue equipment, communication tools, generators, and water purification systems. Such materials and equipment should be stored in an ISO container (such as durable food and medical products) or placed in an ISO container so that they are ready for use at any time (such as water purification systems, communication systems, generators, search, and rescue equipment). Cold chain, refrigerated, ATP-enabled short-lived food products can also be transported in separate ISO containers and stored during operation (for short periods), but this will require additional energy (electricity or diesel fuel). Thus, priority materials can be transported, stored, loaded/unloaded with maximum integration in all types of transportation, and also can be switched between transport modes in a very short time. Furthermore, ISO containers themselves can be used as a micro-level storage area/facility. The system in question will be weatherproof, safe, and resistant to external conditions. When necessary, it will also be able to offer safe and fast storage opportunities in open areas. It will have a simultaneous and secure tracking feature with its built-in GPS-based systems. Thanks to the international serial number of the ISO container, its content, features, and location information will be accessed simultaneously by relevant parties and will bring significant advantages and convenience to logistics operations. However, it should be taken into account that the use of special equipment is required to transfer such a container (ISO Container).

Considering the experience of recent emergencies, 24% of all challenges seen in the field of work in strategic regions (outside the affected country and usually at the headquarters of an agency) outside the affected area of supply chain activities in the first phase (including the supply and transportation of relief materials) of the emergency response are related to "access and transportation." Significant challenges have also been identified in the areas of market capacity, people management and coordination, and storage and storage. Suggested Supply Chain Management solutions to the identified challenges include pre-positioning relief supplies further down the supply chain, in tactical zones (just outside the affected area, but often within the affected country or in a neighboring country). Thus, the efficiency and effectiveness of emergency response can be increased (AusAID, 2007). Additionally, other organizations such as food banks, food pantries, and soup kitchens (DFCS, 2024; State of California, 2024c) can also play important roles in food supply in disaster situations.

While industrial logistics operations and processes currently require the flawless execution of extremely comprehensive and complex global-scale high-level technical operations, logistics operations that can meet the needs and expectations in disaster situations are a fully technical and specific field of logistics. For this reason, all stages and activities of disaster logistics management and operations should be managed and carried out by people and teams with logistics experience who have gained expertise by receiving logistics training. It is deemed necessary to include these logistics experts in disaster-related institutions and organizations of public administrations.

Disclaimer

None.

References

Alpaslanoğlu, A. (2023). *Evaluation on global logistics performance indicators*: 2007-2023. In A. Karaca (Ed.). International research in social, human and administrative sciences XVII. (pp. 89-110). Eğitim Yayınevi.

Arizona Secretary of State. (2023). Title 9. Health Services Chapter 17. *Department of Health Services.*

AusAID. (2007). *Emergency Response Supply Chain Assessment.*

Cabinet Office. (2009). *Logistic Operations for Emergency Supplies, Guidance for Emergency Planners.*

CARE International. (2024a). *1.1 Role of logistics in humanitarian operations.*

CARE International. (2024b). *4.4 Storage of materials and equipment.*

Center for Disaster Philanthropy. (2024a). *The Stafford Act.*

Center for Disaster Philanthropy. (2024b). *Disasters and Long-Term Recovery.*

Department of Homeland Security. (2009). *FEMA's Sourcing for Disaster Response Goods & Services.*

DFCS. (2024). *The Emergency Food Assistance Program Overview.*

Doğan, A. (2014). *Kara Yolu Yük Taşımacılığı.* Beta Basım.

EIS Council. (2024). *The wrath of mother nature: Natural disasters and their impact on critical infrastructure*.

EM-DAT. (2024a). *Home*. https://www.emdat.be/: UNSTATS. (2019). Session 6.3: Disaster-related statistics.

EM-DAT. (2024b). *2023 Disasters in numbers: A significant year of disaster impact*.

EM-DAT. (2024c). *EM-DAT Overview*. https://www.emdat.be/.

EM-DAT. (2024d). *Access all EM-DAT publications*! https://www.emdat.be/publications/.

Federation of American Scientists. (2022). FEMA's Role in Logistics Management for Disaster Response. https://sgp.fas.org/crs/homesec/IF12239.pdf.

FEMA. (2020). *Disaster Emergency Communications*.

Florida Division of Emergency Management. (2024). Florida division of emergency management unified logistics plan.

ITU. (2023). *6 Şubat 2023 Kahramanmaraş ve Hatay Depremleri Ön Inceleme Raporları*.

Jiang, Y. and Yuan, Y. (2019). Emergency Logistics in a Large-Scale Disaster Context: Achievements and Challenges. *Int J Environ Res Public Health*, *16*(5),779. doi: 10.3390/ijerph16050779.

Kundu, Y., Sheu, J-B. and Kuo, H-T. (2022). Emergency logistics management—Review and propositions for future research. *Transportation Research Part E: Logistics and Transportation Review,164,*102789. https://doi.org/10.1016/j.tre.2022.102789.

NASA. (2024). *Disasters data pathfinder*. https://www.earthdata.nasa.gov/learn/pathfinders/disasters.

National Library of Medicine. (2024). *Integrated operations for natural disaster management: A systematic review*. https://www.ncbi.nlm.nih.gov/pmc/articles/PMC9808902/.

Public Safety Canada Search. (2024). *National Emergency Response System*. https://www.publicsafety.gc.ca/cnt/rsrcs/pblctns/ntnl-rspns-sstm/index-en.aspx.

Rodríguez-Espíndola, O., Albores, P. and Brewster, C. (2018). Disaster preparedness in humanitarian logistics: A collaborative approach for resource management in floods. *European Journal of Operational Research*. *264,* 3. 978-993. https://doi.org/10.1016/j.ejor.2017.01.021.

Schulz, F. S. (2009). *Disaster Relief Logistics: Benefits of and Impediments to horizontal cooperation between humanitarian organizations*. Haupt Verlag AG.

Sheu, J-B. (2007). Challenges of emergency logistics management. *Transportation Research Part E 43*. 655–659.

Sphere Association. (2018). *The Sphere Handbook, Humanitarian Charter and Minimum Standards in Humanitarian Response*. Fourth edition. www.spherestandards.org/handbook.

State of California. (2024a). *Logistics Management*.

State of California. (2024b). *Disaster logistics*.

State of California. (2024c). *Emergency Food Assistance Program*. https://www.cdss.ca.gov/emergency-food.

State of Michigan. (2012). *Disaster Logistics and Donations Management Plan*. MSP/EMHSD Pub. 107a October 2012 Final Edition – Planning Template Only

Emergency Management and Homeland Security Division Michigan Department of State Polic.
Strategy and Budget Presidency. (2023). *SBB Deprem sonrası değerlendirme raporu.*
The World Bank. (2024). *What is food security?.*
TMMOB. (2024). 6 Şubat 2023 Kahramanmaraş depremleri raporu.
Transportation Research Board. (2013). *Logistics of disaster response.* https://onlinepubs.trb.org/onlinepubs/trnews/trnews287.pdf.
U.S. Department of Homeland Security. (2010). *FEMA's Logistics management process for responding to catastrophic disasters.*
U.S. Government Publishing Office. (2023). *Federal Food, Drug, and Cosmetic Act.*
UN. (2022). *The Sustainable Development Goals Report 2022.*
UN. (2023). *The Sustainable Development Goals Report, Special edition.*
UNDRR. (2020). *The human cost of disasters: an overview of the last 20 years (2000-2019).*
UNDRR. (2023). *GAR Special Report 2023: Mapping resilience for the Sustainable Development Goals.*
USA Facts. (2023). *Are major natural disasters increasing?*
USAID. (2020). *Tips to optimize storage during emergencies.*
USAID. (2023). *Earthquakes in Türkiye, A Review from 1900 to Today.*
WHO. (2024). *Water sanitation and health.*
World Food Programme. (2002). Emergency Field Operations Pocketbook.
World Food Programme. (2019). *The silent enablers.* https://www.wfp.org/stories/silent-enablers.
World Food Programme. (2024a). *UN humanitarian response depot.* https://www.wfp.org/unhrd.
World Food Programme. (2024b). *Food systems.* https://www.wfp.org/food-systems.
World Health Organization. (2024). *Emergency response framework, Internal WHO procedures.*
World Vision International. (2024). *Disaster management.*
Yan, S. and Shih, Y. L. (2009). Optimal scheduling of emergency roadway repair and subsequent relief distribution. *Computers & Operations Research. 36* (6). 2049–2065. doi: 10.1016/j.cor.2008.07.002.
Yigezu, A. Y., and Sanders, J. H. (2012). Introducing new agricultural technologies and marketing strategies: a means for increasing income and nutrition of farm households in Ethiopia. *African Journal of Food, Agriculture, Nutrition and Development, 12(5), 6365-6384.*

Chapter 15

Emergency Aid and Disaster Management in Hospitals: Crisis Communication

Ayşe Albay*, PhD
Fırat University Hospital, Deputy Hospital Manager, Elazığ, Türkiye

Abstract

The risk of natural disasters occurring at any time in the world has made it mandatory to be prepared for disasters. In Türkiye, plans are made to minimize the damage caused by disasters under the coordination of the Disaster and Emergency Presidency (in Turkish: AFAD). Healthcare organizations, which play an active role, especially in disasters and emergencies, and can even be defined as the main actors in managing the process, prepare a Hospital Disaster and Emergency Plan (HDEP) to effectively manage crises that may arise in emergencies and disasters. Disaster management in hospitals is important to ensure the safety and well-being of both patients and employees in emergency and unexpected situations. With the successful implementation of disaster management in hospitals, patient safety can be ensured, the welfare of healthcare personnel can be sustained, infrastructure can be protected, resources can be distributed in a balanced manner, effective communication can be realized, mass casualties can be responded to quickly, and thus the social impact of disasters can be minimized by maintaining public morale and confidence.

As a classic crisis communication environment, emergencies or disasters create an air of panic in both internal and external stakeholders, and the management of the process becomes difficult due to this panic. In case of panic, proper management of the emergency or disaster by the ten key

* Corresponding Author's Email: aalbay@firat.edu.tr.

In: Disaster Management in the Digital Age and Emerging Technologies
Editors: Ayşe Asiltürk and Ayşegül Sarıoğlu Kemer
ISBN: 979-8-89530-213-2

management teams identified in the Hospital Emergency and Disaster Plan is important to minimize the damage. Among the ten basic management positions, the Public Relations Officer is responsible for communicating information to internal and external stakeholders, including staff, patients, patients' relatives, the media, and the whole world's public opinion, with the approval of the HDEP President, can play a key role in crisis management by providing effective crisis communication. In disasters and emergencies, crisis communication is important to disseminate information in a timely manner, reduce confusion and panic, build trust, coordinate response efforts, facilitate cooperation, prevent misinformation, promote preparedness, and address public concerns. Therefore, the choice of tools to use to inform the public during a disaster or emergency is very important. In today's technological world, mass media tools that enable interactive communication are frequently preferred by the masses to obtain information about disasters and emergencies.

Keywords: crisis communication, disaster, disaster management in hospitals, health organizations, mass media

Introduction

We may encounter many disasters or emergencies in our world. Disasters or emergencies can create serious problems in the places where they occur and can negatively affect both the physical and psychological health of many people, make their living conditions difficult, and even cause people to lose their lives. Natural disasters include hurricanes, severe storms, earthquakes, volcanic eruptions, tornadoes, tropical storms, fires, floods, tsunamis, and droughts, resulting from both geological and meteorological natural processes of the Earth and are often associated with death, trauma, and destruction of property (Saeed and Gargano, 2022). Natural disasters are defined as disruptions of the ecological system that exceed the adaptive capacity of society and therefore require external assistance (Lechat, 1990). Globally, natural disasters can be classified into three main groups (Kouadio et al., 2012);

1. Hydro-meteorological disasters (floods, wave surges, storms, typhoons, hurricanes, and tornadoes).
2. Geomorphological disasters (landslides and avalanches).

3. Geophysical disasters (earthquakes, tsunamis, and volcanic eruptions).

It is possible to classify disasters as natural disasters and artificial disasters. Natural disasters can be of geological origin, such as earthquakes, volcanoes, tsunamis, and landslides, and of meteorological origin, such as lightning, floods, storms, tornadoes, hail, and droughts. On the other hand, artificial disasters can occur due to human actions, such as fires, terrorist incidents, epidemics, and biological attacks (Kadıoğlu, 2017). Whether the cause of the disaster is natural or artificial, a disaster is usually encountered as a result.

Effective and correct management of this process in health institutions that play an active role in the healing process in disasters is very important to minimize the damage that may occur. When we look at the history of disaster management, we see that different aspects are handled and discussed separately. However, among these, the multidisciplinary health systems approach to disaster management suggests that each component of the health system should be resistant to the threats embedded in its own structure. In this way, the entire health system can be strengthened to effectively meet the demands of all kinds of disasters, and coordinated, rapid, and effective response and recovery can be provided (Bayntun et al., 2012).

A health system consists of all organizations, people, and actions whose primary purpose is to promote, improve, or maintain health (WHO, 2007). The health system consists of six building blocks: health service delivery, health workforce, health information management system, medical products, including vaccines and technologies, health financing, and leadership and governance in health (Olu, 2017).

The building blocks listed above must be used, coordinated, and managed appropriately, accurately, and effectively in healthcare organizations, which are the key points in the process of recovery and rehabilitation after emergencies and disasters. Due to their nature, emergencies and disasters create an air of panic in the place where they occur, and this prevents the process from being managed effectively. For this reason, health organizations need to prepare plans for possible emergencies and disaster management. In this context, Hospital Disaster and Emergency Plans (HDEP) are prepared in health institutions in Türkiye.

Hospital Disaster and Emergency Plan (HDEP)

Hospital disaster plans are defined as plans prepared in advance to ensure that hospitals are prepared to respond to disasters, emergencies, and unusual situations in an accurate, timely, and effective manner (AFAD, 2023).

The Hospital Disaster and Emergency Plan (HDEP) covers the pre-disaster, disaster, and post-disaster period and aims to make hospitals safe and prepared for the first 72 hours without interrupting the service and without outside help. In our country, action was taken shortly after the foundation of the republic to effectively manage emergencies and disasters. In 1928, it started to be regulated, and finally on May 29, 2009, "Law No. 5902 on the Organization and Duties of the Disaster and Emergency Management Presidency" was put into force, and the General Directorate of Civil Defense and Provincial/district civil defense directorates were abolished. With this Law, the Prime Ministry Disaster and Emergency Management Presidency and the Provincial Disaster and Emergency Directorates were established. Civil defense services were started to be carried out by the Department of Civil Defense and Provincial Disaster and Emergency (AFAD) directorates within the Presidency (AFAD, 2023).

The multifaceted effects and consequences of disasters and emergencies have made it mandatory for healthcare organizations to be prepared for disasters and emergencies. For this purpose, the Hospital Disaster and Emergency Plan (HDEP) guideline, which is a national guideline on national and international disaster risk reduction and disaster management in the health sector, was first prepared in 2014 under the coordination of the Ministry of Health in Türkiye with participatory teamwork. The first version of the guideline was published in December 2015, and with the feedback received from the training, practices, and drills carried out in the field, the revision was completed in June 2021, and the second version was published and entered into force (MoH, 2021).

According to the Hospital Disaster Plan prepared in Türkiye, an incident management team consisting of ten basic positions has been defined in hospitals to manage emergencies and disasters in an effective, accurate, and planned manner.

Incident Management Team and Duties

The ten basic incident management teams formed in accordance with the Hospital Disaster and Emergency Plan consist of the Pill Chairman, HDEP Secretariat, Occupational Safety Officer, Inter-institutional Coordination Officer, Public Relations Officer, Medical Technical Advisory Board, Operations Chief, Planning Chief, Logistics Chief, and Finance Chief. The duties, authorities, and responsibilities of all defined positions are defined in the HDEP guidelines. According to the MoH HDEP guideline, the duties, authorities, and responsibilities of the ten basic positions are defined as follows (MoH, 2021).

1. Head of HDEP: Organizes and manages the Hospital Incident Management Center. Gives all strategic directives related to hospital disaster management and support activities from emergency response to termination. Approves the Incident Action Plan for each period.
2. Public Relations Officer: Ensures that the information approved by the Head of HDEP is communicated to internal and external stakeholders, including staff, patients, and their relatives, and the press.
3. Inter-Institutional Coordination Officer: Communicates with the representatives of other organizations at the Hospital Incident Management Center and acts as the authorized person who ensures the flow of information about the disaster and emergency.
4. Occupational Safety Officer: Responsible for the safety of the working environment in terms of occupational safety and the supply and control of the use of personal protective equipment. Monitors and identifies existing and potential threats to patients, staff, and visitors. Follow dangerous situations and take the necessary measures to reduce them. In emergency situations that threaten health and life, he/she is authorized to stop any operation with the approval of the HDEP President.
5. HDEP Secretary: Fulfills the instructions given by the personnel assigned in the HDEP office such as activating the disaster and emergency plan, opening the incident management center, and collecting and recording the information requested by the Incident Management Team.
6. Medical Technical Advisory Board: Provides the necessary consultancy services to the HDEP President or department chiefs for

the emergency response according to their areas of expertise on the technical issues needed.

7. Chief of Operations: Develops and implements strategies and tactics to achieve the objectives set by the Head of the HDEP. He/she organizes response areas, manages resources, and assigns medical care, infrastructure, security, hazardous materials, and psycho-social support.
8. Chief Planning Officer: Ensures the collection and analysis of all data related to operations and resource management in disasters and emergencies, the development of alternatives for tactical operations, the initiation of long-term planning, the organization of planning meetings, and the preparation of Incident Action Plan for each operation period.
9. Logistics Chief: Organizes and manages the provision of services, manpower, tools, equipment, food, water, medicines, medical supplies, and other materials required for the activities of the hospital during a disaster and emergency.
10. Chief Finance Officer: Accounts for and monitors the use of financial assets and expenditures. Manages the recording of expenditures and reimbursement of expenses.

The ten basic incident management teams listed above are defined in the HDEP prepared by health institutions. The defined officers are assembled at the request of the HDEP chairperson in case of a possible disaster or emergency. The gathering place of this team is also defined in the HDEP. For this area defined as an incident management center in HDEP, at least two locations should be defined in case of possible damage.

Emergency code call systems are used for rapid communication with the incident management team. There are a total of 11 emergency code call systems defined in the HDEP (MoH, 2021):

1. Code Blue: Adult or child medical emergency (cardiopulmonary arrest), life risk code.
2. Code Gray: Code for aggressive person, armed person, or active shooting or hostage situation.
3. Code Green: Emergency termination code.
4. Code Orange: Code for leakage or release of hazardous material.
5. Code Pink: Code for infant or child abduction.
6. Code Purple: Emergency response plan activation code.

7. Code Red: Fire code.
8. Code White: Attack on employee code.
9. Code Yellow: Evacuation code.
10. Code Turquoise: External mass casualty code.
11. Code Black: Bomb threat code.

Due to these codes defined in the HDEP, rapid communication with employees during an emergency or disaster is established, effective communication is established at the time of risk, the desired message is communicated in a short and clear way, it saves time, helps prevent panic, ensures preparedness for an emergency or disaster, and increases patient and employee safety.

In the event of a possible emergency or disaster, a purple code call is made to gather the incident management team, and the team gathers at the incident management center after this call is made. After this call is made, the incident management team starts to work within the framework of their duties, powers, and responsibilities to take the emergency or disaster under control.

All positions within the incident management team have different roles. The role of the Public Relations Officer, who will manage and direct all internal and external stakeholders in terms of information with the permission of the HDEP chairperson, is very important in the effective management of the process.

The Public Relations Officer takes an active role in crisis communication during emergencies and disasters as he/she informs the press, employees, and the community by using tools such as media, the hospital website, visual announcements on hospital boards, short films on hospital monitors, and printed materials in line with the permission of the HDEP President.

The Crisis Concept and Crisis Communication

The concept of crisis, which is a concept that emerges naturally in disasters or emergencies, is a process in which an unpredictable event that threatens the expectations of internal and external stakeholders can have negative consequences by negatively affecting the performance of the organization (Coombs 2006). According to another definition, a crisis is an important event that affects the organization or the sector, as well as stakeholders, products, services, or image with negative consequences (Fearn-Banks, 2007).

According to Reid (2000), crisis types are categorized under six headings: crises related to natural disasters, crises related to work, crises related to environmental problems, crises related to employee health and safety, crises related to employee relations, and crises related to management problems.

- Crises due to Natural Disasters: Lightning, earthquake, snow or ice conditions, extreme cold, flood or drought, hurricane, tornado, tsunami.
- Operational Crises: Equipment shortages, loss of key subcontractors or suppliers, construction delays, cost overruns, design errors or problems, explosions, fires, essential service failure, environmental or community opposition to a project, collapse or collapse of structures, data or telecommunications failure, and loss of important information.
- Crises due to Environmental Issues: Groundwater contamination, community exposure to toxic chemicals, and release of toxic chemicals into the air and water.
- Crises related to Employee Health and Safety: Chronic safety problems, exposure to carcinogens, employee injuries, or illnesses.
- Employee Relations Crises: Negotiations, unfair labor practices, strikes, and work stoppages.
- Crises due to Management Problems: Bankruptcy, contract dispute with a customer resulting in litigation, hostage, kidnapping, aggressive takeover attempt, key personnel taking jobs with a competitor, management succession, merger or acquisition, murder, negative publicity through rumors, and negative publicity related to political contributions.

Baubion underlines that institutions should be prepared for the crisis in order to develop knowledge and capacities to effectively overcome the crisis. For this, he lists what organizations should do in crisis management as follows (2013);

1. Risk Assessment: Sectoral analysis based on historical events and national risk assessments.
2. Contingency Plans: Creating a scenario- and capability-based information network.
3. Training: Strategic exercises and networking with testing of plans and procedures.

4. Enablement: Development of early warning systems that trigger contingency plans and strategic foresight.

Due to these processes listed by Baubion, organizations will be able to easily take the crisis under control. In this way, recovery and rehabilitation processes can be accelerated, and the crisis can be ended as soon as possible.

The crisis has a multifaceted impact, so managing the crisis correctly and taking the necessary measures for the crisis is important in terms of corporate image, winning internal and external stakeholders, and turning the crisis into an opportunity. Benoit has developed an image restoration approach for use in corporate crises. Benoit states that due to this approach, messages can be prepared by communication experts in times of crisis, and these messages can be critically interpreted and evaluated by critics. Benoit lists the strategies and features of this approach as follows (1997);

1. Denial
   a. Simple Denial: Asserting that no wrongdoing has occurred.
   b. Blame-shifting: Asserting that the mistake was made by someone else.
2. Avoiding Responsibility
   a. Provocation: Attempting to make the error understandable by claiming that the action had to be taken as a result of another action.
   b. Deficiency: Asserting that the error is due to a lack of knowledge or skills.
   c. Accident: Asserting that the error occurred as a result of misfortune.
   d. Good Intentions: Asserting good intentions despite the error.
3. Minimizing the Negatives of Action
   a. Supporting: Emphasizing the good qualities of the activity or the positive aspects of the past, even though there were mistakes.
   b. Minimizing: Asserting that the harm caused by the error is very small.
   c. Distinguishing: Asserting that the mistake caused less harm than similar mistakes made by other firms in the past.
   d. Superiority: Asserting that the benefit obtained as a result of the action is more important than the harm caused by the error.
   e. Attacking the Accusers: Attempting to undermine the credibility of the wrongdoer in society.

f. Compensation: Paying compensation to victims, compensating them for their losses.

4. Corrective Action: Solving the problem and working to prevent the problem from recurring.
5. Acceptance: Apologizing for the mistake.

The planned execution of communication during a crisis is vital for many factors. According to Segger, crisis communication has generally focused on short-term goals, such as doing as little damage as possible to corporate image and resolving the crisis quickly. However, it is important to examine crisis processes over longer periods of time and on a broader scale to apply communication and public relations more comprehensively in a crisis (2002).

Caponigro states that crisis communication processes are carried out in seven stages and lists them as follows (2000):

1. Identifying and assessing weaknesses,
2. Making decisions to prevent a crisis,
3. Creating a crisis plan, identifying worst-case scenarios, and taking precautions,
4. Problem-solving and effective communication,
5. Maintain open, honest, and reliable communication throughout the crisis,
6. Monitoring, evaluating, and adjusting, monitoring the thoughts and behaviors of the target audience during and after the crisis,
7. Protection of the organization to enhance its reputation and credibility.

Successfully managing and controlling the crisis that emerges in disasters or emergencies is important in terms of shortening the recovery process of the crisis. For this reason, the crisis management performance of institutions is very important in the transition to the normalization process. Among these institutions, it is very important that healthcare organizations that play an active role in emergencies and disasters and inform the public about the balance sheet of the disaster or emergency, determine the right crisis communication strategies, and use the right communication technologies.

Crisis Communication in Health Institutions and Technologies Used for Information Transfer in Crisis Communication

There are issues to which health institutions should pay attention in managing the crisis that arises after a disaster or an emergency. Bulduklu lists these issues as follows (2015):

Transparency: As in all institutions and organizations, it is very important for the management to be transparent in health service delivery. Because transparency is the most fundamental element of interaction with the public. It is natural that uncertainties occur during a crisis. However, as long as these uncertainties are not clarified, it may cause more public talk about the event. It should not be forgotten that when uncertainties are not explained and remain silent, different opinions will try to close the gap. For this reason, the evidence should be presented in a concrete and complementary manner in statements or information about the crisis. Furthermore, different opinions about the process should be accepted, evaluated, and used in the management of the crisis process. Since withholding information during crisis communication will have a negative impact, the information should be disclosed rather than withheld. It is the responsibility of both the target audience and other organizations to make clear the source of the information and the target of the communication.

Transmitting the Facts: The explanation style error accounts for a large share of the problems in health communication. It should not be forgotten that the style of explanation itself can create a crisis. For successful health communication in a crisis, it should be known to whom and to what extent the information should be conveyed. According to this understanding, how much of the information will be conveyed to the patient, his/her relatives, or the media? Answering questions such as "What is the relevance of the crisis to this amount or quality of information?" will guide the decisions to be made by crisis managers in the process of information transfer.

Demonstrating Sacrifice: Healthcare organizations are a branch of service that requires sacrifice due to the services they provide. Doing more than they have and conveying this image to the target audience in the right way is very important in crisis communication. As long as healthcare professionals do not love their profession, they cannot perform well in health communication, and therefore they may face a crisis more often. For this, healthcare professionals need to use their empathy skills well, and it is necessary to sacrifice enough without overdoing it.

Not Ignoring Employees: With the emergence of the concept of patient-centered care and communication, employees have started to be prioritized.

However, it is very important to consider the opinions of internal and external stakeholders in times of crisis. In other words, employees should be considered in explanations and communication, as much as the patient is taken into focus. It is an ethical responsibility not to blame colleagues and other healthcare organizations for crisis communication to be carried out at the individual or social level. Similarly, it is also an ethical responsibility to support employees in front of the society and not to make them look bad.

Healthcare organizations constitute a key point in intervening in disasters and explaining the balance sheet of the disaster, as well as in recovery and rehabilitation processes. Although many methods have been used to inform the public about the disasters in Türkiye history, the mass media has been the most preferred among these methods. Thanks to the mass media, information was provided to the public at the time of the disaster. Due to this feature, the correct use of mass media by health institutions will contribute to the acceleration of the healing and rehabilitation process.

In addition to traditional mass media such as newspapers, magazines, television, and radio, with the development of Internet and satellite technologies, Internet journalism, social media platforms such as Twitter, Facebook, Youtube and Blogs, smart mobile phones and tablets have become accessible at almost any time and from any place (Bekiroğlu and Balcı, 2014). In disasters, the masses have turned to traditional or new mass media to fill the information gap.

Today, new media tools, especially social media, which allow interactive communication, are used in crisis communication rather than traditional mass media. Şahinsoy measured the impact of traditional media, which was used as a means of information transfer in the Marmara earthquake, and new media tools, especially Twitter, which was used as a means of information transfer in the Van earthquake. In the study, it was determined that in the Marmara earthquake, which was dominated by traditional media, not enough news was received; in the Van earthquake, which was dominated by new media, especially Twitter, instant information about the region was accessed, aids, campaigns, location notifications were made, and work could be organized instantly thanks to these and spread to large segments quickly (2017).

Proper use of social media in crisis communication plays an important role in controlling the crisis. According to Prentice and Huffman, the advantages of using social media in crisis communication can be listed as follows (2008):

1. It gives credibility to the organization.
2. Social media also guarantees that your message is heard.
3. Another benefit of social media is that it provides crisis communicators with a unique and effective way to defend the organization's reputation.
4. Finally, public blogs are used by authorities to inform on critical issues during a crisis. This balances the scope of the crisis while meeting the demand of stakeholders for timely and accurate information.

In the Covid-19 disaster, which is accepted by the world for its global developments, social media, especially Twitter, was used more by healthcare providers to inform the public. In the Twitter analysis conducted by Ateş and Baran regarding the Covid-19 epidemic for the purpose of effective use of social media in crisis communication; In the posts made from the Twitter account of the Ministry of Health of Türkiye, 74% of Twitter stated that it is an institution for the purpose of providing information and warning, followed by 63% to provide surveillance, monitoring, and status information, and 53% to raise public awareness about risks and crises for its professional purpose. It is similar to Türkiye in terms of effective use in the posts made on the Twitter account of the US Ministry of Health, 59% of Twitter is to provide information and warning, 41% to provide surveillance, monitoring, and status information, and 30% to raise public awareness about risk and crisis. It was operated for the purpose (2020). Similarly, the Ministry of Health shared the secrets of the Kahramanmaraş earthquakes that occurred in 2023 with the public via the social media tool Twitter. The first fall of the severe earthquake in Kahramanmaraş by the Ministry of Health was shared every day for 36 days, made by Alay and Eke in the crisis communication of the Ministry of Health, the tweets increased the most with shares in different sections in the first 20 days, and the tweet sharing was February 9, that is, the earthquake. Today, it has started to increase significantly (2023).

In the study conducted by Çerçi et al. to examine the use of social media as a means of information during the Covid-19 disaster, it was found that social media was used mostly to follow the agenda and get news, and 75.4% of social media users used social media to get information about the Covid-19 epidemic. It has been determined that the information source that social media users use or prefer to use most to get information about the Covid-19 crisis is television, which is a traditional mass media tool, social media is in the second place, participants see social media as a need in times of crisis, and follow the agenda in a crisis environment. They concluded that they use social media to stay informed about the crisis and that they accept social media as a reliable source of information in crisis communication (2020). As can be seen from this study, both traditional and new media tools are of great importance in managing crisis communication and informing the public in times of disaster and emergency.

Since mass media play a leading role in information and awareness-raising activities, the effective use of these tools by the positions responsible for transferring information to internal and external stakeholders in disasters and emergencies is extremely important to overcome the process with minimal damage and to accelerate the recovery and rehabilitation process. Especially in health institutions that play a key role in the management of disasters and emergencies, it is very important that the Public Relations Officer, who is responsible for providing information about the balance sheet of the event to internal and external stakeholders in accordance with HDEP, be conscious in using mass media.

Discussion and Conclusion

Disasters or emergencies result in many negative processes where they occur due to their structure and characteristics. Due to these negative processes, organizations responsible for managing the disaster must implement the correct management policy. It has become necessary to develop professional processes on the subject, especially in our country, which has a high rate of encountering disasters due to its geographical location.

In disasters or emergencies, healthcare institutions undoubtedly have the primary role in the rehabilitation and recovery process. Therefore, health institutions need to be prepared to manage disasters and emergencies. In this preparation process, just making plans is not enough. These plans must be feasible. This is so because executing the prepared plans with the desired effect is very important to manage the process well and quickly return to the old order by reducing chaos and panic.

Health institutions are at a key point not only in reducing the negative outcomes of disasters but also in informing internal and external stakeholders about the event balance sheet. It is the duty authority and responsibility of the Public Relations Officer to inform internal and external stakeholders in the Hospital Disaster Plans prepared by healthcare institutions. It is very important that the Public Relations Officer informs internal and external stakeholders about the consequences of the crisis in a way that does not disrupt the order. While informing internal and external stakeholders, it should be done in line with the principle of transparency, which is one of the characteristics of effective management. The reason is that gaining the public's trust in disasters and emergencies will allow the process to be easily taken under control. For this reason, health institutions must ensure the transfer of information in a way that does not break the trust of the public and does not harm the healing and rehabilitation process.

To convey information to the public at the desired level and in the desired way, the medium in which the information will be presented must be correctly determined. It is necessary to determine the right strategies to inform, especially in social media tools that allow interactive communication and allow the target audience to be informed to intervene immediately in the event.

Disclaimer

None.

References

AFAD(2023). https://www.afad.gov.tr/tanimi-tarihcesi-vegorevleri#:~:text=25%2F05%2F2009%20tarihinde%205902,il%C3%A7e%20sivil%20savunma%20m%C3%BCd%C3%BCrl%C3%BCkleri%20kald%C4%B1r%C4%B1lm%C4%B1%C5%9Ft%C4%B1r. (Accessed: 20.04.2024).

Alay, D., & Eke, E. (2023). A Research on Crisis Communication of the Ministry of Health. *Süleyman Demirel University Social Sciences Institute Journal, 2*(46), 317-337.

Ateş, N. B., & Baran, S. (2020). Effective Use of Social Media in Crisis Communication: Twitter Analysis for the Covid-19 (Coronavirus) Pandemic. *Kocaeli University Faculty of Communication Research Journal 16*, 66-99.

Baubion, C. (2013), "OECD Risk Management: Strategic Crisis Management," OECD Working Papers on Public Governance, No. 23, OECD Publishing. http://dx.doi.org/10.1787/5k41rbd1lzr7-en.

Bayntun, C., Rockenschaub, G., & Murray, V. (2012). Developing a health system approach to disaster management: A qualitative analysis of the core literature to complement the WHO Toolkit for assessing health-system capacity for crisis management. *PLoS Currents*, *4*, e5028b6037259a. https://doi.org/10.1371/5028b6037259a.

Bekiroğlu, O. and Balcı, Ş. (2014). Social Media in Crisis Communication. M. Akdağ and Ü. In Arklan (Ed.), *Crisis Management - Solution-Oriented Strategic Approaches on a Communicative Basis* (pp. 151-181). Konya: Literaturk Academia.

Benoit W L (1997) Image repair discourse and crisis communication. *Public Relations Review, 23*(2), 178-182.

Bulduklu, Y. (2015). Crisis and Crisis Communication in Healthcare. Communication in Health Institutions, (Ed: E. Yüksel) Anadolu University Publications, 203-231.

Caponigro, J. R. (2000). The crisis counselor: A step-by-step guide to managing a business crisis. McGraw-Hill Companies.

Coombs W T (2006) Code Red in the Boardroom: Crisis Management as Organizational DNA, Greenwood Publishing Group.

Çerçi, Ü. Ö., Canöz, N., & Canöz, K. (2020). Use of Social Media as an Information Tool During the Covid-19 Crisis. *Selçuk University Social Sciences Institute Journa l, 44*, 184-198.

Everybody's Business. Strengthening Health Systems to Improve Health Outcomes; WHO's Framework for Action. World Health Organization (WHO) (2007). Available from: http://www.who.int/healthsystems/strategy/everybodys_business.pdf.

Fearn–Banks, K. (2007). Crisis Communications: A Casebook Approach. New Jersey: Lawrence Erlbaum Associates.

Kadıoğlu, M. (2017). Disaster management: Expecting the unexpected, managing the worst. Istanbul: Marmara Municipalities Union Cultural Publications.

Kouadio, I. K., Aljunid, S., Kamigaki, T., Hammad, K., & Oshitani, H. (2012). Infectious diseases following natural disasters: prevention and control measures. *Expert Review of Anti-İnfective Therapy, 10* (1), 95–104. https://doi.org/10.1586/eri.11.155.

Lechat, M. F. (1990). The epidemiology of health effects of disasters. *Epidemiologic Reviews, 12*(1), 192-198.

MoH, (2021). https://www.saglik.gov.tr/TR,1789/hastane-afet-ve-acil-durum-plani-hap-hazirlama-kilavuzu.html.

Olu, O. (2017). Resilient Health System As Conceptual Framework for Strengthening Public Health Disaster Risk Management: An African Viewpoint. *Frontiers in Public Health, 5*, 263. https://doi.org/10.3389/fpubh.2017.00263.

Prentice, S. & Huffman, E. (2008). "Social Media's New Role in Emergency Management Emergency Management and Robotics for Hazardous Environments," *Emergency Management and Robotics for Hazardous Environments,* New Mexico.

Reid, J. L. (2000). Crisis Management: Planning and Media Relations for the Design and Construction Industry. New York: John Wiley &Sons.

Saeed, S. A., & Gargano, S. P. (2022). Natural disasters and mental health. *International review of psychiatry (Abingdon, England), 34*(1), 16–25. https://doi.org/10.1080/09540261.2022.2037524.

Seeger, M. W. (2002). Chaos and crisis: Suggestions for a general theory of crisis communication. *Public Relations Review, 28* (4), 329-337.

Şahinsoy, K. (2017). Traditional and Social Media in Terms of Crisis Management. *Istanbul Aydin University Journal, 9*(4), 1-19.

Chapter 16

Integrated Disaster Management

Sevda Demiröz Yıldırım*, PhD

Department of First and Emergency Aid, Sabuncuoğlu Şerefeddin Health Services Vocational School, Amasya University, Amasya, Türkiye

Abstract

Until recently, disaster management processes were generally focused on healing wounds. In this approach, emphasis was placed on response and recovery processes (crisis management) after disasters occurred. However, disasters have been experienced, and studies have shown that it is not correct to approach disasters only with a focus on crisis management (response and recovery), but pre-disaster risk reduction (mitigation and preparedness) studies are also very important for disaster resilience. The importance of disaster risk reduction strategies began to be emphasized more in international studies that started with The International Decade for Natural Disaster Reduction (IDNDR 1990-2000) and continued with the Yokohama Strategy (1994). In the ongoing process, the Hyogo Framework for Action (2005-2015) and, most recently, the Sendai Framework for Disaster Risk Reduction (2015-2030) have reiterated the importance of risk management for resilient communities within the framework of lessons learned. Although risk management is considered very important for disaster resilience in new disaster management strategies, it is not sufficient alone for successful disaster management. Preparedness, mitigation, response, and recovery processes, which are all stages of disaster management and risk and crisis management, need to be handled. The disaster management system formed by all these processes is called integrated disaster management. Integrated disaster management is based on the focus on preventing

* Corresponding Author's Email: demirozsevda@gmail.com.

In: Disaster Management in the Digital Age and Emerging Technologies
Editors: Ayşe Asiltürk and Ayşegül Sarıoğlu Kemer
ISBN: 979-8-89530-213-2

disasters, reducing their damage, and increasing resilience. This disaster management system increases the resilience of individuals, societies, and countries to disasters.

Keywords: disaster, disaster crisis management, disaster risk management, integrated disaster management, resilience

Introduction

Disasters have profoundly impacted numerous individuals from the past to the present, resulting in death, disability, injury, and/or significant economic losses. Additionally, disasters have led to the collapse of civilizations and mass migrations throughout history. Presently, the frequency and severity of disasters are on the rise, attributed to factors such as rapid population growth and urban concentration, unhealthy and swift urbanization, inappropriate technological utilization, and global climate change.

According to the Emergency Event Database (EM-DAT) report in 2022, 387 natural hazards and disasters were recorded worldwide, resulting in the deaths of 30,704 individuals and affecting 185 million people. These disasters also incurred approximately US$223.8 billion in economic losses (CRED, 2023). Furthermore, it is reported that more than 6.5 million people lost their lives due to the COVID-19 pandemic, classified as a biological disaster, with the actual death toll believed to be higher than the official figure (IFRC, 2023).

Countries conduct numerous local studies to prevent disasters and minimize their potential damage. Additionally, various international studies are undertaken to prevent disasters and mitigate their impact. To prevent disasters and minimize their damage, it is necessary to have a robust disaster management system, as well as a high social resilience and capacity. In the past, due to limited information and technology, disaster response has generally focused on the moment and after the disaster. As we transition from the recent past to the present, with the increase in knowledge about disasters and advancing technology, this understanding has been replaced by a new disaster management approach in which pre-disaster processes are also taken into consideration. This new approach will be discussed in detail in the following section.

Integrated Disaster Management

Societies have grappled with disasters since ancient times due to the damage they inflict on nature and the environment. The approach to disaster management, which previously focused on the immediate aftermath of disasters, has evolved into a comprehensive strategy involving national and international cooperation, encompassing all segments of society within the framework of governance and collaboration among various institutions and organizations. To comprehend this evolution in disaster management, it is essential to first understand the relevant terminology.

The first of these definitions is the concept of disaster management. Disaster management is defined as "The organization, planning, and implementation of measures preparing for, responding to, and recovering from disasters" (UNDRR, 2009). It is a dynamic and continuous process comprising stages such as identifying hazards and their potential impacts on society's ability to maintain a balanced and healthy environment, preventing disasters, mitigating damages, intervening, and subsequent recovery.

Disaster management entails the collaboration of all public institutions, organizations, private sector entities, non-governmental organizations, and all segments of society, starting from individuals, to prevent disasters, reduce damages, respond effectively and swiftly to disasters, and facilitate the recovery of affected communities, creating a safer and more resilient environment post-disaster. It is a multidisciplinary, multi-stakeholder dynamic management process that necessitates cooperation within the framework of good governance.

The disaster management process comprises stages of mitigation and preparedness before a disaster occurs, and response and recovery stages afterward. These stages collectively constitute the disaster management phases.

The success of actions taken in each phase significantly influences subsequent phases. Therefore, considering and evaluating these stages separately may lead to incomplete analyses.

To achieve success as a nation and resilience as a society in the face of disasters, a comprehensive disaster management model encompassing all phases of mitigation, preparedness, response, and recovery is essential. This model is known as Integrated Disaster Management, which incorporates disaster risk management (mitigation, preparedness) and disaster crisis management (response, recovery) processes.

Integrated disaster management is defined as "A management process that acknowledges hazards and undertakes preventive measures, damage reduction, preparedness, response, and recovery activities by mobilizing all societal resources and capabilities to build a resilient society capable of addressing disasters" (AFAD, 2014).

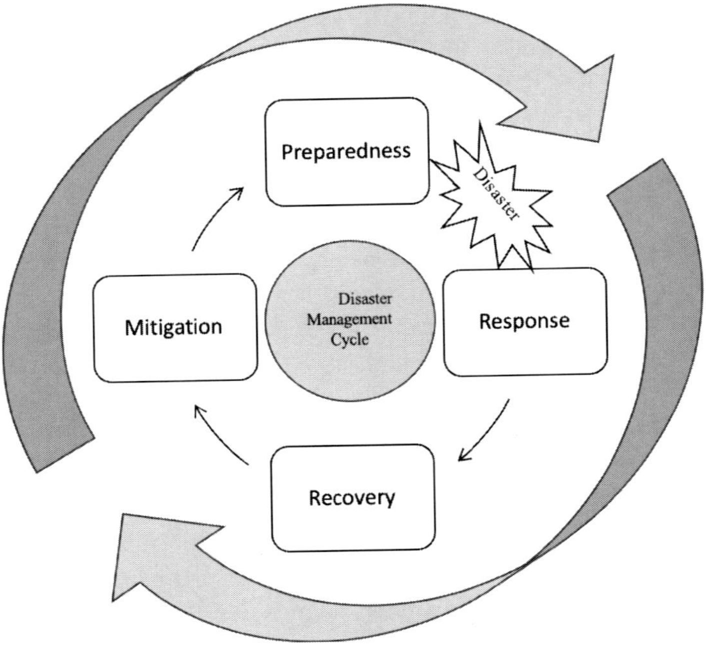

Figure 1. Disaster management cycle.

The integrated disaster management approach consists of four phases, including before, during, and after the disaster. These stages, mitigation, preparation, response, and recovery (Figure 1), constantly follow each other in a cycle. The success of any phase of the cycle greatly affects the success of the next phase. Phases are not entirely separate from each other; two-phase processes can be performed simultaneously. For example, during the response phase, while search and rescue activities are underway, efforts such as establishing shelters, infrastructure, and social rehabilitation can occur concurrently, with the aim of creating more resilient communities before and after the disaster. In the subsequent section of the chapter, the risk and crisis management processes, which are integral to disaster management phases, will be explained in detail.

Disaster Risk Management

Disaster risk refers to the likelihood of losses that a hazard can inflict on people, settlements, and the natural environment in proportion to their vulnerability or damage if it occurs within a certain period of time in the future (AFAD, 2014). Through disaster risk management studies, disaster risks can be reduced and even eliminated. Disaster risk management is defined as "The systematic process of utilizing administrative directives, organizations, and operational skills and capacities to implement strategies, policies, and improved coping capacities in order to lessen the adverse impacts of hazards and the possibility of disaster" (UNDRR, 2009). Its objective is to prevent and mitigate the negative effects of hazards through activities focused on disaster prevention, mitigation, and preparedness.

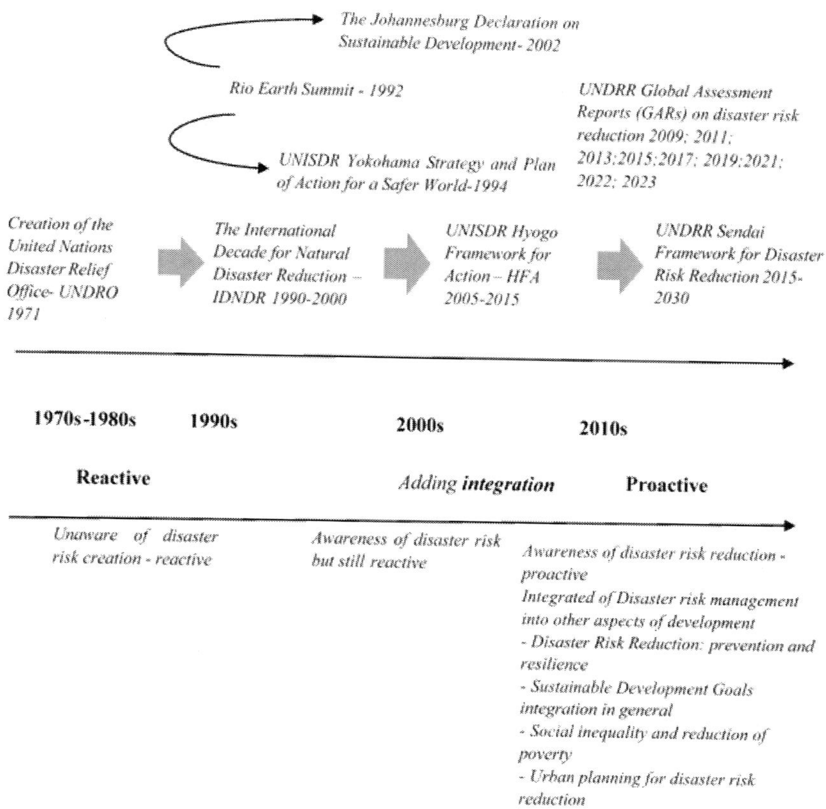

Figure 2. Development of integrated disaster risk management.

The integrated disaster management approach, a contemporary and modern approach, has superseded the previously adopted crisis management approach, which primarily concentrated on response and recovery efforts during and after the disaster. Figure 2 shows the global processes in the development of Integrated Disaster Risk Management.

Today, the disaster management approach aims to prevent disasters and minimize the damage they may cause if prevention is not possible. Within this framework, the impact of disaster risk management studies on sustainable development, individual-based disaster preparedness, and resilient societies has become increasingly significant. Disaster risk management processes encompass mitigation and preparedness activities. Successful disaster risk management is a crucial element that enhances disaster resilience. In the following section, the mitigation and preparedness phases, which are integral to disaster risk reduction processes, will be explained.

Mitigation

The objective of this phase, which encompasses activities conducted before a disaster occurs, is to prevent disasters and to eliminate or minimize hazards and risks that can lead to disasters. This phase encompasses all structural and non-structural measures and activities aimed at averting hazards from escalating into disasters and minimizing damages in the event of their occurrence. Mitigation activities are long-term endeavors that necessitate collaboration among various institutions, organizations, and disciplines, guided by a governance philosophy.

While these activities may appear to be primarily undertaken during the pre-disaster period within the integrated disaster management process, they commence during the recovery phase and persist until a new disaster occurs (AFAD, 2014).

Mitigation constitutes a pivotal stage in integrated disaster management. Prior to the onset of disasters, it is imperative to identify risks, expose vulnerabilities, establish early warning systems for disaster prevention, and devise region-specific disaster plans. At this juncture, it is essential to first ascertain disaster risks. Although disaster risk (DR) is articulated in diverse manners across various sources, it is generally represented as the product of hazard (H), exposure to susceptible elements (E), and vulnerability rate (V) divided by capacity (C). Disaster Risk = (Hazard X Exposure X Vulnerability)/Capacity (AFAD, 2014; UNDRR, 2004; Wisner et al., 2007).

As delineated by the formula, identifying hazards that could culminate in a disaster is paramount to gauging disaster risk. Additionally, it is crucial to pinpoint the potential occurrence locations, magnitudes, frequencies, and affected areas of these hazards. Catalogs detailing the population, environment, infrastructure, structures, and social and economic values susceptible to hazards must also be compiled. This inventory of vulnerable assets helps estimate the physical, economic, and environmental losses that can occur in the event of a disaster (AFAD, 2014).

Mitigation efforts can be implemented through diverse modalities, taking into account factors such as disaster typology, settlement patterns, environmental conditions, and societal structures, as well as cost-benefit analyses. In general, risk reduction strategies encompass risk likelihood reduction, risk consequence reduction, risk avoidance, risk acceptance, and risk transfer, sharing, or spreading. Risk reduction strategies will be briefly elucidated in the subsequent section.

Risk Likelihood Reduction

It is feasible to reduce the likelihood that many hazards evolve into disasters, particularly the potential for technological or human-induced hazards. These hazards, whose existence is dependent on direct human decisions or actions, can be mitigated through preventive measures aimed at avoiding their transformation into disasters, thereby significantly diminishing the associated risk (Coppola, 2015).

Risk Consequences Reduction

This strategy endeavors to mitigate the impact of hazards on individuals, structures, the economy, the environment, or a combination thereof. Anticipating that a hazard can escalate into a disaster, assumptions are made regarding the potential impact, intensity, and magnitude of such a disaster (Coppola, 2015; Lawrence et al., 2020). Precautionary measures are implemented to enhance societal, structural, systemic, and other component resilience against disasters, ensuring that they are either unaffected or less affected.

Risk Avoidance

When certain hazards materialize, they can lead to significant devastation. In such instances, simply mitigating the consequences of the risk is insufficient; instead, risk avoidance becomes imperative to minimize destruction entirely. To avoid risks associated with natural hazards, it is often necessary to relocate entire communities and structures away from endangered areas (Lawrence and Perry, 2017).

Risk Acceptance

Administrators, disaster management experts, legislators, and individuals involved in the disaster management process may deem certain hazards as acceptable risks. The rationale behind accepting these risks includes insufficient funding for risk reduction or avoidance, the potential unintended negative consequences of risk reduction measures, and sociocultural norms hindering relocation to safer areas. Additionally, in certain religious beliefs, risk can be perceived as the will of a higher power rather than as preventable, contributing to the acceptance of risk.

Risk Transfer, Sharing, and Spreading

The objective of risk transfer is to distribute the consequences of risk to a broader audience. The most prevalent method of risk transfer is through insurance systems. Insurance mitigates the financial impact of risk by alleviating the monetary losses associated with damages. The insurance system collects premiums calculated based on the anticipated frequency and potential consequences of the risk (Coppola, 2015). The sharing and spreading of risks involve societal members' efforts to assist disaster victims by sharing limited resources, such as food and other essentials. These efforts are often prevalent in developing countries (Clarke and Dercon, 2015; Lawrence and Perry, 2017; Warner et al., 2009).

During disaster risk management activities, the societal structure, the regional characteristics of the settlements, and the risks of country/international disasters should be taken into account. The hazards and frequencies of the disasters may vary based on geography and societal social factors. While the same disaster threat may not constitute a disaster for a resilient community, it can escalate into a disaster for vulnerable communities due to differences in societal resilience. For successful disaster risk management, communities must actively participate in this process. Social participation is a crucial element that enhances resilience, whereas low or no participation presents a significant obstacle to resilience. Therefore, individuals and all stakeholders in society should be encouraged to participate in disaster risk management activities.

Preparedness

While mitigation processes make significant contributions to disaster prevention, it is unfortunately not feasible to be entirely shielded from disasters today. Therefore, when confronted with a disaster, communities, and countries must be prepared to manage and cope with it effectively.

Disaster preparedness activities are undertaken by various institutions and citizens, including governmental bodies, and public and private organizations. Each of these stakeholders has different responsibilities, and preparedness process studies encompass a wide array of domains. The success of the efforts in this phase directly impacts the effectiveness of the response efforts (Coppola, 2015).

Disaster preparedness can be defined as the state of readiness to respond to a disaster promptly and efficiently. Generally, disaster preparedness activities constitute the human component of pre-disaster hazard management. These activities encompass training the society in disaster response, training disaster workers and volunteers, conducting drills, establishing early warning systems, maintaining emergency response material stocks, developing disaster logistics infrastructure, and organizing infrastructure and logistics for temporary and permanent shelter areas, as well as food reserves (Bullock et al., 2013). Preparedness is not highly effective in preventing disasters from occurring. However, it is a crucial process to improve the capacity to mitigate disaster impacts before they occur and to ensure that communities and teams are well-equipped to respond effectively when disasters strike (Bullock et al., 2013; Sena and Kifle, 2006).

Disaster preparedness is a continuous, integrated process that uses a wide range of activities and resources. From this comprehensive perspective, the goals of disaster preparedness include improving disaster response efficiency and effectiveness at both the community and national levels. Examples of these preparatory efforts include establishing and testing early warning systems, devising evacuation plans and conducting drills, training authorities, society, and disaster response teams, as well as formulating response policies, standards, and organizational and operational plans (IFRC, 2000).

Disaster preparedness is a lengthy process that entails the involvement of numerous stakeholders. Particularly, the full participation of government and individuals is crucial in ensuring success. The primary responsibilities of governments in disaster preparedness include planning, conducting drills, providing training, furnishing equipment, and establishing legal authority. In nearly every country, citizens expect life-saving and sustainable assistance from their governments during times of disaster. Many governments make commitments to their citizens regarding these matters during elections (Bullock et al., 2013). Due to their positions, governments must ensure that disaster preparedness actions are executed comprehensively, and in the event of a disaster, they must promptly address the needs of citizens due to the preparatory work undertaken in advance.

In addition to governments, another vital stakeholder in disaster preparedness is the community. It is recognized that in the immediate aftermath of disasters, existing capacity may be insufficient, leading to significant destruction. Therefore, communities must have the capacity to address their own needs in the initial hours before official disaster assistance arrives. To achieve this, every individual must prepare beforehand. The International Decade for Natural Disaster Reduction has underscored the importance of enhancing disaster risk awareness among societies vulnerable to disasters in disaster risk reduction processes (UN, 1987). Similarly, the Hyogo Framework for Action emphasizes that community hazard awareness and disaster preparedness are pivotal components in bolstering countries' resilience to disasters (UNISDR, 2005a). For disaster resilience to be effective, communities must be prepared for disasters. During these preparatory phases, it is imperative to educate the public, enhance disaster risk awareness within the community through training initiatives, interpret disaster warnings effectively, and foster behavior change by instilling appropriate conduct before, during, and after disasters (Coppola, 2015).

Disaster Crisis Management

Disasters inherently disrupt or halt daily life, often leaving resources such as materials, equipment, and personnel insufficient. Given the urgency of the activities required during disasters, prompt response becomes imperative. Consequently, governments and relevant institutions may enact public policy changes to facilitate swift action (Alexander, 2005). Disaster crisis management entails temporary measures aimed at restoring stability during disasters. It encompasses resilience, response, and recovery activities undertaken by governments, local authorities, voluntary organizations, and other stakeholders (Petak, 1985).

Effective disaster crisis management serves to minimize loss of life, injuries, and environmental damage. Furthermore, the swift restoration of society to its normal rhythm is dependent on the success of crisis management efforts. In cases where disasters cannot be prevented, the subsequent sections will delineate the response and recovery phases of disaster crisis management, crucial for mitigating the impact of disasters to the fullest extent possible.

Response

Despite extensive risk reduction and comprehensive preparation efforts, disasters unfortunately continue to occur worldwide. It is widely acknowledged that the transition from an event to a disaster is directly linked to the response capacity allocated to it. Events that can be managed with local resources remain emergencies rather than disasters. The capability of emergency response teams and the public plays a pivotal role in preventing an event from escalating to disaster proportions. Events where both emergency response teams and the community are inadequate can escalate into disasters necessitating global assistance.

Disaster response encompasses all activities initiated immediately after an event occurs and typically spans a period of 1-2 months, contingent upon the scale of damage and losses incurred. This phase encompasses a wide array of activities aimed at swiftly assessing the disaster-affected area, curtailing the spread of the disaster, minimizing damages, and facilitating rehabilitation. The primary objective of activities during the disaster response phase is to save as many lives as possible within the shortest timeframe, provide treatment for the injured, and ensure continued treatment for those already undergoing medical care. Additionally, this phase aims to promptly address the essential needs of individuals who have lost their homes, including shelter, sustenance, heating, security, communication, evacuation, and social and psychological support, employing the most suitable methods available (AFAD, 2014). Disaster response activities include disaster area and information assessment, evacuation, search and rescue operations, provision of first aid and emergency medical care, distribution of water, food, and shelter, delivery of health and sanitation services, restoration, and activation of critical infrastructure (transportation, communication, etc.), safety and security measures, emergency social services, donation and volunteer management, as well as command, control, and coordination operations (Coppola, 2015).

Response stands as the most intricate phase within the integrated disaster management process. Activities in this phase unfold under high-stress conditions, often with limited resources and information. Eroded trust and unwarranted delays during response can directly contribute to the destruction and increased losses. While training and exercises enhance the capabilities of response teams, disasters, by their very nature, can render even the most well-conceived plans and proficient teams insufficient. Especially in the context of international disaster response, the sudden convergence of diverse groups from disparate regions/countries at the disaster site, each with its own set of expectations, equipment, and missions, poses a challenge to the intervention process (Coppola, 2015; der Heide, 1992).

To achieve success in response efforts, the necessary plans must be established prior to disasters and continually updated. Response plans vary between countries (AFAD, 2014). The quality of response is dependent on robust and comprehensive information about the disaster and effective coordination. The needs of disaster victims and the broader society, the timing of the disaster, the expertise of response teams, and the nature of the response efforts are specific to each disaster. An effective response to disasters necessitates early identification of disasters, swift situational assessment during the disaster, prompt initiation of search, rescue, and emergency aid activities, rapid evacuation when necessary, information management by relevant entities, provision of high-quality and accessible water, food, and shelter, uninterrupted healthcare provision, establishment and accessibility of social service amenities, maintenance of safety and security to avert confusion and disorder, and coordination facilitated by authorities from a singular source (Haddow et al., 2020).

Similar to other phases of integrated disaster management, collaboration between local and central administrations is essential during the response phase. Furthermore, inclusion in all processes, wherever feasible, is imperative. Even in the aftermath of a disaster affecting a prepared society, contributions to response activities such as search and rescue, initial emergency aid, shelter, volunteering, and social support during the acute phase remain feasible.

Recovery

The recovery phase encompasses activities wherein individuals, families, communities, and countries address their losses resulting from a disaster, undertake rebuilding efforts, and implement risk reduction measures against future disasters.

Recovery efforts are aimed at meeting the needs of communities affected by disasters using appropriate methods, restoring normalcy as quickly as possible, enhancing the resilience of communities to potential future disasters, and establishing safer living environments than those existing before the disasters (Chang, 2010), thus minimizing harm. It constitutes a process that entails a comprehensive array of institutional, physical, legal, social, and economic activities (AFAD, 2014; Atsumi et al., 2019). This process commences with the response phase following the disaster in disaster management and extends over an extended period.

Common activities conducted during the recovery phase of disasters include assessment of damage and needs, public communication, provision of temporary or permanent housing, evaluation and repair of damaged structures or their demolition and debris removal, debris clearance, and disposal. Moreover, infrastructure rehabilitation, reconstruction, psychosocial, and cultural rehabilitation initiatives are among the activities falling within the purview of the recovery phase, facilitating the return to normalcy. Measures such as enhancing access to credit for disaster victims, economic support (creating employment opportunities, etc.), reinstating social service activities, rehabilitating the injured, and reassessing disaster hazard risks are also integral to recovery phase efforts (Coppola, 2015). Activities during the recovery phase are diverse and resource-intensive. Rehabilitation endeavors that commence immediately following a disaster may persist for years (Atsumi et al., 2019).

The recovery phase can be delineated into two periods: short-term and long-term recovery phases. In the short-term recovery phase, immediate efforts are made to safeguard the lives of disaster victims and expedite the provision of safe shelter and living spaces. This phase spans the brief interval between the response phase and the commencement of the recovery process. The long-term recovery phase commences subsequent to the conclusion of the response phase and can extend over many years, aimed at rehabilitating individuals, communities, and countries. During this period, activities such as the construction of disaster-resistant structures, the repair of critical infrastructures, undertaking physical, psychological, and social rehabilitation, the promotion of employment and commercial activities, and the implementation of environmental regulations are undertaken (Coppola, 2015). Efforts are also directed to improve the resilience of communities and institutions affected by disasters, preparing them to confront potential future disasters.

One of the primary objectives of the efforts undertaken during the recovery phase is to mitigate the risk of future disasters and prepare society and the country to confront such eventualities. Hence, community participation in recovery, engagement in decision-making processes (Comerio, 2014; Gündüz, 2024), and the fostering of such participation by authorities are crucial. While rectifying the adverse effects of disasters through initiatives in this phase, efforts should also be directed toward cultivating a resilient society. Activities undertaken during the recovery phase should not be viewed solely as post-disaster efforts. The long-term recovery phase commences post-disaster and persists until the occurrence of the subsequent disaster. Consequently, the long-term recovery phase initiated following a disaster may intertwine with the phase focused on reducing risks and harm.

To attain success in disaster management and foster a resilient society, activities across all phases must be implemented effectively. Collaboration among society, local government, government, the private sector, and other stakeholders, guided by a governance philosophy, is imperative at all phases of integrated disaster management. Additionally, the use of technology to prevent disasters and mitigate their impact is essential. In the subsequent discussion, the role, and applications of technology in integrated disaster management will be elucidated.

Discussion and Conclusion

The rapid advancement of technology today has instigated changes and advancements in the field of disaster management, much like in other domains. Compared to the past, technologies are now being used more extensively in all stages of integrated disaster management. The integration of new technologies into disaster management processes has the potential to enhance the effectiveness of disaster risk and crisis management efforts.

The significance of technology is rising in disaster risk management, particularly in the collection, mapping, analysis, and evaluation of crucial information required to assess risks. In disaster preparedness, technology finds widespread application in remote sensing systems (Demirci and Karakuyu, 2004), airflow modeling, and pollution forecasting (Toros and Anbarcı, 2018), as well as in virtual reality-assisted disaster training and scenario simulations (Longo et al., 2019).

During disaster crisis management, Internet of Things (IoT)-supported management platforms facilitate quicker and easier determinations, enabling real-time access to information regarding damage status, affected individuals, and their conditions, and facilitating communication (Küçük et al., 2019). Data and technology play a pivotal role in orchestrating relief efforts and, in certain instances, even in rescuing individuals trapped under debris (Demiröz, 2020). In the recovery phase, rapid and efficient damage detection can be achieved with fewer technical personnel through the use of the Internet of Things (IoT) and Geographic Information Systems (GIS), thereby expediting the commencement of recovery operations (Yeum et al., 2018).

Analyzes, artificial intelligence-supported programs, modeling, and training that use vast amounts of big data from across the globe contribute to mitigating disaster-related vulnerabilities by leveraging them before, during, and after disasters. To exemplify a successful disaster management process, countries must exert greater efforts in harnessing evolving technologies and the data that imbue these technologies with significance.

Disclaimer

None.

References

AFAD. (2014). *Açıklamalı Afet Yönetimi Terimleri Sözlüğü*. Ankara.

Alexander, D. (2005). Towards the development of a standard in emergency planning. *Disaster Prevention and Management: An International Journal, 14*(2). https://doi.org/10.1108/09653560510595164.

Atsumi, T., Seki, Y., & Yamaguchi, H. (2019). The generative power of metaphor: long-term action research on disaster recovery in a small Japanese village. *Disasters, 43*(2). https://doi.org/10.1111/disa.12314.

Bullock, J. A., Haddow, G. D., & Coppola, D. P. (2013). Mitigation, Prevention, and Preparedness. *Introduction to Homeland Security*, 435–494. https://doi.org/10.1016/B978-0-12-415802-3.00010-5.

Chang, S. E. (2010). Urban disaster recovery: A measurement framework and its application to the 1995 Kobe earthquake. *Disasters, 34*(2). https://doi.org/10.1111/j.1467-7717.2009.01130.x.

Clarke, D., & Dercon, S. (2015). Planning for Disasters and the Economics of Disaster Risk Financing and Insurance. In D. D. J. A. Clarke, E. Sadoulet, & E. Skoufias (Eds.), *Disaster Risk Financing and Insurance: Issues and results*. FERDi.

Comerio, M. C. (2014). Housing Recovery Lessons from Chile. *Journal of the American Planning Association, 80*(4), 340–350. https://doi.org/10.1080/01944363.2014.968188.

Coppola, D. P. (2015). *Introduction to international disaster management* (Third Edition). Butterworth Heinemann.

CRED. (2023). *2022 Disasters in Numbers*. https://cred.be/sites/default/files/2022_EMDAT_report.pdf.

Demirci, A., & Karakuyu, M. (2004). The Role of Geographic Information Technologies on Disaster Management. *Doğu Coğrafya Dergisi*, *9*(12).

der Heide, E. A. (1992). Disaster Response: Principles of Preparation and Coordination. *Public Productivity & Management Review*, *15*(3), 381. https://doi.org/10.2307/3380618.

Gündüz, F. (2024). A phenomenological study on women's disaster experience in Türkiye. *International Journal of Disaster Risk Reduction*, *100*, 104196. https://doi.org/10.1016/J.IJDRR.2023.104196.

Haddow, G. D., Bullock, J. A., & Coppola, D. P. (2020). The Historical Context of Emergency Management. In *Introduction to Emergency Management* (7th Edition). Elsevier. https://doi.org/10.1016/b978-0-12-817139-4.00001-4.

IFRC. (n.d.). *What is a disaster?* Retrieved February 15, 2024, from https://www.ifrc.org/our-work/disasters-climate-and-crises/what-disaster.

IFRC. (2000). *Introduction to Disaster Preparedness - Disaster Preparedness Training Programme Participant resource & learning module*.

IFRC. (2023). *Trust, Equity and Local Action Lessons from the COVID-19 pandemic to avert the next global crisis*. IFRC.

Küçük, K., Bayılmış, C., Sönmez, A. F., & Kaçar, S. (2019). IoT Teknolojilerini Kullanan Afet Sonrası Yönetim Sistemi. *Academic Platform Journal of Engineering and Science*, *7*(2). https://doi.org/10.21541/apjes.474822.

Lawrence, J. F., & Perry, M. A. (2017). Assessing Risk and Vulnerabilities. In *Physical Security: 150 Things You Should Know* (pp. 79–96). Butterworth-Heinemann. https://doi.org/10.1016/B978-0-12-809487-7.00002-4.

Longo, F., Nicoletti, L., & Padovano, A. (2019). Emergency preparedness in industrial plants: A forward-looking solution based on industry 4.0 enabling technologies. *Computers in Industry*, *105*. https://doi.org/10.1016/j.compind.2018.12.003.

Petak, W. J. (1985). Emergency Management: A Challenge for Public Administration. *Public Administration Review*, *45*. https://doi.org/10.2307/3134992.

Sena, L., & Kifle, M. (2006). Disaster Prevention and Preparedness. In *Ethopia Public Health Training Initiative*.

Toros, H., & Anbarcı, K. (2018). Hava Kirliliğine Dirençli Kentler. In *Şehir ve Toplum*.

UN. (1987). *International Decade for Natural Disaster Reduction*. https://www.un.org/depts/dhl/resguide/r42_resolutions_table_eng.html.

UN. (1994). *Yokohama Strategy and Plan of Action for a Safer World*.

UN. (2002). *Report of the World Summit on Sustainable Development: Johannesburg, South Africa, 26 August-4 September 2002*. United Nations.

UNDHA. (1992). *Internationally agreed glossary of basic terms related to disaster management*.

UNDRR. (2004). Risk awareness and assessment. *Living with Risk: A Global Review of Disaster Reduction Initiatives*.

UNDRR. (2009). *UNISDR Terminology on Disaster Risk Reduction*. https://www.undrr.org/terminology/disaster.

UNDRR. (2019). *GAR on DRR 2019*. ISBN:9789210041805.

UNDRR. (2021). *GAR on DRR 2021: Special Report on Drought.* ISBN: 9789212320274.
UNDRR. (2022). GAR on DRR 2022: Our World at Risk: Transforming Governance for a Resilient Future. ISBN: 9789212320281.
UNDRR. (2023). *GAR on DRR 2023 Mapping Resilience for the Sustainable Development Goals.* ISBN:9789210028301.
UNISDR. (2005a). *Hyogo Framework for Action 2005-2015: Building the Resilience of Nations and Communities to Disasters.* www.unisdr.org/wcdr.
UNISDR. (2005b). *National Report of Japan on Disaster Reduction for the World Conference on Disaster Reduction.*
UNISDR. (2009). *GAR on DRR. Risk and poverty in a changing climate Invest today for a safer tomorrow Invest today for a safer tomorrow.* ISBN:978-92-1-132028-2.
UNISDR. (2011). *GAR 2011 - Revealing Risk, Redefining Development.* ISBN: 9789211320305.
UNISDR. (2013). *GAR on DRR. From Shared Risk to Shared Value: The Business Case for Disaster Risk Reduction.* ISBN: 9789211320381.
UNISDR. (2015a). GAR on DRR 2015. ISBN:9789211320428.
UNISDR. (2015b). *Sendai Framework for Disaster Risk Reduction 2015 - 2030.*
Warner, K., Ranger, N., Surminski, S., Arnold, M., Linnnerooth-Bayer, J., Michel-Kerjan, E., Kovacs, P., & Herweijer, C. (2009). *Adaptation to Climate Change: Linking Disaster Risk Reduction and Insurance.* www.unisdr.org.
Wisner, B., Blaikie, P., Cannon, T., & Davis, I. (2007). *At Risk: natural hazards, people's vulnerability and disasters* (Second edition). Routledge.
Yeum, C. M., Dyke, S. J., & Ramirez, J. (2018). Visual data classification in post-event building reconnaissance. *Engineering Structures, 155.* https://doi.org/10.1016/j.engstruct.2017.10.057.

Chapter 17

Organizational Adaptations in Disaster Management: The Use of Digital Innovation

Tuğba Dikbaş[1,*]
and Tuba Büyükbeşe[2]
[1]Social Sciences Institute, Department of Business Administration, Hasan Kalyoncu University, Şahinbey, Gaziantep, Türkiye
[2]Department of Business Administration, School of Business Administration, Hasan Kalyoncu University, Gaziantep, Şahinbey, Türkiye

Abstract

Due to the negative social and economic impacts of the increasing frequency and severity of disasters, countries, and international communities have made improving methods for managing disasters a key priority. To overcome these difficulties, institutions/organizations need to be resilient by developing strategies regarding the devastating effects of post-disasters. The critical task in designing digital adaptive systems is to identify the foundations for effective organizational adaptation. Organizational adaptation is one of the most common concepts in organizational theory and strategic management. Many perspectives, dating from the early 20th century to the emergence of scientific management (Taylor, 1911) and industrial management (Fayol, 1916, 1949), have highlighted the role of adaptation in explaining organizational success (Aldrich and Ruef, 2006). There is an assumption in the literature that managers have the capacity to evaluate their environment, take action, and take advantage of opportunities and new technologies to be successful and increase the chances of the survival of organizations.

* Corresponding Author's Email: tugbaranadikbas@gmail.com (Tuba Büyükbeşe, Professor).

In: Disaster Management in the Digital Age and Emerging Technologies
Editors: Ayşe Asiltürk and Ayşegül Sarıoğlu Kemer
ISBN: 979-8-89530-213-2

Awareness of the concepts of innovation or novelty is becoming increasingly widespread in both business and daily life. There is a need to understand and assimilate innovation in social life and to increase the effects of innovations on the efficiency of organizational and social outputs. In this context, it is important to carry out studies on innovation and the assimilation of digital innovation in its new form in the public sector, the private sector, and the social base.

Keywords: disaster management, organizational alignment, digital innovation, technology integration

Introduction

With the increasing social and economic destruction caused by disasters around the world, the international community and national-level Disaster Management (DM) authorities have determined that taking precautions, being prepared, intervening, and improving recovery processes are goals of the highest priority. Technological advances and the Fourth Industrial Revolution are vital to achieving these goals. However, they also present new challenges by changing the fundamental operational, organizational, and social dynamics of traditional disaster management systems. Currently, there is only a paucity of studies that examine these aspects beyond technology and investigate the impact of digital innovation on the entire lifecycle of disaster management at the national level. This research aims to fill this gap by aiming to understand the effects and determinants of digital innovation in LM systems by combining interdisciplinary concepts from different fields such as Disaster Management, Information Technologies, Management, and Organization.

Although disaster management seems to be limited to emergency aid, rescue, recovery, and reconstruction activities generally carried out after a disaster, it actually has a much broader scope (Alsamhi et al., 2018). Post-disaster work is only aimed at repairing the damage and is not a comprehensive disaster management strategy. Modern understanding defines disaster management as the identification of disaster risks before they occur, the elimination of these risks, or the minimizing of their damage. Disaster management is a professional discipline and management approach that aims to effectively respond to disasters by improving existing systems using the experiences gained from each event.

Disaster management covers crisis and risk management activities carried out in the period between disasters. In other words, disaster management is the process of reducing and preventing disaster losses, responding to disasters quickly, effectively, and on time, and creating a safer social environment. This process includes planning, coordination, support, and implementation of activities before, during, and after the disaster and brings together the resources and opportunities of all institutions/organizations in the society (Alexandru et al., 2019). International documents such as the Yokohama Strategy Document, the Hyogo Framework Action Plan, and the Sendai Framework have determined strategic approaches and action plans for disaster management. These documents include policies and strategies to be implemented at the national and local levels to reduce disaster risks and increase the resilience of societies against disasters. Information technologies play an important role in disaster management. Data collected through mobile devices and other digital tools enable disaster managers to make faster and more accurate decisions. These technologies have the potential to optimize emergency responses by collecting and analyzing real-time data from disaster areas and transmitting this information to coordination centers (Atzori et al., 2010). In summary, disaster management systems require an integrated approach to understand disaster risks, prepare for disasters, respond effectively to emergencies, and manage recovery processes.

It is now clear that digital innovation, as the cornerstone of innovative processes, plays a critical role in shaping competition. Advances in digital technologies have not only facilitated access to necessary data and information but have also provided significant improvements in factors such as time and cost in the information processing process. For example, analyzing millions of data quickly and at low cost by supercomputers or producing a new product as a low-cost prototype with 3D printers immediately after the CAD data are completed has been a great innovation for manufacturing organizations (Christensson, 2014).

The radical changes brought about by digital technology create a rapid transformation that continues in all processes in organizations. They see the spread of digitalization in industry ecosystems as the basis for this change (Markus and Loebbecke, 2013). Each ecosystem consists of a series of sub-ecosystems, and these sub-ecosystems interact with each other and with the parent ecosystem. Sometimes a parent ecosystem can change all sub-ecosystems within it, and sometimes a strong sub-ecosystem can transform the parent ecosystem. This type of change in the communications technology ecosystem is an important example of how it can affect entire sub-ecosystems. Digital technology has brought about two fundamental innovations in digital business ecosystems. First, the effective and efficient management of all processes. Second, allowing the establishment of new integrated structures by creating new platform structures between the upper and lower ecosystems (Markus and Loebbecke, 2013). These innovations have enabled organizations to be effective in innovative product and service concepts.

Organizations must create strategic action plans to achieve their mission and vision. Lee and Berente (2012) state that organizations' first steps towards digital technology began with the establishment of IT departments starting in the 1990s. This situation, which is considered the beginning of digitalization in businesses, first started to be used effectively in internal operation processes. The success of the strategies to be implemented by the business depends on effective management and compliance with other sub-strategies of the organization. In the 2000s, IT strategy began to play an effective and decisive role in achieving the main strategic goals of organizations with the multifaceted development of digital technology. Kulatilaka and Venkatraman (2001) emphasize that the IT strategies of organizations should be well synchronized with the upper strategic structure of the organization. After the success of digitalization technology in the operational processes of organizations, its importance is increasing to achieve the strategic goals of the organization.

According to Yoo et al. (2012), the tools of digital technology have gone beyond improving the operational processes of organizations and started to play an active role in product and service processes. For example, websites that were once created solely for promotional purposes can today turn into powerful marketing and sales channels when used correctly and provide organizations with a competitive advantage. Advances in digital technology are no longer limited to intra-organizational processes but have also begun to be used effectively in non-organizational processes. This shows that digitalization plays an important role in shaping competition in the market as a strategic tool in both the product structure and the processes.

The increasing trend of digital technology has pushed organizations both in the same sector and in other sectors that interact with each other to make significant changes in their strategies regarding the new products and services offered. Organizations develop strategies that will provide a competitive advantage in the market and turn these strategies into opportunities by using new technological opportunities. This has brought digital technology to a central position in organizations' product and service processes and has raised the question of how to effectively manage digital innovation. Organizations must have a dynamic management structure to monitor their digital innovation performance.

Organizational Adaptation and the Use of Digital Innovation Against Disasters

Organizational Adaptation

Changeis to make things different. Organizational change is a phenomenon that concerns the entire organization, including its environment (Dikbaş and Büyükbeşe, 2020). Chakravarthy (1982) defines adaptation as "the main purpose of strategic management" (Sarta, Durand, and Vergne, 2020). According to Cameron (1984); "Organizational adaptation" refers to changes and modifications made to the organization or its components to adapt to changes in the external environment. Its purpose is to restore balance in an unstable situation. Adaptation generally does not refer to an event, but rather to a process by which changes are initiated in organizations (Cameron, 1984). According to Dikbaş and Büyükbeşe, 2020, changes that occur in the environment of the organization begin to affect the organization through the organization's inputs and disrupt the balance of the organization. When changes in the organization's environment reach large dimensions, the organization is forced to change in line with the demands of the environment in order to continue its vital activities. Every change leads to an interaction, and as a result of this interaction, the organization may have to make significant changes in existing relationships, habits, ways, and methods, as well as changes in business, technology, and group structures (Dikbaş and Büyükbeşe, 2020). Organizations adopt innovations to make changes in their results, structures, and processes to maintain or improve their performance or effectiveness levels (Damanpour and Gopalakrishnan, 1999; Dikbaş and Büyükbeşe, 2020).

Organizations use digital technologies to develop new products and services, implement new business processes, or operate innovative business models (Legner et al., 2017; Nambisan et al., 2017). In this context, they invest in flexible IT infrastructures or governance structures that will strategically handle their digital transformation in order to evaluate the advantages obtained from digital technologies.

Digital transformation is the process of rebuilding traditional structures, processes, and models with new digital technologies in order to increase the competitiveness of organizations and respond to constantly changing needs in the digital economy (Hapuarachchi, 2008). Innovation is associated with digital transformation as both a driving force and an outcome leading to better organizational performance. Innovation is about the creation and adoption of new ideas, which means that digital transformation is about innovation, and innovation is about newness and creativity (Sheridan, 2020; Solis et al., 2014). Innovation for organizations can occur in the form of creating or adopting new management systems, technologies, processes, policies, structures, products, or services (Somech and Drach-Zahavy, 2013). In terms of disaster management and related systems, the impact of technological innovation on all stages of disaster management is well known. However, more information about the nature and determinants of this effect is needed to increase effectiveness, efficiency, and competitiveness outcomes (Stolterman and Fors 2004).

Digital Innovation

Innovation derives from the Latin word "innovatus" and means "the introduction of new methods in social, cultural, and administrative fields" (Fichman et al., 2014). Innovation can occur in various forms, such as product and process innovation, management, and technology innovation. An organization's capacity to create such innovations is seen as one of the key factors of its sustainability and success (Wang and Ahmed, 2004; Ellonen, 2008). The ability to innovate is considered a fundamental quality that enables an organization to survive and gain an advantage over its competitors. Organizational innovation can be defined as the process of identifying information sources, integrating information created with new ideas into the organization, and responding to new approaches from various sources (Zhu et al., 2006; Dikbaş and Büyükbeşe, 2020). Organizational innovation can also be defined as an organization's tendency and ability to adopt new ideas to develop new products and services (Ruber and Kirca, 2012). Kanter (1983) defines innovation as "the introduction, acceptance, and implementation of new ideas, processes, products, or services."

Digital innovation refers to the effective use of digital technologies in innovation activities. Digital innovation has fundamentally transformed the character and structure of products and services, developed new methods of creating and sharing value, created innovation communities with the participation of actors with different goals and skills, created a new type of innovative processes, and, in general, led to the transformation of many industries (Boudreau and Lakhani 2013; Hui 2014; Iansiti and Lakhani 2014; Porter and Heppelmann 2015).

The increasing impact of digitalization has encouraged academics to re-evaluate the validity and value of existing innovation theories and organizational studies. Some researchers, such as Benner and Tushman (2015), have suggested that innovation and organizational sciences may be in a process of transformation, with changes in the focus of innovation and the idea that the digital revolution may shake or change basic organizational principles.

Digital innovations cause significant changes in organizations' product and service ranges, business processes, and business models. This type of digital transformation occurs when organizations are adequately prepared. In this context, there is a third group of research focusing on factors that enable digital innovations. This branch of research examines the contribution of organizational capabilities (Levallet and Chan 2018; Nwankpa and Datta 2017; Stoeckli et al., 2018) and digital resources (Nambisan 2013; Yoo et al., 2010), for example, digital infrastructures (Blaschke et al., 2010). 2016; Henfridsson and Bygstad 2013) or digital platforms and ecosystems (Karimi and Walter 2015; Lyytinen et al., 2016; Parker et al., 2017).

It is particularly important to realize four key factors at the organizational level (Hess and Barthel, 2017). The (re)combination of digital technologies leads to the development of digital solutions and, as a result, the development of digital business concepts. For digital innovation to emerge and create value, these digital solutions and business concepts must be integrated into the organization. This integration concerns the integration of digital solutions and business concepts into both existing IT structures (Bygstad 2017; Tilson et al., 2010) and existing organizational structures (Hess et al., 2016). While the first case is often represented through an organization's IT application portfolio, the second case typically involves three key elements: organizational culture (Hartl and Hess 2017), organizational structures (Matt et al., 2015), and organizational capabilities (Grant 1996).

Hawkes et al. (2013) investigated the effect of short visual instruction right before performing a newborn intubation and found that this approach was effective in improving both knowledge and procedural skills for novice learners. Medical students performed better with patient encounters in an emergency department setting when they had immediate access to an instructional video right before seeing a patient in the emergency room (Tews et al., 2011). A similar effect was found in another study. According to Noguera et al. (2013), when second-year students in physical therapy had access to two- or three-dimensional images of anatomy on their mobile devices during a practice session, their knowledge gain was higher than that of those who did not have such learning environments. With resources on their mobile devices, the students were able to visualize the specific anatomy of the knee, ankle, and pelvic zone as they practiced physical therapy. This allowed participants to deeply engage in learning how physical therapy works in relation to a specific anatomy structure.

Digitalization and Organizational Adaptation

The use of digital technologies plays an important role in improving the performance of institutions. Digitalization is the process of evaluating the effects of these technologies on the organization and using digital tools effectively. The effects of digitalization on organizations include organizational flexibility, learning capacity, digital innovations, and the development of business ecosystems (Kuusisto, 2017; Büyükbeşe et al., 2022).

Organizational Flexibility

In today's rapidly changing world, organizational flexibility is not just a goal but a necessity. According to a study by the Economist Intelligence Unit, the majority of executives (88%) view flexibility as a key factor for global success. The two main advantages of resilience are: The first is the ability to respond quickly to business threats. The second is the capacity to quickly identify and take advantage of emerging opportunities. Organizational flexibility can be considered a unique and inimitable benefit that supports the long-term performance of organizations (Alavi et al., 2014).

Organizational Learning

Organizational learning is of vital importance for organizations as it increases the innovation and efficiency of processes. This can be defined as a dynamic process of knowledge creation at the core of the organization, through individuals and groups, contributing to the creation of unique capabilities that improve the performance and results of the organization (Real et al., 2006; Dikbaş and Büyükbeşe, 2020).

Digital Innovation

Digital innovations are emerging more frequently and faster today. While new technologies were seen every ten years in the 1980s, various innovations have been encountered every moment since the millennium. According to Fichman et al. (2014), digital innovation is a phenomenon that is broadly accepted as new, requires significant changes for adopters, and emerges as a concrete product, process, or business model (Büyükbeşe et al., 2022).

Business Ecosystems

Business ecosystem refers to a community of organizations that come together and interact with each other, usually around the same market or product. This term has biological roots and is defined by The New Oxford English Dictionary (1993) as "a community of living species that occupy a habitat and interact with the environment in which they live" (Kuusisto, 2017).

Discussion and Conclusion

Organizations need to develop the organizational competencies necessary to implement digital innovations and incorporate them into their corporate structures. First, it is important for organizations to have sufficient IT capabilities to manage digital technologies and create a basis for digital innovations (Nwankpa and Datta 2017). Additionally, organizations need specific digital capabilities that will enable them to use digital resources for innovation purposes (Li et al., 2018). To develop these capabilities, organizations can resort to various methods. They can develop these skills through organic means, such as training, job changes, or online seminars, or they can acquire them through inorganic means, such as mergers and acquisitions, or outsourcing.

Businesses act in such a way that they will not be left out of the Industry 4.0 revolution by monitoring digital technological advances in their own fields or in different sectors and incorporating them into their value chains. The effectiveness and efficiency of such integrations depend on a flexible and solid organizational structure. Due to the rapid advancement in digital technologies, the value of organizations capable of learning is increasing. Businesses must make the appropriate arrangements in their organizational structures to successfully manage their digital innovations. While making these changes, they must consider the interactions in internal and external environmental conditions.

Researchers in the field of Information Technology (IT) have paid special attention to the conceptual development of digital innovations in the last decade (Kohli and Melville, 2018; Vial 2019; Yoo et al., 2012). These innovations are considered to be a fundamental element of the digital transformation of organizations. Additionally, studies on the factors that make digital innovations possible, such as; new potentials mentioned by Li et al. (2018), discussion of cultural change (Hartl and Hess, 2017; Henfridsson and Bygstad, 2013), digital infrastructures (Gawer and Cusumano, 2014), and digital platforms (Karimi et al., 2009), are on the increase.

Researchers such as Bharadwaj et al. (2013), Hess and Barthel (2017), and Kohli and Melville (2018) are aware of the challenges associated with digital innovations and emphasize the value of a structured approach to the governance of these innovations. To cope with challenges such as volatile markets, disruptive technologies, and accelerating innovation cycles stated by Chan et al. (2018) and Karimi and Walter (2015), organizations need to establish dedicated digital units, and as stated by Fuchs et al. (2019), employ multi-functional teams.

Zhu et al. (2006) examined the concept of innovation in the context of digital transformation and evaluated corporate digital transformation as an example of technology-supported innovations. They have created a comprehensive model to identify the factors that determine the subsequent stages of innovation adoption. In this model, four innovation characteristics (relative advantage, compatibility, cost, and security concern) and four contextual factors (technological competence, organizational size, competitive pressure, and partnership readiness) are considered.

Vial G. (2019) conducted a study covering the existing literature on digital transformation and sought an answer to the question: "What do we know about digital transformation?." This study aimed to compile the knowledge on the subject using an inductive approach.

Verhoef et al. (2021) examined digital transformation from a multidisciplinary perspective and examined the facts and literature in various fields. The three main objectives of the study are to identify external factors that increase the need for digital transformation; discuss the strategic imperatives brought about by digital transformation, namely the required digital resources, appropriate organizational structure, growth strategies, and metrics; and present a research agenda that will guide future digital transformation research. As a result, the importance of companies focusing on digital resources, organizational structure, growth strategies, and metrics is emphasized.

When organizations operate in uncertain and variable environmental conditions, they can improve their performance with knowledge-centered strategies. In situations of environmental uncertainty, obtaining accurate and timely information is vital to the success of organizations. Research shows that, under uncertainties in social, economic, and political factors, organizations that act with IT approaches are more successful.

The increasing use of technology and an information-oriented structure enables organizations to continue their activities more effectively and efficiently, leading to a positive increase in their performance. Organizations can achieve their goals more quickly and effectively by using technology more effectively and continuing their knowledge-based work. Increasing the use of technology is of strategic importance for organizations that want to reach large audiences and carry out their activities more effectively.

Disclaimer

None.

References

Alavi, S., Abd. Wahab, D., Muhamad, N., & Arbab Shirani, B. (2014). Organic structure and organisational learning as the main antecedents of workforce agility. *International Journal of Production Research,* *52*(21), 6273-6295. https://doi.org/10.1080/00207543.2014.919420.

Aldrich, H. E., & Ruef, M. (2006). Organizations evolving (2nd ed.). London: Sage.

Alexandru, A., Ianculescu, M., Marinescu, I. A., & Popescu, T. D. (2019, May). Shaping the digital citizen into a smart citizen on the basis of IoT capabilities. In 2019 22nd International Conference on Control Systems and Computer Science (CSCS) (pp. 707-714). IEEE. doi: 10.1109/CSCS.2019.00126.

Alsamhi, S. H., Ansari, M. S., & Rajput, N. S. (2018). Disaster coverage prediction for the emerging tethered balloon technology: Capability for preparedness, detection, mitigation, and response. *Disaster Medicine and Public Health Preparedness, 12*(2), 222-231. doi:10.1017/dmp.2017.54.

Atzori, L., Iera, A., & Morabito, G. (2010). The internet of things: A survey. *Computer Networks, 54*(15), 2787-2805. https://doi.org/10.1016/j.comnet.2010.05.010.

Benner, M. J., & Tushman, M. L. (2015). Reflections on the 2013 Decade Award—'Exploitation, Exploration, and Process Management: The Productivity Dilemma Revisited' ten years later. *Academy of Management Review, 40*(4), 497-514. https://doi.org/10.5465/amr.2015.0042.

Bharadwaj, A., El Sawy, O. A., Pavlou, P. A., & Venkatraman, N. (2013). Digital business strategy: Toward a next generation of insights. *MIS Quarterly, 37*(2), 471–482.

Blaschke, M., Haki, K., Riss, U., Winter, R., & Aier, S. (2016). Digital infrastructure: A service-dominant logic perspective. Paper presented at the 37th International Conference on Information Systems, Dublin, Ireland.

Boudreau, K. J., & Lakhani, K. R. (2013). Using the crowd as an innovation partner. *Harvard Business Review, 91*(4), 60-69.

Büyükbeşe, T., Dikbaş, T., Klein, M., & Ünlü, S. B. (2022). A study on digital leadership scale (DLS) development. *Kahramanmaraş Sütçü İmam University Journal of Social Sciences, 19*(2), 740-760. https://doi.org/10.33437/ksusbd.1135540.

Bygstad, B. (2017). Generative innovation: A comparison of lightweight and heavyweight IT. *Journal of Information Technology, 32*(2), 180–193.

Cameron, K. S. (1984). Organizational adaptation and higher education. *The Journal of Higher Education, 55*(2), 122-144. https://doi.org/10.2307/1981182.

Chakravarthy, B. S. (1982). Adaptation: A promising metaphor for strategic management. *Academy of Management Review, 7*, 35-44.

Chan, C. M., Teoh, S. Y., Yeow, A., & Pan, G. (2018). Agility in responding to disruptive digital innovation: Case study of an SME. *Information Systems Journal, 28*(1), 1–21.

Christensson, P. (2014). 3D printer definition. Retrieved March 12, 2024, from: https://techterms.com.

Damanpour, F., & Gopalakrishnan, S. (1999). Organizational adaptation and innovation: The dynamics of adopting innovation types. In K. Brockhoff, J. Hauschildt, & A. K. Chakrabarti (Eds.), *The Dynamics of Innovation* (pp. 53-80). Berlin: Springer.

Dikbaş, T., & Büyükbeşe, T. (2020). The mediating role of psychological capital in the effect of authentic leadership on employee performance: A field study in a public institution. *Anemon Muş Alparslan University Journal of Social Sciences, 8*(4), 1213-1224. https://doi.org/10.18506/anemon.645661.

Ellonen, R. (2008). The role of trust in organizational innovativeness. *European Journal of Innovation Management, 11*(2), 160-181.

Fichman, R. G., Dos Santos, B. L., & Zheng, Z. (2014). Digital innovation as a fundamental and powerful concept in the information systems curriculum. *MIS Quarterly, 38*(2), 329–343.

Fuchs, C., Barthel, P., Herberg, I., Berger, M., & Hess, T. (2019). Characterizing approaches to digital transformation: Development of a taxonomy of digital units. Paper presented at the 14th International Conference on Wirtschaftsinformatik, Siegen, Germany.

Gawer, A., & Cusumano, M. A. (2014). Industry platforms and ecosystem innovation. *Journal of Product Innovation Management, 31*(3), 417–433.

Hapuarachchi, N. P. (2008). The usage of geo-information and technology in disaster management by sub-national level organizations: Case study—Ratnapura in Sri Lanka.

Hartl, E., & Hess, T. (2017). The role of cultural values for digital transformation: Insights from a Delphi study. Paper presented at the 23rd Americas Conference on Information Systems, Boston, USA.

Henfridsson, O., & Bygstad, B. (2013). The generative mechanisms of digital infrastructure evolution. *MIS Quarterly, 37*(3), 907–931.

Hess, T., & Barthel, P. (2017). Wieviel digitale Transformation steckt im Informationsmanagement? Zum Zusammenspiel eines etablierten und eines neuen Managementkonzepts. HMD Praxis der Wirtschaftsinformatik, 54(3), 313–323.

Hess, T., Matt, C., Benlian, A., & Wiesböck, F. (2016). Options for formulating a digital transformation strategy. *MIS Quarterly Executive, 15*(2), 123–139.

Hui, G. (2014, July 29). How the Internet of Things changes business models. Harvard Business Review. Retrieved from: https://hbr.org.

Iansiti, M., & Lakhani, K. R. (2014). Digital ubiquity: How connections, sensors, and data are revolutionizing business. *Harvard Business Review, 92*(11), 91-99.

Kanter, R. M. (1983). Frontiers for strategic human resource planning and management. *Human Resource Management, 22*(1-2), 9-21.

Karimi, J., & Walter, Z. (2015). The role of dynamic capabilities in responding to digital disruption: A factor-based study of the newspaper industry. *Journal of Management Information Systems, 32*(1), 39–81.

Karimi, J., Somers, T. M., & Bhattacherjee, A. (2009). The role of ERP implementation in enabling digital options: A theoretical and empirical analysis. *International Journal of Electronic Commerce, 13*(3), 7–42.

Kohli, R., & Melville, N. P. (2018). Digital innovation: A review and synthesis. *Information Systems Journal, 29*(1), 200–223.

Kulatilaka, N., & Venkatraman, N. (2001). Strategic options in the digital era. *Business Strategy Review, 12*(4), 7-15.

Kuusisto, M. (2017). Organizational effects of digitalization: A literature review. *International Journal of Organization Theory and Behavior, 20*(03), 341-362.

Lee, J., & Berente, N. (2012). Digital innovation and the division of innovative labor: Digital controls in the automotive industry. *Organization Science, 23*(5), 1428-1447.

Legner, C., Eymann, T., Hess, T., Matt, C., Böhmann, T., Drews, P., … & Zimmermann, H.-D. (2017). Digitalization: Opportunity and challenge for the business and information systems engineering community. *Business & Information Systems Engineering, 59*(4), 301–308.

Li, L., Su, F., Zhang, W., & Mao, J. Y. (2018). Digital transformation by SME entrepreneurs: A capability perspective. *Information Systems Journal, 28*(6), 1129–1157.

Lyytinen, K., Yoo, Y., & Boland Jr., R. J. (2016). Digital product innovation within four classes of innovation networks. *Information Systems Journal, 26*(1), 47–75.

Markus, M. L., & Loebbecke, C. (2013). Commoditized digital processes and business community platforms: New opportunities and challenges for digital business strategies. *MIS Quarterly, 37*(2), 649-653.

Matt, C., Hess, T., & Benlian, A. (2015). Digital transformation strategies. *Business and Information Systems Engineering, 57*(5), 339–343.

Nambisan, S. (2013). Information technology and product/service innovation: A brief assessment and some suggestions for future research. *Journal of the Association for Information Systems, 14*(4), 215–226.

Nambisan, S., Lyytinen, K., Majchrzak, A., & Song, M. (2017). Digital innovation management: Reinventing innovation management research in a digital world. *MIS Quarterly, 41*(1), 223–238.

Nwankpa, J. K., & Datta, P. (2017). Balancing exploration and exploitation of IT resources: The influence of digital business intensity on perceived organizational performance. *European Journal of Information Systems, 26*(5), 469–488.

Parker, G., Van Alstyne, M. W., & Jiang, X. (2017). Platform ecosystems: How developers invert the firm. *MIS Quarterly, 41*(1), 255–266.

Porter, M. E., & Heppelmann, J. E. (2014). How smart, connected products are transforming competition. *Harvard Business Review, 92*(11), 64-88.

Porter, M. E., & Heppelmann, J. E. (2015). How smart, connected products are transforming companies. *Harvard Business Review, 93*(10), 96-114.

Real, J. C., Leal, A., & Roldán, J. L. (2006). Information technology as a determinant of organizational learning and technological distinctive competencies. *Industrial Marketing Management, 35*(4), 505-521.

Rubera, G., & Kirca, A. H. (2012). Firm innovativeness and its performance outcomes: A meta-analytic review and theoretical integration. *Journal of Marketing, 76*, 130-147. https://doi.org/10.1509/jm.10.0494.

Sarta, A., Durand, R., & Vergne, J.-P. (2020). Organizational adaptation. *Journal of Management, 20*(10), 1-33. https://doi.org/10.1177/0149206320929088.

Sheridan, C. (2020). Fast, portable tests come online to curb coronavirus pandemic. *Nature Biotechnology, 38*(5), 515-518.

Solis, B., Lieb, R., & Szymanski, J. (2014). The 2014 state of digital transformation–How companies are investing in the digital customer experience. San Francisco, CA.

Somech, A., & Drach-Zahavy, A. (2013). Translating team creativity to innovation implementation: The role of team composition and climate for innovation. *Journal of Management, 39*(3), 684-708.

Stoeckli, E., Dremel, C., & Uebernickel, F. (2018). Exploring characteristics and transformational capabilities of InsurTech innovations to understand insurance value creation in a digital world. *Electronic Markets, 28*(2), 1–19.

Stolterman, E., & Fors, A. C. (2004). Information technology and the good life. In *Information systems research: Relevant theory and informed practice* (pp. 687-692).

Taylor, F. W. (1911). Principles and methods of scientific management. *Journal of Accountancy, 12*(3), 3.

Tilson, D., Lyytinen, K., & Sørensen, C. (2010). Research commentary - Digital infrastructures: The missing IS research agenda. *Information Systems Research, 21*(4), 748–759.

Verhoef, P. C., Broekhuizen, T., Bart, Y., Bhattacharya, A., Dong, J. Q., Fabian, N., & Haenlein, M. (2021). Digital transformation: A multidisciplinary reflection and research agenda. *Journal of Business Research, 122*, 889-901.

Vial, G. (2019). Understanding digital transformation: A review and a research agenda. *The Journal of Strategic Information Systems, 28*(2), 118–144.

Wang, C., & Ahmed, P. (2004). The development and validation of the organizational innovativeness construct using confirmatory factor analysis. *European Journal of Innovation Management, 7*(4), 303-313.

Yoo, Y., Boland Jr., R. J., Lyytinen, K., & Majchrzak, A. (2012). Organizing for innovation in the digitized world. *Organization Science, 23*(5), 1398–1408.

Yoo, Y., Henfridsson, O., & Lyytinen, K. (2010). Research commentary - The new organizing logic of digital innovation: An agenda for information systems research. *Information Systems Research, 21*(4), 724–735.

Zhu, K., Dong, S., Xu, S. X., & Kraemer, K. L. (2006). Innovation diffusion in global contexts: Determinants of post-adoption digital transformation of European companies. *European Journal of Information Systems, 15*(6), 601-616.

Chapter 18

Artificial Intelligence in Disaster Management

Ezgi Atalay*, PhD
Department of Medical Services and Techniques, Bolu Abant İzzet Baysal University, Bolu, Türkiye

Abstract

Disasters are events that can cause loss of life, economic loss, or major damage to the environment. Disaster management is a discipline that includes coordination and implementation processes to minimize damages caused by natural or man-made disasters. The disaster management process covers the stages of anticipation, preparation, response, and recovery. The main objective is to minimize the effects of disasters, increase the resilience of society against disasters, and ensure their rapid recovery. Today, with the development of technology, applications of artificial intelligence can provide various advantages in chaotic situations, such as disasters. In particular, risk predictions can be made by quickly analyzing large data sets before a disaster. In the time of a disaster, it can help to make quick decisions. Due to artificial intelligence, it can detect changes in disaster areas using data from various sensors or cameras. It can also be used to improve communication and coordination in disaster situations. In this way, providing fast and reliable information to the public, directing aid requests, and effective communication between aid teams are ensured. In the post-disaster recovery process, it is used in damage assessment, resource management, and reconstruction processes. There are still significant concerns about the use of artificial intelligence in disaster management in terms of social acceptance, ethics, data security, and privacy. Today, with increasing technological developments, it is thought

* Corresponding Author's Email: ezgiatalay@ibu.edu.tr.

In: Disaster Management in the Digital Age and Emerging Technologies
Editors: Ayşe Asiltürk and Ayşegül Sarıoğlu Kemer
ISBN: 979-8-89530-213-2

that the developments regarding the use of artificial intelligence in disasters will increase at the same rate.

Keywords: disaster, technology, disaster management, artificial intelligence

Introduction

A disaster is a natural, technological, or human-induced event that causes physical, economic, and social losses for the whole or certain segments of the society, stops or interrupts normal life and human activities, and the coping capacity of the affected society is not sufficient. At the same time, disaster does not refer to the event itself but to the consequences it brings about (AFAD, 2014:23). Disasters are events that can negatively affect a large number of people on a global scale. Disasters, which have caused significant losses worldwide from past periods to the present, are still continuing to be experienced (Usta, 2023). Disasters are events that can affect society in the short and long term, whether they occur under the influence of natural or human factors. Disasters are events that cause great material and moral losses, sometimes at the local level and sometimes at the national level. Disasters have existed in all periods of history, and human beings have tried to produce different defense mechanisms against these events (Yılmaz and Gülüm, 2024). Starting from an early age, people have developed policies and strategies to prevent loss of life, property, and environmental losses, especially loss of life, in their struggle against disasters of increasing type and severity (Mavi, 2020).

To cope with disasters and reduce their damages, it is necessary to take a series of actions before, during, and after disasters. The disaster management cycle is a dynamic planning and implementation framework covering the process from pre-disaster to post-disaster. In general, the disaster management cycle consists of four different phases: "mitigation," "preparedness," "response," and "recovery." Disaster management is a management field that includes many different disciplines and develops preparedness and response strategies against disasters using fields such as health, engineering, architecture, economy, management, business, social sciences, psychology, and geography. The disaster management process requires the identification of hazards and risks in advance, taking pre-disaster measures to prevent or mitigate damages, ensuring effective response and coordination, and carrying out post-disaster recovery activities with a holistic approach (Kadıoğlu, 2011:48; Karaman, 2016:3).

In general, the disaster management process has a complex and dynamic structure. For this structure, which is difficult to control, the possibilities provided by technology are needed for storing, processing, mapping, and evaluating data from systems (Partigöç, 2022). Technology can bring significant benefits at different stages of integrated disaster management, which is now accepted and implemented all over the world (Memiş and Babaoğlu, 2020a). Technological advances have led to a significant improvement in the monitoring and management of natural disasters. Remote sensing technologies make it possible to monitor natural disasters, such as volcanic eruptions, landslides, and forest fires in real-time, allowing the determination of important parameters such as the impact and speed of the spread of disasters. Analyzing the data allows the development of strategic plans for more effective evacuation and rescue operations. In addition, technologies such as the Internet of Things (IoT), smartphones, and social media make significant contributions to disaster management through rapid data collection and dissemination mechanisms. Technology can provide real-time data to first responders in disaster areas and improve emergency response. Smartphones and social media platforms can support disaster management by alerting citizens in disaster-affected areas and providing the necessary information. Integrating innovative technologies into disaster management can make disaster management processes more effective and societies more prepared for disasters (Krichen et al., 2023). Additionally, developments in the technological field bring new opportunities and can neutralize the previous ones. In this context, applications such as satellite systems, television, and the short message system (SMS) have been effective in disaster management (Genç, 2012: 360). Today, innovations such as the Internet of Things, artificial intelligence, robots, and smart systems are gaining importance. Although technology has some limitations and negatives, it provides significant benefits in identifying risks and realizing preparations before disasters, during disasters when it is very important to act quickly, and in post-disaster recovery efforts. In other words, the use of innovative technologies will provide significant gains in terms of risk reduction before the disaster as well as after the disaster. To summarize, technology can function with its capacity to ensure communication between parties in the disaster management process and to obtain data when used (Memiş and Babaoğlu, 2020b: 171; Partigöç, 2022; Atalay and Kaya, 2023:192; Yıldırım and Gündüz, 2023: 216).

Artificial Intelligence (AI)

In the world of science, researchers and technologists are trying to unravel the complexity of the human brain, a labyrinth of neurons that are interconnected and transmit signals to the whole body. In the scientific community, efforts to design a model to mimic the human brain have led to the concept of "Artificial intelligence (AI)" (Tandon et al., 2020; Akdoğan and Özdemir, 2024). In general, AI is an information technology that simulates and imitates human behaviors such as learning, reasoning, and decision-making. The basis of this system is to act independently of humans by imitating human cognitive characteristics (Pamuklu and Fındıkcı, 2023:178). AI is a multifaceted and interdisciplinary field of study that brings together different disciplines. AI is a comprehensive subject combining computer science, mathematics, philosophy, neuroscience, psychology, cybernetics, linguistics, and other professional fields (Tan et al., 2021; Rezk, 2023:2). Although AI attracts a great deal of attention in society, there is no single widely accepted definition. There are different definitions of artificial intelligence (McCarthy, 1988; Poole et al., 2010:3-5; Russell et al., 2016:1-5; Kaplan et al., 2019). AI is to provide a computer or machine with abilities such as reasoning like a human, making sense, making generalizations, and learning from past experiences to model human intelligence (Yılmaz, 2022:1). Despite varying definitions among academics, AI is expressed as a general term that refers to performing human tasks with the help of machines and technology (Ahmed et al. 2021).

Alan Turing, the famous British logician, and mathematician who first introduced the philosophy of AI, presented the first assumptions of AI with his article "Computing Machinery and Intelligence" published in 1950 about the possibility of intelligent machines (Pirim, 2006; 89). Alan Turing developed a test that measures whether a computer or other system has the same mental abilities as humans. This test measures whether an expert can distinguish between the performance of a machine and that of a human. If not, then the machine is as mentally capable as humans. In this test, a human and a computer are hidden from the experimenter. The experimenter communicates with both of them without knowing which one (Pirim, 2006; 90; Pamuklu and Fındıkcı, 2023:179). This study, also known as the Turing test, tests whether a machine can be understood as a human during communication between a human and a computer. It was measured whether the performance of a computer and the performance of a human mind could be the same. This test forms the basis of AI (Benli and Şenel, 2020: 300; Kocaman, 2021: 3005). The term artificial intelligence was first used at

Dartmouth College in the United States. In 1956, the foundations of artificial intelligence were laid at a meeting attended by major researchers at Dartmouth (Karaküçük and Eker, 2020:124; Arslan, 2020). The term "artificial intelligence" was first used by John McCarthy, who is now recognized as the father of the field. In 1956, he organized the first academic meeting on the subject of "The science and engineering of producing intelligence robots" at Dartmouth College in Hanover, New Hampshire (Rezk, 2023:2).

In the present era, AI is defined as any machine or technology that can mimic the cognitive abilities of human intelligence. Deep Learning (DL), Machine Learning (ML), Artificial Neural Network (ANN), Recurrent Neural Network (RNN), Convolutional Neural Network (CNN), Big Data (BD), Internet of Things (IoT), and Relationship Network (RN) are among the frequently used AI tools (Sun et al., 2020; Ayaydın et al., 2022; Şimşek et al., 2023; Pamuklu and Fındıkcı, 2023:178). To better understand the concept of AI, it is important to know the basic concepts such as machine learning, neural networks, and deep learning (Khanagar et al., 2021; Ossowska et al., 2022; Akdoğan and Özdemir, 2024).

Disasters and Artificial Intelligence

AI technologies can assist humans in the management and communication of crises (Özgür, 2024). AI models are playing a remarkable role in the processing of large and non-linear data, demonstrating extremely impressive success and superiority. These developments offer excellent tools for critical processes such as disaster monitoring and management and provide a significant advantage in combating disasters (Tan et al., 2021). AI systems can be used for many different purposes in disaster management. AI is used to automate tasks and improve decision-making, especially in this aspect, it can have important areas of use in disaster management. Thanks to its advanced analytical capabilities, AI can increase the effectiveness of disaster preparedness, response, and recovery processes. "Disaster prediction and early warning, damage assessment, logistics and resource management, public communication, and public information, disaster risk assessment, training and simulation, rescue operations, post-disaster recovery" are some of the fundamental disaster management tasks that AI can perform (Şimşek et al., 2023; Karaca, 2023:1321). AI can analyze weather and geological data to determine the potential impacts of natural disasters. It can be used to determine evacuation routes, create emergency plans, and direct aid teams in disaster situations. It can distribute resources effectively by identifying post-disaster human needs and directing requests for assistance. In this way, it can accelerate decision-making processes in combating disasters, enable us to use resources more effectively, and help us manage the recovery process more effectively (Demirel and Arıkan, 2023).

Examples of the Use of Artificial Intelligence in Disaster Maganement

There are many applications and methods in the literature regarding the use of AI in disasters. There are many systematic reviews focusing especially on pre-disaster, disaster response, and post-disaster situations (Nunavath and Goodwin, 2019, Tan et al., 2021; Abid et al., 2021; Linardos et al., 2022; Farazmehr and Wu, 2023; Bari et al., 2023; Ghaffarian et al., 2023). Since it is not possible to cover all applications in this section, some of the most prominent use cases are mentioned.

The use of AI in natural disaster risk reduction can be listed as seismic building damage prediction, earthquake prediction, earthquake-induced built-in damage detection, seismic hazard prediction, landslide prediction, and multi-hazard susceptibility assessment (Kıbrıs, 2022: 369). For example, "DeepShake" is an early warning system developed by Daniel J. Wu from Stanford University and his team and adopts an approach based on deep neural networks. This method was developed to predict how the shaking of a new earthquake will act using previous earthquakes (Datta et al., 2022). There are also studies focusing on prediction systems for hurricane (Dawood et al., 2020) and volcanic (Rudi et al., 2020) activities. Flood warning systems can reduce disaster risks and damage by providing life-saving effects. The AI model called "Flood Hub" developed by Google Research is specially designed to be used in flood forecasting. It was developed to provide accurate and effective warnings against flooding in large-scale rivers with dense populations. The model records river basins in more than 80 countries and analyzes local river flood movements and water trends. Flood Hub uses an improved modeling method to provide real-time flood forecasts and warnings (Nevo et al., 2022). Human-input supervised data sets can be used for urban flood forecasts. Social media and crowdsourced data can be processed using Natural Language Processing and Computer Vision techniques. By analyzing social media data, the reality and location of the disaster are obtained (Wang et al., 2018).

It is important that buildings are resistant to disasters before disasters occur. While architectural design is being realized, designs should be realized with the awareness of earthquake architecture. For this purpose, deep learning and image processing methods were used, and an Irregularity Control Assistant was created. Due to this system, general information will be given to architects about the compliance of carrier system decisions with earthquake regulations in the early stages of architectural design (Bingöl et al., 2020).

Casualty assembly centers are response centers that have the capacity to provide emergency health services to a large number of injured people at the same time in earthquakes, hurricanes, floods, or many disasters. The locations of these centers should be determined before disasters. Using AI methods, it can be predetermined where the casualty gathering centers will be deployed (Arslan, and Yıldız, 2016).

The increasing popularity of social media platforms has changed the dynamics of information sharing during disasters and mass emergencies. Social media information can be an important resource for emergency responders and decision-makers, especially in the early stages of disasters when little or no information is available through other traditional channels (Ofli et al., 2020). Powers et al. (2023) analyzed "calls for help" shared on Twitter using machine learning methods. They identified how Twitter was used in disasters such as Hurricane Harvey. They classified the tweets shared during Hurricane Harvey as urgent or non-urgent by creating neural and non-neural machine learning models. It is a sample study to classify individuals' requests for help in life-threatening disaster situations.

Effective crisis management is needed to minimize the loss of life and property in the region after disasters. The quick and accurate determination of road damage in the disaster area is very important for effective crisis management. Using Convolutional Neural Networks (CNN), road damage after earthquakes can be detected as soon as possible and new routes can be determined with the shortest path algorithm. This system can be used to optimize the algorithm for the shortest path in real-time (Ünlü and Şimşek, 2021). During and after a disaster, transferring the data collected through AI applications very quickly and directly from the scene facilitates disaster management. Using the mobile application, it can be determined whether the buildings are damaged or not with the device camera. This system is based on a system based on similarity rate. As a result, damaged buildings will be recorded in the database with location information, and support will be provided for all processes that will take place after the disaster (Tarhan et al., 2022: 127).

AI can also be used in biological disasters (Su et al., 2021). In particular, it is an innovative technology that has helped fight the COVID-19 pandemic (Vaishya et al., 2020). AI contributes promisingly to mitigating the effects of pandemics and shows high performance in areas such as diagnosis, prognosis assessment, outbreak prediction, and drug discovery (Wang et al., 2021; Hasan et al., 2023; Ulaşan, 2023). Robots can help prevent and manage the virus by providing information to humans and healthcare workers by recognizing COVID-19 symptoms (Ashique et al., 2024). AI is being used in the early diagnosis and treatment of infection and drug and vaccine development processes, as well as in helping to reduce the workload of healthcare workers (Vaishya et al., 2020).

Technological developments not only affect natural disaster measures but also affect the preparation process of information designs used to increase disaster literacy. It is thought that artificial intelligence technology can also be used in information design aiming to increase disaster literacy. In a study, natural disaster information designs produced by five expert designers and five information designs produced by the DALL-E artificial intelligence model were analyzed. After the evaluations, it was revealed that the designs of expert designers and the designs produced by the DALL-E artificial intelligence model were acceptable and functional informative infographic designs to increase earthquake awareness in the context of graphic design. The study found that AI technology can be used in the preparation of information designs to increase disaster literacy, and alternative information designs can be prepared with this technology. As a result, it was understood that artificial intelligence technology can be used to improve disaster literacy (Kahraman and Gülaçtı, 2023).

In disaster situations, thanks to technologies such as unmanned vehicles, automation systems, and the Internet of Things, loss of property and life can be prevented, especially in logistics processes. Problems such as the transportation of wrong, inadequate, and missing materials in disaster logistics can be prevented. Managing warehouses with AI, quickly tracking which materials and medicines are in which warehouse in a disaster, reporting deficiencies to the center, and tracking orders will be carried out in a healthy way. AI technology prevents the transportation of wrong, unnecessary, inadequate, and missing materials (Adıgüzel, 2022).

Discussion and Conclusion

Disasters are increasingly common and affect communities around the world. To address these challenges, advances in AI technology offer unique opportunities. AI applications for disasters can increase the success of risk assessment, preparedness, and disaster response. AI technologies can also be used to improve disaster reconstruction processes. In other words, it has qualities that can play an important role before and after disasters. Considering that technological developments are gaining momentum, more governments and organizations will benefit from AI technology in disaster management in the future. However, this technology may also present some challenges and risks that need to be carefully evaluated and managed. When AI is used in consultation with ethical experts and emergency management professionals, it can prove to be an invaluable tool in disaster response. As AI technology continues to evolve and become more widely adopted, its applications in disaster management are likely to become even more powerful and widespread.

Disclaimer

None.

References

Abid, S. K., Sulaiman, N., Chan, S. W., Nazir, U., Abid, M., Han, H., ... & Vega-Muñoz, A. (2021). Toward an integrated disaster management approach: how artificial intelligence can boost disaster management. *Sustainability, 13* (22), 12560. https://doi.org/10.3390/su132212560.

Adıgüzel, S. (2022). Afet durumlarında yapay zeka teknolojisi ile lojistik yönetimi örnekleri. *Akademik İzdüşüm Dergisi, 7* (1), 47-70.

AFAD (2014). Açıklamalı Afet Terimleri Sözlüğü. Retrieved March 29, 2024, fromhttps://www.afad.gov.tr/kurumlar/afad.gov.tr/3495/xfiles/sozluk.pdf.

Ahmed, N., Abbasi, M. S., Zuberi, F., Qamar, W., Halim, M. S. B., Maqsood, A., & Alam, M. K. (2021). Artificial intelligence techniques: analysis, application, and outcome in dentistry—a systematic review. *BioMed Research International,* 2021, *15.* https://doi.org/10.1155/2021/9751564.

Akdoğan, C., & Özdemir, H. (2024). Yapay Zekâ'nın Diş Hekimliği Pratiğine Kazanımları. *Türk Diş Hekimliği Araştırma Dergisi,* 2(3), 278-287. https://doi.org/10.58711/turkishjdentres.vi.1296215.

Arslan, H. M., & Yıldız, M. S. (2016). Yapay Zekâ Optimizasyon Yöntemi İle Yaralı Toplanma Merkezlerinin Konuşlandırılması. *Bartın Üniversitesi İktisadi ve İdari Bilimler Fakültesi Dergisi, 7*(14), 165-188.

Arslan, K. (2020). Eğitimde yapay zekâ ve uygulamaları. *Batı Anadolu Eğitim Bilimleri Dergisi, 11*(1), 71-88.

Ashique, S., Mishra, N., Mohanto, S., Garg, A., Taghizadeh-Hesary, F., Gowda, B. J., & Chellappan, D. K. (2024). Application of artificial intelligence (AI) to control COVID-19 pandemic: Current status and future prospects. *Heliyon. 10* (2024) e25754 https://doi.org/10.1016/j.heliyon.2024.e25754.

Atalay, E. & Kaya, N. (2023). Afetlerde Exoskeleton Kullanımı. S. Şahinöz (Eds.), *Afet Yönetimi ve Teknoloji* (pp. 189-199). İstanbul Tıp Kitabevi.

Ayaydın A, Akcayol MA (2022) Deep learning based forecasting of delay on flights. *Journal of Information Technologies, 15*(3):239-249. https://doi.org/10.17671/gazibtd.1060646.

Bari, L. F., Ahmed, I., Ahamed, R., Zihan, T. A., Sharmin, S., Pranto, A. H., & Islam, M. R. (2023). Potential use of artificial intelligence (AI) in disaster risk and emergency health management: a critical appraisal on environmental health. *Environmental Health Insights, 17*, 11786302231217808. https://doi.org/10.1177/11786302231217808.

Benli, E., & Şenel, G. (2020). Yapay zekâ ve haksız fiil hukuku. *Ankara Sosyal Bilimler Üniversitesi Hukuk Fakültesi Dergisi, 2* (2), 296-336. https://doi.org/10.47136/asbuhfd.713190.

Bingöl, K., Aslı, E. R., Örmecioğlu, H. T., & Arzu, E. R. (2020). Depreme dayanıklı mimari tasarımda yapay zeka uygulamaları: Derin öğrenme ve görüntü işleme yöntemi ile düzensiz taşıyıcı sistem tespiti. *Gazi Üniversitesi Mühendislik Mimarlık Fakültesi Dergisi, 35* (4), 2197-2210. https://doi.org/10.17341/gazimmfd.647981.

Datta, A., Wu, D. J., Zhu, W., Cai, M., & Ellsworth, W. L. (2022). DeepShake: Shaking intensity prediction using deep spatiotemporal RNNs for earthquake early warning. *Seismological Society of America, 93* (3), 1636-1649. https://doi.org/10.1785/0220210141.

Dawood, M., Asif, A., & Minhas, F. U. A. A. (2020). Deep-PHURIE: deep learning based hurricane intensity estimation from infrared satellite imagery. *Neural Computing and Applications, 32* (13), 9009-9017. https://doi.org/10.1007/s00521-019-04410-7.

Demirel, Y. T. & Arıkan, N. İ. (2023). Yapay Zekânın Afet Bölgelerinde Kullanımı. *IJESOS International Journal Of Educational and Social Sciences*, 2 (2), 77-82. https://doi.org/10.5281/zenodo.10429545.

Farazmehr, S., & Wu, Y. (2023). Locating and deploying essential goods and equipment in disasters using AI-enabled approaches: A systematic literature review. *Progress in Disaster Science, 100292*. https://doi.org/10.1016/j.pdisas.2023.100292.

Genç, N. (2012). Afet yönetimi ve iletişim teknolojileri. M. Z. Sobacı & M. Yıldız (Eds.), *E-Devlet: Kamu yönetimi ve teknoloji ilişkisinde güncel gelişmeler* (pp. 353-380). İstanbul: Nobel Yayıncılık.

Ghaffarian, S., Taghikhah, F. R., & Maier, H. R. (2023). Explainable artificial intelligence in disaster risk management: Achievements and prospective futures. *International*

Journal of Disaster Risk Reduction, 98, 104123. https://doi.org/10.1016/j.ijdrr.2023.104123.

Hasan, M. M., Islam, M. U., Sadeq, M. J., Fung, W. K., & Uddin, J. (2023). Review on the evaluation and development of artificial intelligence for COVID-19 containment. *Sensors, 23* (1), 527. https://doi.org/10.3390/s23010527.

Kadıoğlu, M. (2011). *Afet yönetimi: Beklenilmeyeni beklemek, en kötüsünü yönetmek.* İstanbul: Marmara Belediyeler Birliği.

Kahraman, M. E., & Gülaçtı, İ. E. (2023). Doğal Afet Okuryazarlığı İçin Bilgilendirici İnfografik Tasarımlarında Yapay Zekâ Kullanımı. *Türk Deprem Araştırma Dergisi, 5* (2), 166-182. https://doi.org/10.46464/tdad.1296530.

Kaplan, A., & Haenlein, M. (2019). Siri, Siri, in my hand: Who's the fairest in the land? On the interpretations, illustrations, and implications of artificial intelligence. *Business horizons, 62* (1), 15-25. https://doi.org/10.1016/j.bushor.2018.08.004.

Karaca, M. (2023). Yapay Zekâ Tabanlı Stratejik Afet Yönetimi: Verilerin Tam Kullanımı. *Afet ve Risk Dergisi, 6* (4), 1312-1331. https://doi.org/10.35341/afet.1324323.

Karaküçük, Y., & Eker, S. (2020). Oftalmolojide Yapay Zeka ve Derin Öğrenme Uygulamaları. A. R. Şahin, K. Doğan & S. Sivri (Eds.), *Sağlık Bilimlerinde Yapay Zeka* (pp. 123-144). Ankara: Akademisyen Kitabevi.

Karaman, Z. T. (2016). Afet yönetimine giriş ve Türkiye'de örgütlenme. Z. T. Karaman ve A. Altay, (Eds.), *Bütünleşik Afet Yönetimi* (pp. 1-36). İzmir: İlkem Yayınları.

Khanagar, S. B., Al-Ehaideb, A., Maganur, P. C., Vishwanathaiah, S., Patil, S., Baeshen, H. A., ... & Bhandi, S. (2021). Developments, application, and performance of artificial intelligence in dentistry–A systematic review. *Journal of Dental Sciences, 16* (1), 508-522. https://doi.org/10.1016/j.jds.2020.06.019.

Kıbrıs, M. E. (2022, 20-22 April). Doğal Afet Risklerinin Azaltılmasında Yapay Zeka Uygulamaları. TMMOB Afet Sempozyumu bildiriler kitabı (pp. 357-372) Ankara.

Kocaman, Şeref (2021). Grafik Tasarım Endüstrisinde Yapay Zekâ. *Journal of Social and Humanities Sciences Research, 8* (77), 3000-3015. https://doi.org/10.26450/jshsr.2843.

Krichen, M., Abdalzaher, M. S., Elwekeil, M., & Fouda, M. M. (2023). Managing natural disasters: An analysis of technological advancements, opportunities, and challenges. *Internet of Things and Cyber-Physical Systems, 4* (2024) 99-109. https://doi.org/10.1016/j.iotcps.2023.09.002.

Linardos, V., Drakaki, M., Tzionas, P., & Karnavas, Y. L. (2022). Machine learning in disaster management: recent developments in methods and applications. *Machine Learning and Knowledge Extraction, 4* (2), 446-473. https://doi.org/10.3390/make4020020.

Mavi, E. E. (2020). Afet kriz yönetiminde medya: 30 Ekim 2020 İzmir depremi. *Karadeniz Teknik Üniversitesi İletişim Araştırmaları Dergisi*, 10(2), 31-53.

McCarthy, J. (1988). Mathematical logic in artificial intelligence. *Daedalus,* 297-311.

Memiş, L. & Babaoğlu, C. (2020a). Acil durum ve afet yönetiminde süreç yaklaşımı ve teknoloji. *Ömer Halisdemir Üniversitesi İktisadi ve İdari Bilimler Fakültesi Dergisi, 13* (4), 776-791. https://doi.org/10.25287/ohuiibf.731103.

Memiş, L. & Babaoğlu, C. (2020b). Afet yönetimi ve teknoloji. M. Yaman &E. Çakır (Eds.), *Farklı boyutlarıyla afet yönetimi* (pp. 163-178), Ankara: Nobel Yayıncılık.

Nevo, S., Morin, E., Gerzi Rosenthal, A., Metzger, A., Barshai, C., Weitzner, D., ... & Matias, Y. (2022). Flood forecasting with machine learning models in an operational framework. *Hydrology and Earth System Sciences, 26* (15), 4013-4032. https://doi.org/10.5194/hess-26-4013-2022.

Nunavath, V., & Goodwin, M. (2019, December). *The use of artificial intelligence in disaster management-a systematic literature review.* In 2019 International Conference on Information and Communication Technologies for Disaster Management (ICT-DM) (pp. 1-8). *IEEE.* https://doi.org/10.1109/ICT-DM47966.2019.9032935.

Ofli, F., Imran, M., & Alam, F. (2020). Using artificial intelligence and social media for disaster response and management: an overview. T. V. V. Kumar & K. Sud (Eds.), *AI and Robotics in Disaster Studies* (pp. 63-81). Singapore: Palgrave Macmillan. https://doi.org/10.1007/978-981-15-4291-6_5.

Ossowska, A., Kusiak, A., & Świetlik, D. (2022). Artificial intelligence in dentistry—Narrative review. *International Journal of Environmental Research And Public Health, 19* (6), 3449. https://doi.org/10.3390/ijerph19063449.

Özgür, Ö. F. (2024). Türkiye'de DergiPark Sistemindeki Yapay Zekâ ve Kriz Konulu Akademik Çalışmalar Üzerine Bir Değerlendirme. *AJIT-e: Academic Journal of Information Technology, 15* (1), 49-70. https://doi.org/10.5824/ajite.2024.01.003.x.

Pamuklu, A. G., & Fındıkcı, M. B. (2023). Grafik Tasarımın Geleceği: Yapay Zekâ ve İnsan. *Bilim Eğitim Sanat ve Teknoloji Dergisi, 7* (2), 177-191. https://doi.org/10.46328/bestdergi.100.

Partigöç, N. S. (2022). Afet risk yönetiminde yapay zekâ kullanımının rolü. *Bilişim Teknolojileri Dergisi, 15* (4), 401-411. https://doi.org/10.17671/gazibtd.1067831.

Pirim, H. (2006). Yapay zeka. *Yaşar Üniversitesi E-Dergisi, 1* (1), 81-93.

Poole, D. L., & Mackworth, A. K. (2010). *Artificial Intelligence: foundations of computational agents.* Cambridge University Press. https://doi.org/10.1017/CBO 9780511794797.

Powers, C. J., Devaraj, A., Ashqeen, K., Dontula, A., Joshi, A., Shenoy, J., & Murthy, D. (2023). Using artificial intelligence to identify emergency messages on social media during a natural disaster: A deep learning approach. *International Journal of Information Management Data Insights, 3* (1), 100164. https://doi.org/10.1016/ j.jjimei.2023.100164.

Rezk, S. M. M. (2023). The role of artificial intelligence in graphic design. *Journal of Art, Design and Music, 2* (1), 1-12. https://doi.org/10.55554/2785-9649.1005.

Rudi, J., Malossi, A. C. I., Isaac, T., Stadler, G., Gurnis, M., Staar, P. W., ... & Ghattas, O. (2015, November). *An extreme-scale implicit solver for complex PDEs: highly heterogeneous flow in earth's mantle.* In Proceedings of the international conference for high performance computing, networking, storage and analysis (pp. 1-12). https://doi.org/10.1145/2807591.2807675.

Russell, S. J., & Norvig, P. (2016). *Artificial intelligence: a modern approach.* Pearson.

Su, Z., McDonnell, D., Bentley, B. L., He, J., Shi, F., Cheshmehzangi, A., ... & Jia, P. (2021). Addressing Biodisaster X threats with artificial intelligence and 6G technologies: literature review and critical insights. *Journal of Medical Internet Research,* 23 (5), e26109. https://doi.org/10.2196/26109.

Sun W, Bocchini P, Davison BD (2020) Applications of artificial intelligence for disaster management. *Natural Hazards, 103* (3):2631-2689. https://doi.org/10.1007/s11069-020-04124-3.

Şimşek, D., Kutlu, İ., & Şık, B. (2023). *The role and applications of artificial intelligence (AI) in disaster management.* In 3rd International Civil Engineering and Architecture Congress.137-150. https://doi.org/10.31462/icearc.2023.arc992.

Tandon, D., Rajawat, J., & Banerjee, M. (2020). Present and future of artificial intelligence in dentistry. *Journal of Oral Biology and Craniofacial Research, 10* (4), 391-396. https://doi.org/10.1016/j.jobcr.2020.07.015.

Tarhan, C., Ozgur, A. S., Teke, I. & Komesli, M. (2022). Görüntü İşleme Entegre Afet Yönetiminde Yapay Zeka Yöntemi Olarak Kullanılabilir mi?. *Journal of Research in Business, 7* (1), 116-131. https://doi.org/10.54452/jrb.1025382.

Ulaşan, F. (2023). *Koronavirüsle Mücadelede Yapay Zekânin Yerinin Kamu Yönetimi Temelinde Değerlendirilmesi.* In I. Global (Eds.), 3. International Mediterranean Congress, (pp. 109-123). Mersin.

Usta G. (2023). Dünya'da meydana gelen afetlerin istatistiksel olarak analizi (1900-2022). *Gümüşhane Üniversitesi Sosyal Bilimler Dergisi, 14* (1), 172-186.

Ünlü, R., & Şimşek, N. (2021). *Deprem sonrası hasarlı yolların yapay zekâ ile tespit edilerek en kısa yol güzergâhlarının anlık optimizasyonu.* In 1st International Congress on Artificial Intelligence and Data Science Proceeding Book (pp. 119-124).

Vaishya, R., Javaid, M., Khan, I. H., & Haleem, A. (2020). Artificial Intelligence (AI) applications for COVID-19 pandemic. *Diabetes & Metabolic Syndrome: Clinical Research & Reviews, 14* (4), 337-339. https://doi.org/10.1016/j.dsx.2020.04.012.

Wang, L., Zhang, Y., Wang, D., Tong, X., Liu, T., Zhang, S., ... & Clarke, M. (2021). Artificial intelligence for COVID-19: a systematic review. *Frontiers in Medicine, 8,* 1457. https://doi.org/10.3389/fmed.2021.704256.

Wang, R. Q., Mao, H., Wang, Y., Rae, C., & Shaw, W. (2018). Hyper-resolution monitoring of urban flooding with social media and crowdsourcing data. *Computers & Geosciences, 111,* 139-147. https://doi.org/10.1016/j.cageo.2017.11.008.

Yıldırım, S., D. & Gündüz, F. (2023). Afetlerde Veri Kullanım Sistemleri. S. Şahinöz (Eds.), Afet Yönetimi ve Teknoloji (pp. 200-224). İstanbul Tıp Kitabevi.

Yılmaz, A. (2022). Yapay zekâ. İstanbul: Kodlab.

Yılmaz., H. & Gülüm, E. (2024). Afet Acil Durum Yönetimi ve Afet Müdahale Planı Çerçevesinde Uygulamalar: Pazarcık – Elbistan Depremi. *Akademik Yaklaşımlar Dergisi, 15* (1-Deprem Özel Sayısı-), 1-25. https://doi.org/10.54688/ayd.1412307.

Chapter 19

Human Resources in Disaster Management

Samet Dinçer[*]**, PhD**

Department of Medical Services and Techniques, Durağan Vocational School, Sinop University, Sinop, Türkiye

Abstract

Disasters represent complex challenges with profound implications for individuals, communities, and nations, necessitating robust preparedness strategies to mitigate their impact and accelerate recovery efforts. This paper delves into the essential components of disaster preparedness, emphasizing the critical roles of human resource management and technological use. Drawing upon an extensive array of scholarly sources, it thoroughly examines key aspects, including education and awareness, emergency planning, drills and practices, communication strategies, cooperation and solidarity, resilience, and psychosocial support in disaster preparedness efforts. Furthermore, it elucidates the intricate interplay between human resource management and technology in disaster response, highlighting the pivotal contributions of technological innovations such as warning systems, satellite imaging, communication networks, and social media to improve disaster management capabilities. The abstract underscores the imperative to integrate human resources and technology to effectively mitigate the impact of disasters, underscoring the significance of ongoing training initiatives, technological infrastructure enhancement, and strategic planning in fortifying communities against the multifaceted challenges posed by disasters.

Keywords: disasters, human resources, using technology

---

[*] Corresponding Author's Email: sdincer@sinop.edu.tr.

In: Disaster Management in the Digital Age and Emerging Technologies
Editors: Ayşe Asiltürk and Ayşegül Sarıoğlu Kemer
ISBN: 979-8-89530-213-2

Introduction

Disasters have a profound impact on individuals, societies, and even the global community. It is important to note that this impact can be felt at all levels, from the individual to the global scale. They result in physical, economic, and social losses, often disrupting public order and daily life (Gündüz, 2024). The distinction between the terms 'disaster' and 'emergency' is based on the coping capacity of the society affected by the crisis. In an emergency, a society can manage a crisis situation within its capacity, while in a disaster, external assistance is required to deal with the crisis (Al-Dahash, Thayaparan, & Kulatunga, 2016).

Dealing with disasters, which have become more destructive and varied in recent years, requires more than just process management led by a single actor. Combating disasters and emergencies requires not only economic resources and technological equipment but also human resources (Duchek, 2020). Communities exposed to disaster-inducing situations will be able to respond to the extent of their available resources and the community's coping capacity until external assistance arrives. The effectiveness of this community reaction is determined by various components (Makwana, 2019). These components include social dimensions such as social capital, socio-economic status, level of general education, and social trust of the community exposed to disasters. It is important to have an awareness of the risks and hazards in the region where a society lives, as well as a strong social fabric, to effectively cope with disasters (Pilatin and Ayaydın, 2022). In the event of a potential disaster, it is crucial to involve the community in response and recovery efforts to mitigate risks and prevent losses. Raising awareness of potential risks among existing human resources and mobilizing society can help achieve this goal (Pramono et al., 2020).

It is crucial for communities and institutions to be prepared for disasters and emergencies caused by unexpected events such as natural disasters, industrial accidents, epidemics, and terrorist attacks. Human resources play a critical role in dealing with disasters and emergencies, and their preparedness is based on factors such as training, awareness raising, and planning. Our study has identified important components for the preparedness of human resources in disasters and emergencies. These components include training and awareness, emergency plans, practices and drills, communication plans, cooperation and solidarity, resilience and mobilization, and psychosocial support. In the final part of the study, we will examine the relationship between human disaster preparedness and technology.

Education and Awareness

Tackling disasters with public power and capacity alone is very difficult. Despite the size of the capacities and opportunities, they may not be sufficient to eliminate the damages caused by disasters that affect a large part of society (Muttarak and Pothisiri, 2013). For this reason, individuals and society should be able to cope with the negative effects of disasters without external assistance for a certain period of time (Tan et al., 2017). Education and awareness raising are key to a community's ability to cope with disasters and how it should behave in the face of disasters. Access to information sources is essential for a community to be aware of the nature of the risks it faces and their possible consequences (Mishra and Suar, 2012).

Organizing training sessions on disaster response, implementation of emergency plans, and communication methods is vital to the preparation of human resources. Specifically, providing prior knowledge of various types of disasters and their effects through training will enable the correct action to be taken in the event of a disaster. Training can facilitate behavioral change by increasing the knowledge of the person concerned about an unknown situation (Mishra and Suar, 2012). In addition to education and training, information on emergency plans drawn up by the authorities before the disaster and information on their role in these plans will help individuals prepare for adverse situations (Nix-Stevenson, 2013). Furthermore, it is vital to determine communication methods for emergencies before the adverse situation. This includes determining how to communicate with family members, relatives, and all other individuals who need to be contacted during emergencies. By being prepared in the pre-disaster period, communication-related problems, which can be one of the biggest issues after a disaster, can be minimized (Johnston et al., 2020). When all these components are strong together, awareness can be expected to increase.

Emergency Plans

The term 'emergency' typically refers to sudden adverse situations that can be managed with local resources. It denotes a state of crisis (Herstein et al., 2021). An emergency plan is essential to prevent major accidents and injuries, minimize damage to the workplace, surrounding buildings, and equipment, protect society and the environment from the effects of emergencies, and ensure a swift return to normal working order while minimizing material and moral damage (İytemür and Tekeli Yeşil, 2020). To ensure the least possible material and moral damage during an emergency, it is crucial to plan. This involves determining the necessary measures that need to be taken and ensuring the ease of implementation when necessary. It is important that all members of society, including families, workplaces, schools, and local administrations, have disaster and emergency plans in place (Alexander, 2002).

Emergency plans should include details such as identifying the emergency situation, officials and their tasks, evacuation routes, communication methods, and meeting points. The determination of the emergency situation should result from risk assessment processes. All risks that may be encountered should be measured and analyzed with scientific methods to identify emergency situations that may pose a danger (Alexander, 2002). To ensure effective emergency management, it is crucial to assign specific tasks to individuals based on identified risks. Acting without control during an emergency can lead to increased losses (Bowen, 2008). Determining how and from which directions to evacuate homes or workplaces before the event against events such as a fire, earthquake, or CBRN threat will greatly prevent the loss of life that may occur (Alexander, 2015). In the event of an emergency, it is crucial to establish alternative communication methods in case of interruptions. Radio or other similar alternatives are essential to maintain uninterrupted communication during disasters and emergencies (Wang et al., 2023). In case of an emergency, having safe meeting points for individuals identified prior to the incident will help minimize problems such as getting lost due to communication breakdowns.

Practices and Drills

One of the most important prerequisites for individuals to be prepared for disasters and emergencies is provided by drills. Drills are generally training and preparation activities used in many fields. These activities can cover various scenarios such as emergencies, fire evacuation, disaster situations, workplace safety, and military operations (McAlister, 2011). When designing an exercise, it involves several stages. First, it is essential to determine the purpose of the exercise, as it should always serve a specific goal. Clearly defining this purpose helps participants develop the necessary skills. Second, it is crucial to create realistic scenarios that allow the modeling of real risks and threats that may be encountered (Rokach et al., 2008). Third, it is important to clarify the roles and responsibilities of all participants involved in the exercises. Additionally, it may be beneficial to ensure that professional diversity is maximized during exercises. Fourth, prior to the exercise, it is crucial to train and inform all participants about the scenario, tasks, and protocols. This will increase the efficiency of the exercise (Zavar and Nelan, 2020). Fifth, evaluate the performance of the participants and provide feedback at the end of the training. Sixth, review any elements that may pose a risk or danger before exercising and take the necessary precautions. This will facilitate the recognition of positive behaviors and the identification of areas for improvement. Seventh, document and report the exercise results and participants' performance when required. This text provides feedback for future improvements. Continuous improvement activities are essential. Exercise should be included in the continuous improvement process of the organization. Feedback and exercise results must be continuously evaluated, and safety measures updated (Bowen, 2008).

Drills are a crucial tool to improve the preparedness of an organization or community and to be ready for potential crises or incidents. Well-designed and implemented drills, in line with the principles outlined above, enable participants to learn and develop effectively. Drills and exercises should be based on disaster scenarios, allowing people to put their plans into practice and gain valuable experience in how to react in the event of a disaster (Alexander, 2015).

Communication Plans

The purpose of communication plans in disaster situations is to ensure fast, effective, and coordinated communication. There are some basic elements of a communication plan that can be used in disaster situations. First, it is necessary to determine who will assume which role in communication in emergencies. A communication chain should be established, and the roles and responsibilities of everyone in this chain should be clearly defined. This chain should include everyone, from managers to staff and community members (Hansson et al., 2020). Second, a list should be created containing the contact details of the people to be contacted in case of an emergency. This list should include contact details of employees, emergency teams, local authorities, the media, and other interested parties. Additionally, alternative means of communication should be identified in cases where traditional means of communication, such as telephone lines or Internet connections, may be disabled. Alternative means of communication, radio, walkie-talkies, social media, etc. may be considered preferred methods (Liu and Mehta, 2021). There is a need to establish a strategy to inform the public in the case of a disaster. The strategy should aim to ensure the rapid dissemination of accurate and up-to-date information about emergencies. Various communication tools, such as press releases, social media updates, brochures, or newsletters for the community, can be used. Additionally, communication teams should be formed for emergency information, and each team should have specific roles and responsibilities. During emergencies, these teams should be responsible for coordinating communication and gathering and disseminating information (Johnston et al., 2020).

When preparing communication plans, it is essential to ensure that individuals are capable of applying the plan during emergencies. Regular drills and training sessions can help individuals become familiar with the communication plan and implement it effectively. Additionally, individuals should have access to current information and resources for emergencies. To achieve this, it is crucial to use technologies that incorporate an up-to-date database of local authorities, public institutions, and other relevant organizations. Additionally, the communication plan should undergo a continuous review, with improvements made in areas that may be lacking. Feedback mechanisms should be established to evaluate the effectiveness of the communication plan during disaster situations, and the plan should be updated accordingly based on this feedback (Johnston et al., 2020). The preparation and implementation of communication plans in disaster situations are critical to ensure a rapid and effective response. Regular review and update of these plans will ensure better preparedness for disaster situations. The continuity of communication in emergencies is of critical importance. In the pre-disaster period, individuals should have a plan explaining how they can communicate with their families and relatives. At the same time, local governments, public institutions, and emergency teams should also create communication strategies to inform the community (Özer, 2017).

Cooperation and Solidarity

Social capital is a key concept in disasters and emergencies. Solidarity of society and mutual aid are key components of social capital. The presence of organizations and social networks that facilitate support is essential to mitigate the impact of disaster situations (Dany, 2018).

In disaster situations, solidarity is vital because it is essential for society to act in unity and solidarity to minimize the damage and speed up the healing process of those affected. Providing solidarity is subject to certain conditions, and it is important to identify and prioritize needs during such times. This stage involves determining and prioritizing the basic needs of those affected, such as shelter, food, water, and medical aid (Bendell, 2022). Effective coordination and communication are crucial for aid and solidarity initiatives to be successful in disaster situations. It is essential to maintain regular and effective communication between emergency management units, non-governmental organizations, volunteers, and other relevant stakeholders. Ensuring security and maintaining order in disaster-stricken areas is of utmost importance. Ensuring public safety and protecting victims enhances the effectiveness of relief and rescue operations (Kotzur and Schmalenbach, 2014).

The involvement of volunteers and the community is crucial in disaster situations. Their contributions, according to their abilities, strengthen solidarity and support those affected. Sensitivity to the emotional and psychological needs of those affected is paramount. It is important to understand the trauma experienced by everyone and to act respectfully (Spade, 2020). In addition to rescue and emergency relief, it is important to provide long-term support in the aftermath of disasters. Long-term projects and programs should be developed to support the recovery of those affected and help communities rebuild. Relief and solidarity initiatives must ensure equal access to all segments of society and take into account the needs of all. It is crucial to prevent discrimination and distribute resources fairly to build social trust. According to these principles, solidarity can be achieved in disaster situations, effectively supporting those affected and aiding in the rebuilding of society (Kotzur and Schmalenbach, 2014).

Resilience and Mobilization

Given the unpredictable nature of disasters and emergencies, it is imperative that human resources possess the necessary resilience and capability to mobilize and respond to such situations. The ability to remain resilient and mobilize quickly is crucial for individuals and communities to be well-prepared, respond effectively, and accelerate the recovery process (Marien, 2012). Being prepared for the potential impacts of disasters is considered essential for building resilience. This involves creating disaster plans, keeping emergency preparedness kits, knowing safe evacuation routes, and taking preventive measures. Raising awareness and educating the community about disasters is believed to increase the capacity to respond quickly and effectively. It is important to conduct regular training on emergency procedures, disaster risks, and prevention methods (Norris, Stevens, Pfefferbaum, Wyche, and Pfefferbaum, 2008). Effective communication and coordination are important in disaster situations to ensure that aid and resources are directed efficiently, increasing resilience capacity. Flexibility is also important to adapt to changing conditions (Masys, 2014). Strengthening neighborhood relations and mutual support can enhance resilience and facilitate collective action. In disaster situations, it is important to promote social solidarity and unity. Additionally, it is essential to react promptly and carry out immediate rescue operations. The rapid mobilization of rescue teams and swift assistance to those affected are of vital importance (Kundak, 2017). Long-term recovery and reconstruction efforts are considered crucial in aiding communities to recover after disasters. These efforts are aimed at reducing post-disaster stress, minimizing damage, and helping society return to normal. Resilience and mobilization in disaster situations enable societies and individuals to be stronger and better prepared against disasters. In this way, the effects of disasters are minimized, and societies can recover more quickly (Kaye-Kauderer et al., 2021).

Psychosocial Support

Disasters and emergencies can be stressful and traumatic. It is important to have resources and experts available to provide psychosocial support to affected individuals. This support can help them cope with the emotional and psychological effects of the disaster, including trauma, anxiety, fear, and stress, and support the recovery process (Schonfeld et al., 2015). The provision of fast and accessible assistance to affected individuals following a disaster is crucial to effective psychosocial support. This support can be provided by a range of professionals, including psychologists, social workers, volunteers, and aid organizations (Kar, 2009). The creation of safe and supportive environments for those affected is of utmost importance. This can be achieved through the establishment of temporary shelters, collection centers, or aid points. It is crucial that those providing assistance to demonstrate sensitivity and empathy toward affected individuals, to comprehend and address their emotional needs (Everly and Lating, 2021).

Providing accurate information about the effects of the disaster and educating those affected on coping strategies can be helpful in reducing anxiety and fears, and improving their ability to cope. Emotional support can also be obtained through support groups and individual counseling services, which allow individuals to share their experiences and support each other through trauma. In disaster situations, it is important to ensure that all individuals receive the necessary support, including children and those with special needs. This may involve providing appropriate support to children, such as play therapy, or ensuring access to the necessary medical or psychosocial services for those with special needs (Schonfeld et al., 2015). Post-disaster psychosocial support can be a long-term process. Therefore, long-term recovery and recovery plans for those affected should be established and continuous support should be provided when necessary. Psychosocial support in disaster situations plays a critical role in ensuring the emotional well-being of those affected. This support helps people cope with post-traumatic stress and accelerates the community recovery process (Everly and Lating, 2021).

Human Resource Management and Technology in Disasters

Disasters and emergencies require preparedness not only from state institutions but also from individuals and society as a whole. Therefore, it is important to provide training, increase awareness, and prepare human resources, especially using digital developments. This will enhance society's capacity to cope with disasters and emergencies (Adıgüzel, 2022).

The preparedness of human resources and the use of technology are important in minimizing the impact of disasters and responding promptly. Awareness and training against disasters are among the essential aspects of preparedness and the use of technology. Disaster preparedness training helps individuals understand disaster risks and prepares them to act appropriately in the case of a disaster. The implementation of disaster plans and regular drills is crucial to ensure that individuals are well-prepared to act in the event of a disaster. Additionally, maintaining emergency preparedness kits in homes and workplaces is essential to meet basic needs during disasters. Furthermore, the participation and training of volunteers and professional groups in disaster response teams is vital to ensure a prompt and effective response to disasters (Memiş and Babaoğlu, 2020).

The use of technology in disaster management can assist in predicting disaster risks, improving emergency communications, and coordinating rescue efforts. For example, tools such as pre-disaster warning systems, weather forecasts, satellite imaging technology, and communication networks are considered crucial in disaster management (Youb and Yousfi, 2021). Pre-disaster warning systems have the ability to predict the potential effects of natural disasters such as earthquakes, tsunamis, hurricanes, floods, and fires. These systems provide timely warnings to individuals and authorities about emergencies, allowing them to take appropriate action (Hermans et al., 2022). Weather forecasts can be helpful in monitoring and predicting potential disasters, which can enable the implementation of preventive measures. For example, they can help determine the path of a hurricane, and identify flood-prone areas, or areas at risk of fire. Satellite imaging technology is crucial to monitoring disaster areas, assessing their impact, and providing aid. Satellite imagery serves many useful purposes, such as determining the extent of damage, directing post-disaster rescue operations, and identifying affected areas (Jiang et al., 2022). Communication networks enable the communication and dissemination of emergency information during a disaster. They help people receive information about emergencies, communicate with their loved ones, and coordinate rescue teams. These technological tools are used to reduce the impact of disasters and respond more effectively. However, the effectiveness of these tools is directly related to the ability to obtain accurate data, process it quickly, and share it with people. Therefore, it is important to continuously develop and improve them (Haresco and San Jose, 2021).

The use of the Internet and social media for communication and coordination in disaster situations helps to inform society quickly and strengthen disaster response. Social media platforms, in particular, provide an instant flow of information that quickly informs society and strengthens the response to disasters. People can get information about what is happening in the disaster area, be directed to safe areas, and send their requests for help (Ogie et al., 2022).

Satellite and drone technology plays an important role in damage assessment and rescue operations in disaster areas. These technologies help identify the impact of a disaster more quickly and intervene more effectively. Satellite imaging systems can quickly identify damage in large areas and guide rescue teams. Similarly, drones can provide detailed damage assessment in confined areas and facilitate rescue operations (Agrawal et al., 2021).

The use of these technologies helps to identify the impact of a disaster more quickly and respond more effectively. This, in turn, can reduce post-disaster losses and make rescue operations more efficient. However, the effective use of technology requires good planning, training, and coordination. Issues such as privacy and data security should also be considered (Arora et al., 2015).

Discussion and Conclusion

The preparedness of human resources and the effective use of technology are essential pillars in the effort to mitigate the impact of disasters and accelerate society's recovery. Continuous training of individuals and the reinforcement of technological infrastructure are imperative measures to achieve this goal. Leveraging today's technological advancements becomes paramount in safeguarding disaster-affected communities. Societies must optimize the utilization of existing technological assets to confront the risks they encounter while simultaneously embarking on capacity-building initiatives aimed at enhancing these assets further. Technological opportunities for disaster reduction represent a cornerstone in securing the future of humanity and safeguarding the progress attained thus far. By harnessing these opportunities, societies can significantly diminish the detrimental effects of disasters and fortify their resilience in the face of future challenges. It is through a comprehensive approach that combines human resource readiness with cutting-edge technology that societies can forge a path toward a safer and more resilient future, ensuring the preservation of hard-won gains and the continued advancement of civilization.

Disclaimer

None.

References

Adıgüzel, S. (2022). Afet durumlarinda yapay zeka teknolojisi ile lojistik yönetimi örnekleri. *Akademik İzdüşüm Dergisi, 7*(1), 47-70. Retrieved from https://dergipark.org.tr/en/pub/beuiibfaid/issue/69148/1057267.

Agrawal, T., Suraj, & Meleet, M. (2021). *Classification of Natural Disaster Using Satellite & Drone Images with CNN Using Transfer Learning*.

Al-Dahash, H., Thayaparan, M., & Kulatunga, U. (2016). *Understanding the terminologies: Disaster, crisis and emergency*. Paper presented at the Proceedings of the 32nd annual ARCOM conference, ARCOM 2016.

Alexander, D. (2015). Disaster and emergency planning for preparedness, response, and recovery. *Oxford Research Encyclopedia of Natural Hazard Science*.

Alexander, D. E. (2002). *Principles of emergency planning and management*: Oxford University Press on Demand.

Arora, P., Gunaseelan, R., & Bhardwaj, J. R. (2015). Evaluation of an international disaster response team's perspective post 2011 Japan multiple disasters: Pivotal learnings in human resource management. *Journal of Strategic Human Resource Management, 4*(3), 39-56.

Bendell, J. (2022). Replacing Sustainable Development: Potential Frameworks for International Cooperation in an Era of Increasing Crises and Disasters. *Sustainability, 14*(13), 8185.

Bowen, A. A. (2008). Are We Really Ready? The Need for National Emergency Preparedness Standards and the Creation of the Cycle of Emergency Planning. *Politics & Policy, 36*(5), 834-853. https://doi.org/10.1111/j.1747-1346.2008.00137.x.

Dany, C. (2018). Failed cooperation in times of natural disasters: explaining the rejection of humanitarian aid. *International Relations of the Asia-Pacific, 20*(2), 193-223. https://doi.org/10.1093/irap/lcy025.

Duchek, S. (2020). Organizational resilience: a capability-based conceptualization. *Business Research, 13*(1), 215-246. doi:10.1007/s40685-019-0085-7.

Everly, G. S., & Lating, J. M. (2021). Psychological first aid (PFA) and disasters. *International Review of Psychiatry, 33*(8), 718-727. doi:10.1080/09540261.2021.2016661.

Gündüz, F. (2024). A phenomenological study on women's disaster experience in Türkiye. *International Journal of Disaster Risk Reduction, 100*, 104196. https://doi.org/10.1016/j.ijdrr.2023.104196.

Hansson, S., Orru, K., Siibak, A., Bäck, A., Krüger, M., Gabel, F., & Morsut, C. (2020). Communication-related vulnerability to disasters: A heuristic framework. *International Journal of Disaster Risk Reduction, 51*, 101931. https://doi.org/10.1016/j.ijdrr.2020.101931.

Haresco, A. A., & San Jose, A. E. (2021). Communication skills and disaster preparedness of junior high school students. *Asian Journal of Education and Social Studies*, 20 (1), 21-31.

Hermans, T. D. G., Šakić Trogrlić, R., Van Den Homberg, M. J. C., Bailon, H., Sarku, R., & Mosurska, A. (2022). Exploring the integration of local and scientific knowledge in early warning systems for disaster risk reduction: a review. *Natural Hazards, 114*(2), 1125-1152. https://doi.org/10.1007/s11069-022-05468-8.

Herstein, J. J., Schwedhelm, M. M., Vasa, A., Biddinger, P. D., & Hewlett, A. L. (2021). Emergency preparedness: What is the future? *Antimicrobial Stewardship & Healthcare Epidemiology, 1*(1), e29. https://doi.org/10.1017/ash.2021.190.

İytemür, A., & Tekeli Yeşil, S. (2020). Bir Üniversite Hastanesinde Çalışan Hemşirelerin Hastane Afet ve Acil Durum Planları ile İlgili Görüşlerinin İncelenmesi. *Hacettepe*

Üniversitesi Hemşirelik Fakültesi Dergisi, 7(2), 138-148. https://doi.org/10.31125/hunhemsire.763162.

Jiang, Z., Yang, S., Liu, Z., Xu, Y., Xiong, Y., Qi, S., . . . Xu, T. (2022). Coupling machine learning and weather forecast to predict farmland flood disaster: A case study in Yangtze River basin. *Environmental Modelling & Software, 155*, 105436. https://doi.org/10.1016/j.envsoft.2022.105436.

Johnston, K. A., Taylor, M., & Ryan, B. (2020). Emergency management communication: The paradox of the positive in public communication for preparedness. *Public Relations Review, 46*(2), 101903. https://doi.org/10.1016/j.pubrev.2020.101903.

Kar, N. (2009). Psychological impact of disasters on children: review of assessment and interventions. *World Journal of Pediatrics, 5*(1), 5-11. https://doi.org/10.1007/s12519-009-0001-x.

Kaye-Kauderer, H., Feingold, J. H., Feder, A., Southwick, S., & Charney, D. (2021). Resilience in the age of COVID-19. *Bjpsych Advances, 27*(3), 166-178. https://doi.org/10.1192/bja.2021.5.

Kotzur, M. T., & Schmalenbach, K. (2014). Solidarity Among Nations. *Archiv des Völkerrechts, 52*(1), 68-91. Retrieved from http://www.jstor.org/stable/24260786.

Kundak, S. (2017). Dirençliliğin Temeli. *Resilience, 1*(1), 55-69. https://doi.org/10.32569/resilience.358081.

Liu, B. F., & Mehta, A. M. (2021). From the periphery and toward a centralized model for trust in government risk and disaster communication. *Journal of Risk Research, 24*(7), 853-869. https://doi.org/10.1080/13669877.2020.1773516.

Makwana, N. (2019). Disaster and its impact on mental health: A narrative review. *J Family Med Prim Care, 8*(10), 3090-3095. https://doi.org/10.4103/jfmpc.jfmpc_893_19.

Marien, M. (2012). Resilient People, Resilient Planet: A Future Worth Choosing. In *Report of the United Nations Secretary-General's High-Level Panel on Global Sustainability* (pp. 109). NY: United Nations.

Masys, A. J. (2014). *Disaster Management: Enabling Resilience*: Springer.

McAlister, V. C. (2011). Drills and exercises: the way to disaster preparedness. *Can J Surg, 54*(1), 7-8. https://doi.org/10.1503/cjs.036910.

Memiş, L., & Babaoğlu, C. (2020). Acil durum ve afet yönetiminde süreç yaklaşimi ve teknoloji. *Omer Halisdemir Universitesi Iktisadi ve Idari Bilimler Fakültesi Dergisi, 13*(4),776-791.

Mishra, S., & Suar, D. (2012). Effects of Anxiety, Disaster Education, and Resources on Disaster Preparedness Behavior. *Journal of Applied Social Psychology, 42*(5), 1069-1087. https://doi.org/10.1111/j.1559-1816.2011.00853.x.

Muttarak, R., & Pothisiri, W. (2013). The role of education on disaster preparedness case study of 2012 indian ocean earthquakes on thailand's andaman coast. *Ecology and Society, 18*(4), 51.

Nix-Stevenson, D. (2013). Human Response to Natural Disasters. *3*(3), 2158244013489684. https://doi.org/10.1177/2158244013489684.

Norris, F. H., Stevens, S. P., Pfefferbaum, B., Wyche, K. F., & Pfefferbaum, R. L. (2008). Community resilience as a metaphor, theory, set of capacities, and strategy for disaster readiness. *American Journal of Community Psychology, 41*(1-2), 127-150. https://doi.org/10.1007/s10464-007-9156-6.

Ogie, R. I., James, S., Moore, A., Dilworth, T., Amirghasemi, M., & Whittaker, J. (2022). Social media use in disaster recovery: A systematic literature review. *International Journal of Disaster Risk Reduction, 70*, 102783. https://doi.org/10.1016/j.ijdrr.2022.102783.

Özer, Y. E. (2017). Afet konusundaki algı ve yerel aktörlerin sorumlulukları. [Perception about disaster and responsibilities of local actors]. *Sayıştay Dergisi,* 106, 1-34.

Pilatin, A., & Ayaydın, H. (2022). Social capital measurement in turkiye: creating an index by province. *Journal of Economy Culture and Society, 66,* 235-259.

Pramono, J., Kusumastuti, D., Sekarwangi, M., Choerudin, A. J. J. o. T. D., & Excellence. (2020). The Community Participation in Disaster Mitigation to Managing the Impact of Natural Disasters in Indonesia. *Journal of Talent Development and Excellence, 12,* 2396-2403.

Rokach, A., Pinkert, M., Nemet, D., Goldberg, A., & Bar-Dayan, Y. (2008). Standards in collaborative international disaster drills: a case study of two international search and rescue drills. *Prehospital and Disaster Medicine, 23*(1), 60-62. https://doi.org/10.1017/S1049023X00005586.

Schonfeld, D. J., Demaria, T., Krug, S. E., Chung, S., Fagbuyi, D. B., Fisher, M. C., . . . Wissow, L. S. (2015). Providing psychosocial support to children and families in the aftermath of disasters and crises. *Pediatrics, 136*(4), e1120-e1130. https://doi.org/10.1542/peds.2015-2861.

Spade, D. (2020). Solidarity not charity: mutual aid for mobilization and survival. *Social Text, 38*(1), 131-151. https://doi.org/10.1215/01642472-7971139.

Tan, Y., Liao, X., Su, H., Li, C., Xiang, J., & Dong, Z. (2017). Disaster Preparedness Among University Students in Guangzhou, China: Assessment of Status and Demand for Disaster Education. *Disaster medicine and public health preparedness, 11*(3), 310-317. https://doi.org/10.1017/dmp.2016.124.

Wang, Q., Li, W., Yu, Z., Abbasi, Q., Imran, M., Ansari, S., . . . Zhu, T. (2023). An overview of emergency communication networks. *Remote Sensing, 15*(6), 1595.

Youb, A., & Yousfi, R. (2021). Digitizing Human Resources Management as a Mechanism to Predict and Manage Disasters. *International Journal of Inspiration & Resilience Economy, 5*(1), 13-16.

Zavar, E., & Nelan, M. (2020). Disaster drills as experiential learning opportunities for geographic education. *Journal of Geography in Higher Education, 44*(4), 624-631. https://doi.org/10.1080/03098265.2020.1771684.

Chapter 20

Digital Volunteering and Management in Disasters

Galip Usta*, **PhD**
Department of Medical Services and Techniques, Trabzon University, Trabzon, Türkiye

Abstract

Disasters are defined as a situation that usually occurs suddenly, cause great damage and destruction in the places where they occur, cause damage to people, cannot be handled with local means, and require external assistance at the national or international level. To minimize the negative effects of disasters, it is important to increase the resilience of societies against disasters. Increasing social resilience can be possible through the adoption of community-based disaster management, minimization of structural risks, the establishment of an applicable disaster management strategy, and the successful implementation of disaster management processes. Volunteer-based approaches are of critical importance in the effective implementation of these processes. Volunteerism is the participation of individuals in activities for the benefit of society voluntarily and without coercion. In this context, digital volunteering in disasters has emerged as a new concept with technological developments. Digital volunteering refers to individuals offering their time, skills, and tasks for the benefit of society through the Internet, without any financial expectation. Digital volunteers can contribute to improving response mechanisms in disaster-prone countries by bringing together key information from data sources such as emergency numbers, websites, and social media accounts. Involving digital volunteers in disaster management can be effective in reducing

* Corresponding Author's Email: galipusta@trabzon.edu.tr.

In: Disaster Management in the Digital Age and Emerging Technologies
Editors: Ayşe Asiltürk and Ayşegül Sarıoğlu Kemer
ISBN: 979-8-89530-213-2

the effects of unavoidable disasters, and in avoidable disasters, the negative effects of disasters can be minimized, or disasters can be completely prevented with the contributions of digital volunteers. Therefore, it can be said that digital literacy and the resulting conscious digital volunteer organizations are important in building a resilient society against disasters and increasing the success of disaster interventions. These organizations can ensure the rapid transmission of accurate information to disaster managers and response teams. As a result, integrating digital volunteering into disaster management processes affects the success of the disaster management process. Organizations and authorities responsible for disaster management can successfully manage information in disasters by benefiting from the activities of digital volunteers. In this way, relevant institutions and organizations can improve their disaster response capacities and skills.

Keywords: disaster, disaster management, digital volunteering in disasters, volunteering

Introduction

Disasters are defined as a situation that usually occurs suddenly, causes great damage and destruction in the places where it occurs, causes harm to people and cannot be handled by local means, and requires external assistance at the national or international level (Guha-Sapir et al., 2016, p. 13). Disasters generally occur as a result of exposure to hazards, existing vulnerabilities, and inadequate capacity or measures to cope with or minimize the potential negative consequences of these hazards (United Nations, 2009, p. 9). In recent years, the frequency of disasters has increased due to factors such as pollution, overpopulation, deforestation, and climate change. This increase has also led to an increase in the number of people affected, economic losses, and disaster-related damages (IFRC, 2018, p. 168; Yigitcanlar and Kankanamge, 2022, p. 51). In this context, it is important to increase the resilience of societies against disasters. Increasing social resilience can be possible through the adoption of community-based disaster management, the minimization of structural risks, the establishment of an applicable disaster management strategy, and the successful implementation of disaster management processes. Volunteer-based approaches are of critical importance in the effective implementation of these processes (Fernandez, 2006, p. 9; Alexander, 2010, p. 151). Volunteerism is the participation of individuals in activities for the benefit of society, without financial interest, voluntarily, and without coercion (Yönten

Balaban and Çoban İnce, 2015, p. 150-152; Cnaan et al., 1996, p. 365-367). Individuals generally participate in volunteer activities for reasons such as developing their own skills, gaining experience, helping people, and feeling satisfaction from the work done (Borst, 2010, p. 8; Güngör and Çölgeçen, 2014, p. 183; Köçer and Aslan, 2023, p. 834). Volunteers make significant contributions to many areas such as fighting poverty, gender equality, education, health, environmental protection, and disaster risk reduction (AFAD, 2014, p. 80; Palabiyik, 2011, p. 91). Volunteers play critical roles in various fields such as search and rescue, logistics, transport, nutrition, shelter, and health services during disasters or extraordinary situations (Alexander, 2010, p.151). Technological advances have enabled volunteering activities to accelerate and diversify. Recent disasters emphasize the importance of technological advances and social media in volunteering (Reuter, 2022, p. 191). Technological advances in recent years have facilitated citizens' access to critical information in disaster situations through online platforms. This has increased the effectiveness of volunteer activities and contributed significantly to the success of rescue operations (Yigitcanlar and Kankanamge, 2022, p. 51-53). The evolution of digital media platforms and tools has fostered digital volunteering that addresses emergency needs online using social media and mapping software, accelerating the ability of these volunteers to play critical roles through their ability to manage and analyze crisis data by collaborating in open networks (Estellés-Arolas and González-Ladrón-de-Guevara, 2012; Park and Johnston, 2017; Starbird and Palen, 2011, p. 1071). The advancement of technology and social media platforms is transforming how crises are responded to and communicated, enabling citizens and authorities to take active roles in disaster management (Hornmoen et al., 2018). The integration of volunteers with training and social networks provides critical support to communities in times of disaster, enabling individuals to gain experience and contribute to society. Thus, every individual can reinforce social solidarity and solidarity as a potential volunteer in disaster management (Ekşi, 2017, p. 165). Encouragement of disaster volunteerism can support the survival of disaster victims by increasing the effectiveness of response processes in disasters and can significantly contribute to the successful management of the post-disaster recovery phase (Şahin et al., 2018, p. 157-158). Therefore, it can be stated that volunteer participation is important in disaster preparation and response processes, and volunteering activities should be encouraged. In this context, it has been observed that literature studies emphasize volunteer participation in various events such as earthquakes (Quarantelli, 1993, p. 29; Schilderman, 1993, p. 415; Shaw and Goda, 2004, p. 18; Yurttagüler, 2013. p. 18-22;

O'Brien and Mileti, 1992, p.71; Jalali, 2002, p.120), flood (Harris et al., 2017: Albahari and Schultz, 2017; Bier et al., 2023), explosion/attack (Adams and Boscarino, 2015; John and Fuchs, 2002), and the importance of volunteering activities in disaster response. In this study, digital volunteering practices in disasters, which constitute the new face of volunteering activities that have an important place in disaster preparation and response processes, have been examined.

Digital Volunteering in Disasters

Integration of the community into disaster management processes and the use of technological developments support the efficient use of resources, contribute to minimizing losses, and can increase the efficiency of crisis management operations by facilitating rapid access to information for managers (Schoch-Spana et al., 2007, p. 8). Access to fast and reliable information in disasters is considered an important factor in the successful realization of rescue and humanitarian aid operations (Kumar et al., 2022, p. 582-583). Thanks to technological developments, digital volunteers can assist disaster management organizations in organizing, raising awareness, and sharing real-time information by managing big data in disasters and contributing to accelerating disaster response processes (Jaeger et al., 2007, p. 602; Hughes and Palen, 2009, p. 248). In this context, it is argued that relevant institutions and authorities should support digital volunteers who can make significant contributions to every stage of disaster management processes and give more importance to good practice examples (Haworth and Bruce, 2015, p. 245). Digital volunteering means individuals offering their time, skills, and tasks for the benefit of society through the Internet without any financial expectation (Mukherjee, 2011, p. 253; Volunteer Canada, 2020, p. 1-2; Whittaker et al., 2015, p. 364). Digital volunteering transcends geographical boundaries, enabling people to have a global impact and making it important to use technology for social good (Amichai-Hamburger, 2008, p. 544, 551). Digital volunteers can take part in various tasks such as transferring information, analyzing information, organizing social media campaigns, mapping, and determining logistic needs to support post-disaster response efforts (Coleman et al., 2009, p. 332; Abdulhamid et al., 2019, p. 1). In disaster situations, analyzing and rapidly sharing important data received from digital volunteers and technological tools with rescue teams increases the probability of survival of disaster victims by ensuring the effective execution of tasks (Hu and Janowicz, 2016, p. 137; Liu and Nejat, 2016, p. 512). Digital volunteers can contribute to crisis management by helping to establish communication between citizens and authorities or by taking part in other information-oriented tasks (Smith et al., 2021, p.127). Digital volunteers can contribute to improving response mechanisms in disaster-prone countries by bringing together key information from data sources such as emergency numbers, websites, and social media accounts (Abdulhamid et al., 2019, p. 11-12). Such activities conducted by digital volunteers can contribute to increasing situational awareness, obtaining and structuring reliable information from the incident site, planning assistance, enabling authorities to make effective decisions, and organizing rescue teams (Saura et al., 2020, p. 11; Cobb et al., 2014, p. 888-889; Starbird, 2013, p. 801; Latonero and Shklovski, 2011, p. 1). Even more people can be motivated to volunteer in disasters through digital volunteers (Nahkur et al., 2022, p. 9). Involving digital volunteers in disaster management can be effective in reducing the effects of unavoidable disasters, and in avoidable disasters, the negative effects of disasters can be minimized or disasters can be prevented completely with the contributions of digital volunteers (Badarudin et al., 2020, p. 1). In this context, virtual support teams also play an important role. By evaluating the data sent by digital volunteers, virtual support teams undertake tasks such as creating crisis maps, analyzing social media trends, conducting sentiment analysis, ensuring the dissemination of accurate information, filtering false news, and establishing a digital information network (Fathi et al., 2020, p. 2). However, disaster management authorities recognize that it is difficult to take full advantage of the information and data provided by digital volunteers (Fathi et al., 2020, p. 2). Because digital volunteers can unintentionally enable the spread of misinformation (Nahkur et al., 2022, p. 10). It is recognized that there are various reasons for this situation. For example, lack of adoption of digitalization, age factor, lack of digital literacy, geographical barriers, and difficulties in adapting to technology are among the factors that prevent participation in digital volunteering (Lim et al., 2022, p. 5-6; Sun et al., 2021, p. 957). Furthermore, factors such as lack of training and awareness, the unclear scope of tasks, and conditions of participation constitute some of the barriers to participation in digital volunteering, leading to a lack of understanding of digital volunteering (West and Pateman, 2016, p. 5; Vahidi et al., 2021, p. 25). In this context, it can be argued that digital literacy has an important role to play in promoting digital volunteering. The level of digital literacy plays a critical role in the preparation of societies for disasters and the creation of resilient communities, in preventing the spread of false information, and in individuals acting based on scientific data (Asteria et al., 2023, p. 1,6; Syam et al., 2023; Eshet, 2004, p. 93-94). Developing individuals' digital competencies is important in the fight against fake news, misinformation, and disinformation. Therefore, it can be said that digital literacy and the resulting conscious digital volunteer organizations are important in building a resilient society against disasters and increasing the success of disaster responses. These organizations can ensure the rapid transmission of accurate information to disaster managers and response teams. Furthermore, they can help create a conscious society in the fight against disinformation and encourage society to act consciously in this regard.

Application Examples and Future Trends Regarding Digital Volunteering in Disasters

It is known that being prepared for disasters has a significant impact on the disaster response process. The willingness for voluntary participation in the preparation process is generally satisfactory; however, it is observed that the actual participation may not be at the same level at the time of the disaster (Ma et al., 2021, p. 1-3). In this context, it is observed that digital volunteer applications have become widespread in the disaster response process with the development of technology. Literature studies emphasize the importance of social media platforms and digital volunteers in the disaster response process (Chatfield and Brajawidagda, 2014; Chernobrov, 2018: 940; Mehta et al., 2022, p. 180). In the study by Kaufhold and Reuter (2016, p. 137), it was determined that social media tools were used effectively in coordinating aid activities between affected citizens and volunteers and organizing people. Chatfield and Brajawidagda (2014, p. 2237-2240) analyzed the Oklahoma Hurricane and pointed out that many tweets were sent about the hurricane through social media. In the same study, it was emphasized that citizens play an important role in the crowdsourcing of disaster management institutions. Chernobrov (2018, p. 940) argues that the development of digital volunteering and digital communication in humanitarian crises provides authorities and responders with new sources of information and helps to reshape the roles and powers of traditional actors. In the study conducted by Park and Johnston (2019, p. 155), it was mentioned that digital volunteers collect, analyze, and visualize disaster data and share these data with response teams. In this case, it is emphasized that it increases the capacity of response teams to cope with the disaster. Mehta et al. (2022, p. 180) stated that during the COVID-19 pandemic, many community-oriented organizations benefited from digital volunteers to improve their capacities and provide access to more citizens. Tan et al. (2023, p. 184) reported that virtual volunteering activities during the COVID-19 outbreak enabled information sharing by organizing and establishing social networks. In the study conducted by Lachance (2021, p. 104), it was evaluated that digital volunteering practices can be an effective strategy to respond to organizations in outbreaks such as COVID-19, which require social distancing and isolation measures. In the study conducted by Peng (2017, p. 335), it was stated that the Yushu earthquake was effective in revealing the importance of crowdsourcing. In the same study, it was seen that emphasis was placed on the promotion of digital volunteering and the integration between non-governmental organizations and digital volunteers. In a study conducted in Turkey, it was stated that various activities were carried out to meet the needs of earthquake victims through the WhatsApp communication lines at many stages of the earthquake response process (Çelebi et al., 2024). In different studies in the literature, the importance of digital volunteering in disasters, the tasks undertaken by these volunteers, and their effects on the disaster response process have been evaluated (Heinzelman and Waters, 2010; Shanley et al., 2013). However, despite these developments, digital volunteering may have some drawbacks. The ambiguity of the roles of digital volunteers in disasters, lack of coordination, and ambiguity of duties and authority between authorities and volunteers can negatively affect digital volunteering activities (Nahkur et al., 2022, p. 11). A clear inclusion of digital volunteering in disaster plans can contribute to the solution of this problem. The lack of guidelines for digital volunteering activities and outdated institutional policies affect digital volunteering activities (Mehta et al., 2022, p. 180). Therefore, updating organizational policies by considering digital volunteering activities can contribute to the effectiveness of volunteering activities. Reliability of data collected through volunteers in disasters, lack of policies and directives, and personnel competence are considered as other factors that cause the digital volunteering practice to be questioned (Munkvold et al., 2015, p. 1-5). In this context, giving importance to volunteer training, strengthening data management systems, and cooperating with professional teams involved in disaster response can contribute to digital volunteering practices (Hughes and Tapia, 2015). Pre-planning the role of digital volunteers in disaster management is especially important in situations such as pandemics where physical interactions need to be limited. However, relevant guidelines need to be developed to gain more benefit from such practices (Nahkur et al., 2022, p.10).

To develop digital volunteering in disasters, it is considered that more work should be done to ensure the integration between decision-makers and volunteers and to improve this integration (Rogstadius et al., 2013, pp. 874-875).

Discussion and Conclusion

In this study, in which digital volunteering activities in disasters are examined, it is evaluated that digital volunteering practices are gaining momentum. It has been observed that technological developments affect this situation. The inclusion of digital volunteers in disaster management processes can be effective in reducing the effects of unavoidable disasters, and in preventable disasters, the negative effects of disasters can be minimized, or disasters can be completely prevented with the contributions of digital volunteers. Therefore, it has been evaluated that the development of technology affects digital volunteering practices, and these activities affect the disaster management process. As mentioned in the content of the study, it has been determined that digital volunteers are used to obtain information from the field in many areas, from epidemics to earthquakes, and to share this information with the authorities responsible for disaster management. However, despite these developments and successful practices, it has been evaluated that digital volunteering activities may cause some problems. Uncertain roles for digital volunteers in disasters, lack of coordination, and ambiguity of duties and authorities between authorities and volunteers can negatively affect digital volunteering activities. It could be argued that defining the guidelines for the management of digital volunteers in disaster plans and making clear the boundaries of their responsibilities and powers will help to resolve this issue. It has been determined that out-of-date institutional regulations and the absence of standards for digital volunteering activities have an impact on these initiatives as well. The effectiveness of volunteering activities can therefore be increased by revising institutional regulations to include consideration for digital volunteering activities. Therefore, the success of the crisis management process is impacted by incorporating digital volunteers while considering strengths and shortcomings. The efforts of digital volunteers can help organizations and authorities in charge of disaster management handle information during disasters. Institutions and organizations that are pertinent can enhance their capabilities and expertise in disaster response in this way.

Disclaimer

None.

References

Abdulhamid, N. G., Perry, M., & Kashefi, A. (2019). Towards an Understanding of Digital Disaster Response Workflow in Humanitarian Emergencies. *The Journal of Web Science*, *7*, 16-16.

Adams, R. E., & Boscarino, J. A. (2015). Volunteerism and well-being in the context of the World Trade Center terrorist attacks. *International journal of emergency mental health*, *17*(1), 274. https://doi.org/10.4172/1522-4821.1000158.

AFAD. (2014). Açıklamalı Afet Yönetimi Terimleri Sözlüğü. Ankara. https://www.afad.gov.tr/kitaplar.

Albahari, A., & Schultz, C. H. (2017). A qualitative analysis of the spontaneous volunteer response to the 2013 Sudan floods: changing the paradigm. *Prehospital and disaster medicine*, *32*(3), 240-248. https://doi.org/10.1017/S1049023X17000164.

Alexander, D. (2010). The voluntary sector in emergency response and civil protection: review and recommendations. *International Journal of Emergency Management*, *7*(2), 151-166. https://doi.org/10.1504/IJEM.2010.033654.

Amichai-Hamburger, Y. (2008). Potential and promise of online volunteering. *Computers in Human Behavior*, *24*(2), 544-562. https://doi.org/10.1016/j.chb.2007.02.004.

Asteria, D., Surpi, N. K., Brotosusilo, A., & Suja'i, I. S. (2023). Integration of Local Capacity Building in Countering False Information about Disaster Into Community-based Disaster Risk Management. *In IOP Conference Series: Earth and Environmental Science*, *1275* (1), 012028. IOP Publishing. doi:10.1088/1755-1315/1275/1/012028.

Badarudin, P. H. A. P., Wan, A. T., & Phon-Amnuaisuk, S. (2020). A blockchain-based assistance digital model for first responders and emergency volunteers in disaster response and recovery. In *Proceedings of the 8th International Conference on Information and Communication Technology* (ICoICT) (pp. 1-5). https://doi.org/10.1109/ICoICT49345.2020.9166389.

Bier, M., Fathi, R., Stephan, C., Kahl, A., Fiedrich, F., & Fekete, A. (2023). Spontaneous volunteers and the flood disaster 2021 in Germany: Development of social innovations in flood risk management. *Journal of Flood Risk Management*, 1-20. https://doi.org/10.1111/jfr3.12933.

Borst, I. (2010*). Understanding Crowdsourcing: Effects of motivation and rewards on participation and performance in voluntary online activities* (No. EPS-2010-221-LIS).

Chatfield, A. T., & Brajawidagda, U. (2014). Crowdsourcing hazardous weather reports from citizens via twittersphere under the short warning lead times of EF5 intensity tornado conditions. In *2014 47th Hawaii International Conference on System Sciences* (pp. 2231-2241). IEEE. https://doi.org/10.1109/HICSS.2014.281.

Chernobrov, D. (2018). Digital volunteer networks and humanitarian crisis reporting. *Digital Journalism, 6* (7), 928-944. https://doi.org/10.1080/21670811.2018.1462666.

Cnaan, R. A., Handy, F., & Wadsworth, M. (1996). Defining who is a volunteer: Conceptual and empirical considerations. *Nonprofit and Voluntary Sector Quarterly, 25*(3), 364-383. https://doi.org/10.1177/0899764096253006.

Cobb, C., McCarthy, T., Perkins, A., Bharadwaj, A., Comis, J., Do, B., & Starbird, K. (2014). Designing for the deluge: understanding & supporting the distributed, collaborative work of crisis volunteers. *In Proceedings of the 17th ACM conference on Computer supported cooperative work & social computing* (pp. 888-899). https://doi.org/10.1145/2531602.2531712.

Coleman, D., Georgiadou, Y., & Labonte, J. (2009). Volunteered geographic information: The nature and motivation of produsers. *International Journal of Spatial Data Infrastructures Research, 4*(4), 332-358.

Çelebi, İ., Tuna, A., & Çalişkan, C. (2024). Evaluation of incoming messages to WhatsApp communication network created by volunteers in major disasters: the case of Turkey earthquake. *Disaster Medicine and Public Health Preparedness, 18,* e53. https://doi.org/10.1017/dmp.2024.65.

Ekşi A., (2017). *Demokrasinin temsil krizi bütünleşik afet yönetimi pratiğiçözümleyici* mi?. İzmir: Kitapana.

Eshet, Y. (2004). Digital literacy: A conceptual framework for survival skills in the digital era. *Journal of Educational Multimedia and Hypermedia, 13*(1), 93-106.

Estellés-Arolas, E., & González-Ladrón-de-Guevara, F. (2012). Towards an integrated crowdsourcing definition. *Journal of Information Science, 38*(2), 189–200. https://doi.org/10.1177/0165551512437638.

Fathi, R., Thom, D., Koch, S., Ertl, T., & Fiedrich, F. (2020). VOST: A case study in voluntary digital participation for collaborative emergency management. *Information Processing & Management, 57*(4), 102174. https://doi.org/10.1016/j.ipm.2019.102174.

Fernandez, L. S., Barbera, J. A., & Van Dorp, J. R. (2006). Strategies for managing volunteers during incident response: A systems approach. *Homeland Security Affairs, 2*(3), 1-15.

Guha-Sapir, D., Hoyois, P., Wallemacq, P., & Below, R. (2016). Annual disaster statistical review 2016. *The Numbers And Trends*, 1-91.

Güngör, F., & Çölgeçen, Y. (2014). STK'larda gönüllü yönetimi ve motivasyonun performansa etkisi. *Akademik İncelemeler Dergisi, 8*(3), 163-187.

Harris, M., Shaw, D., Scully, J., Smith, C. M., & Hieke, G. (2017). The involvement/exclusion paradox of spontaneous volunteering: New lessons and theory from winter flood episodes in England. *Nonprofit and Voluntary Sector Quarterly, 46*(2), 352-371. https://doi.org/10.1177/0899764016654222.

Haworth, B., & Bruce, E. (2015). A review of volunteered geographic information for disaster management. *Geography Compass, 9*(5), 237-250. https://doi.org/10.1111/gec3.12213.

Heinzelman, J., & Waters, C. (2010). *Crowdsourcing crisis information in disaster-affected Haiti*. Washington, DC: US Institute of Peace.

Hornmoen, H., Backholm, K., Frey, E., Ottosen, R., Reimerth, G., & Steensen, S. (2018). *Social media in risks and crises.* In R. Andersen, P. L. Silva, & L. Purnaka (Eds.), The Routledge companion to media and humanitary action. Routledge.

Hu, Y., & Janowicz, K. (2016). Understanding the mapping sequence of online volunteers in disaster response. *In International Conference on GIScience Short Paper Proceedings, 1*(1). https://doi.org/10.21433/B3118N57H6ZV.

Hughes, A. L., & Palen, L. (2009). Twitter adoption and use in mass convergence and emergency events. *International journal of emergency management*, 6(3-4), 248-260. https://doi.org/10.1504/IJEM.2009.031564.

Hughes, A. L., & Tapia, A. H. (2015). Social media in crisis: When professional responders meet digital volunteers. *Journal of Homeland Security and Emergency Management, 12*(3), 679-706. https://doi.org/10.1515/jhsem-2014-0080.

IFRC. (2018). *World disasters report.* International Federation of Red Cross and Red Crescent.

Jaeger, P. T., Shneiderman, B., Fleischmann, K. R., Preece, J., Qu, Y., & Wu, P. F. (2007). Community response grids: E-government, social networks, and effective emergency management. *Telecommunications Policy, 31*(10-11), 592-604. https://doi.org/10.1016/j.telpol.2007.07.008.

Jalali, R. (2002). Civil society and the state: Turkey after the earthquake. *Disasters, 26*(2), 120-139. https://doi.org/10.1111/1467-7717.00196.

Kaufhold, M. A., & Reuter, C. (2016). The self-organization of digital volunteers across social media: The case of the 2013 European floods in Germany. *Journal of Homeland Security and Emergency Management, 13*(1), 137-166. https://doi.org/10.1515/jhsem-2015-0063.

Köçer, M. S., & Aslan, R. (2023). Gönüllü arama kurtarma ekiplerinin orman yangınlarındaki tahliye deneyimleri: 2021 Akdeniz orman yangınları. *Afet ve Risk Dergisi, 6*(3), 829-851. https://doi.org/10.35341/afet.1236021.

Kumar, A., Joshi, S., Sharma, M., & Vishvakarma, N. (2022). Digital Humanitarianism and Crisis Management: An empirical study of antecedents and consequences. *Journal of Humanitarian Logistics and Supply Chain Management, 12*(4), 570-593. https://doi.org/10.1108/JHLSCM-02-2022-0020.

Lachance, E. L. (2021). COVID-19 and its impact on volunteering: Moving towards virtual volunteering. *Leisure Sciences, 43*(1-2), 104-110. https://doi.org/10.1080/01490400.2020.1773990.

Latonero, M., & Shklovski, I. (2011). Emergency management, Twitter, and social media evangelism. *International Journal of Information Systems for Crisis Response and Management (IJISCRAM), 3*(4), 1-16. https://doi.org/10.4018/jiscrm.2011100101.

Lim, H. A., Lee, J. S. W., Lim, M. H., Teo, L. P. Z., Sin, N. S. W., Lim, R. W., ... & Tang, K. W. A. (2022). Bridging connectivity issues in digital access and literacy: reflections on empowering vulnerable older adults in Singapore. *JMIR Aging, 5*(2), e34764. https://doi.org/10.2196/34764.

Liu, Y., & Nejat, G. (2016). Multirobot cooperative learning for semiautonomous control in urban search and rescue applications. *Journal of Field Robotics*, 33(4), 512-536. https://doi.org/10.1002/rob.21597.

Ma, Y., Zhu, W., Zhang, H., Zhao, P., Wang, Y., & Zhang, Q. (2021). The factors affecting volunteers' willingness to participate in disaster preparedness. *International Journal of Environmental Research and Public Health*, *18*(8), 4141. https://doi.org/10.3390/ijerph18084141.

Mehta, D., Prasad, V., Chitta, T., Naik, N. S., Prakash, A., & Vashistha, A. (2022, June). Mobilizing Digital Volunteers to Support Underserved Communities in India During COVID-19 Lockdowns. In *Proceedings of the 5th ACM SIGCAS/SIGCHI Conference on Computing and Sustainable Societies* (pp. 180-194). https://doi.org/10.1145/3530190.3534827.

Mukherjee, D. (2011). Participation of older adults in virtual volunteering: A qualitative analysis. *Ageing International, 36,* 253-266. https://doi.org/10.1007/s12126-010-9088-6.

Munkvold, B. E., Flaten, M., & Nguyen, R. P. (2015). Volunteers' perceptions of the use of social media in emergency management. In Palen, Büscher, Comes, Hughes (Eds.), *Proceedings of the ISCRAM 2015 Conference Short Paper – Practitioner Cases and Practitioner-Centered Research*, Kristiansand, 24–27 May 2015.

Nahkur, O., Orru, K., Hansson, S., Jukarainen, P., Myllylä, M., Krüger, M., ... & Rhinard, M. (2022). The engagement of informal volunteers in disaster management in Europe. *International Journal of Disaster Risk Reduction, 83,* 103413. https://doi.org/10.1016/j.ijdrr.2022.103413.

O'Brien, P., & Mileti, D. S. (1992). Citizen participation in emergency response following the Loma Prieta earthquake. *International Journal of Mass Emergencies & Disasters*, *10*(1), 71-89. https://doi.org/10.1177/028072709201000104.

Palabiyik, H. (2011). Gönüllülük ve yerel hizmetlere gönüllü katilim üzerine açıklamalar. *Çanakkale Onsekiz Mart Üniversitesi Yönetim Bilimleri Dergisi, 9*(1), 82-113.

Park, C. H., & Johnston, E. (2019). Determinants of collaboration between digital volunteer networks and formal response organizations in catastrophic disasters. *International Journal of Organization Theory & Behavior,* *22*(2), 155-173. https://doi.org/10.1108/IJOTB-07-2018-0088.

Park, C. H., & Johnston, E. W. (2017). A framework for analyzing digital volunteer contributions in emergent crisis response efforts. *New Media and Society, 19*(8), 1308–1327. https://doi.org/10.1177/1461444817706877.

Peng, L. (2017). Crisis crowdsourcing and China's civic participation in disaster response: Evidence from earthquake relief. *China Information*, *31*(3), 327-348. https://doi.org/10.1177/0920203X17735668.

Quarantelli, E. L. (1993). Organizational response to the Mexico City earthquake of 1985: Characteristics and implications. *Natural Hazards*, *8*(1), 19-38. https://doi.org/10.1007/BF00596233.

Reuter, C. (2022). Self-Organisation of digital volunteers across social media. In *A European Perspective on Crisis Informatics: Citizens' and Authorities' Attitudes Towards Social Media for Public Safety and Security* (pp. 191-235). Wiesbaden: Springer Fachmedien Wiesbaden. https://doi.org/10.1007/978-3-658-39720-3_6.

Rogstadius, J., Teixeira, C., Karapanos, E., & Kostakos, V. (2013). An introduction for system developers to volunteer roles in crisis response and recovery. In T. Comes, F. Fiedrich, S. Fortier, J. Geldermann, & T. Müller (Eds.), *Proceedings of the 10th*

International ISCRAM Conference (pp. 874-883). Baden-Baden, Germany, May 2013.

Saura, J. R., Palos-Sanchez, P., & Velicia-Martin, F. (2020). What drives volunteers to accept a digital platform that supports NGO projects?. *Frontiers in Psychology, 11*, 520300. https://doi.org/10.3389/fpsyg.2020.00429.

Schilderman, T. (1993). Disasters and development. A case study from Peru. *Journal of International Development, 5*(4), 415-423. https://doi.org/10.1002/jid.3380050407.

Schoch-Spana, M., Franco, C., Nuzzo, J. B., & Usenza, C. (2007). Community engagement: leadership tool for catastrophic health events. *Biosecurity and Bioterrorism, 5*(1), 8-25. https://doi.org/10.1089/bsp.2006.0036.

Shanley, L., Burns, R., Bastian, Z., & Robson, E. (2013). Tweeting up a storm: the promise and perils of crisis mapping. *Photogrammetric Engineering & Remote Sensing, 79* (10): 865–879. https://doi.org/10.2139/ssrn.2464599.

Shaw, R., & Goda, K. (2004). From disaster to sustainable civil society: the Kobe experience. *Disasters*, 28(1), 16-40. https://doi.org/10.1111/j.0361-3666.2004.00241.x.

Smith, W. R., Robertson, B., Stephens, K., & Murthy, D. (2021). A different type of disaster response digital volunteer: Looking behind the scenes to reveal coordinating actions of disaster knowledge workers. *Journal of Contingencies and Crisis Management*, 29(2), 116–130. https://doi.org/10.1111/1468-5973.12352.

John, C., & Fuchs, J. (2002). The heartland responds to terror: Volunteering after the bombing of the Murrah Federal Building. *Social Science Quarterly, 83*(2), 397-415. https://doi.org/10.1111/1540-6237.00091.

Starbird, K. (2013). Delivering patients to sacré coeur: collective intelligence in digital volunteer communities. In *Proceedings of the SIGCHI Conference on Human Factors in Computing Systems* (pp. 801-810). https://doi.org/10.1145/2470654.2470769.

Starbird, K., & Palen, L. (2011). "Voluntweeters": Self-organizing by digital volunteers in times of crisis. *In Proceedings of the SIGCHI conference on human factors in computing systems*. https://doi.org/10.1145/1978942.1979102.

Sun, P. C., Morrow-Howell, N., Pawloski, E., & Helbach, A. (2021). Older adults' attitudes toward virtual volunteering during the COVID-19 pandemic. *Journal of Applied Gerontology, 40*(9), 953-957. https://doi.org/10.1177/07334648211006978.

Syam, B. M., Sutisna, S., Kurniadi, A., & Kusuma, K. (2023). The role of digital literacy capability of Babinsa Kodim 0705/Magelang to improve community preparedness for natural disasters. In *E3S Web of Conferences* (Vol. 464, p. 04001). EDP Sciences. https://doi.org/10.1051/e3sconf/202346404001.

Şahin, Y., Lamba, M., & Öztop, S. (2018). Üniversite öğrencilerinin afet bilinci ve afete hazırlık düzeylerinin belirlenmesi. *Medeniyet Araştırmaları Dergisi, 3*(6) 149-159.

Tan, W., Zhu, Y., Wu, L., Hou, J., Yi, J., Qi, T., & Zhang, W. (2023). Virtual volunteering, community support, and self-care in Chinese communities in Canada. *Interdisciplinary Nursing Research, 2*(3), 180-187. https://doi.org/10.1097/NR9.0000000000000030.

United Nations. (2009). *UNISDR Terminoloji on Disaster Risk Reduction*. United Nations International Strategy for Disaster Reduction (UNISDR) Geneva, Switzerland.

Vahidi, H., Taleai, M., Yan, W., & Shaw, R. (2021). Digital citizen science for responding to COVID-19 crisis: Experiences from Iran. *International Journal of Environmental Research and Public Health, 18*(18), 9666. https://doi.org/10.3390/ijerph18189666.

Volunteer Canada (2020). Virtual volunteering. https://volunteer.ca/vdemo/ResearchAndResources_DOCS/Virtual%20Volunteering%20Mar2020.pdf.

West, S. E., & Pateman, R. M. (2016). Recruiting and retaining participants in citizen science: What can be learned from the volunteering literature? *Citizen Science: Theory and Practice, 1*, 1–10. https://doi.org/10.5334/cstp.8.

Whittaker, J., McLennan, B., & Handmer, J. (2015). A review of informal volunteerism in emergencies and disasters: Definition, opportunities and challenges. *International Journal Of Disaster Risk Reduction, 13*, 358-368. https://doi.org/10.1016/j.ijdrr.2015.07.010.

Yigitcanlar, T., & Kankanamge, N. (2022). Volunteer crowdsourcing and social media. *In Urban Analytics with Social Media Data* (pp. 51-82). CRC Press. https://doi.org/10.1201/9781003278986-3.

Yönten Balaban, A., & Çoban İnce, İ. (2015). Gençlerin sivil toplum kuruluşlarındaki gönüllülük faaliyetleri ve gönüllülük algısı : Türkiye Eğitim Gönüllüleri Vakfı (TEGV) Örneği. *Dokuz Eylül Üniversitesi İktisadi İdari Bilimler Fakültesi Dergisi, 30*(2), 149-169.

Yurttagüler, L. (2013). *Gönüllülük ve Vatandaşlık Kavramı*. UN Volunteers. (Ed.) Türkiye'de Gönüllülük, (26-36) Ankara: Tasarımhane.

Chapter 21

An Evaluation of Disasters from a Microbiology Perspective

Mehtap Usta[1,*], PhD
and Seda Biryol[2], PhD
[1]Department of Health Care Services, Tonya Vocational School, Trabzon University, Trabzon, Türkiye
[2]Department of Medical Services and Techniques, Tonya Vocational School, Trabzon University, Trabzon, Türkiye

Abstract

Disasters are characterized as the result of natural and technological / human-induced events that cause loss of life and property in the places where they occur and exceed the coping capacity of society. It has been evaluated that disasters affect the microbial population where they occur as much as they affect humans. Changes in the microbial population can become a potential threat to other living species, especially humans, and can cause changes in ecological balance. Since the term disaster microbiology is current and open to interpretation, it is not clear which event will be evaluated within this scope, but the American Society for Microbiology defines disaster microbiology as a field of study that focuses on the microbial effects caused by disasters. Disaster microbiology allows experts to make microbiological evaluations of disaster-induced microbiological evaluations in line with a common purpose by bringing together many scientific disciplines. The main purpose of disaster microbiology is to detect microorganisms after disasters and examine their interaction with humans. It is important to evaluate the effects of disaster microbiology in societies where disasters

* Corresponding Author's Email: mehtapyakupoglu@trabzon.edu.tr.

In: Disaster Management in the Digital Age and Emerging Technologies
Editors: Ayşe Asiltürk and Ayşegül Sarıoğlu Kemer
ISBN: 979-8-89530-213-2

are frequently experienced. In this context, this study aims to evaluate disasters from the perspective of microbiology.

Keywords: biological threats in disasters, disaster, disaster microbiology, microbiology

Introduction

Disaster microbiology is a new field of study focusing on the impact of natural and human-induced disasters on microbial populations and their ecological interactions (Smith and Casadevall, 2022). Disasters can alter the microbial population in affected areas, leading to new biological processes and potential consequences for species, including humans (Viswanathan et al., 2021). Microorganisms play an important role in disasters, whether bioterrorism events, pandemics such as COVID-19, or mass disasters, where identification by microbiological methods becomes crucial (Wagar, 2016; Noviana et al., 2022; Iacono and Nichols, 2017). Disaster microbiology emphasizes the significance of planning, risk assessment, and safety procedures in clinical microbiology laboratories to respond effectively to biological disasters. (Wagar, 2016). Understanding the influence of the environment on infectious diseases and the transmission pathways of pathogens has been a very important breakthrough in the microbiological era, highlighting the necessity of microbiological research in disaster management. (Iacono and Nichols, 2017). Microbiologists play a critical role in the post-disaster response by contributing to laboratory diagnosis, situation analysis, assessment, planning, prevention, and control of infectious diseases (Garbern, 2024). They assess the risk of disease outbreaks, establish laboratory services, train personnel in field settings, and address biosecurity and waste management (Bhattacharyya et al., 2020). Infectious diseases are a major concern in disaster areas, with increased risks due to overcrowding, interrupted access to safe water and sanitation, and limited health services (Dhar et al., 2022). Proper planning, sample collection, disease diagnosis, and biosecurity measures are essential for the management of infectious diseases during disasters. Overall, disaster microbiology is a very important area of research given the increasing frequency of disasters and their potential impact on microbial populations and public health.

Microbiological technologies are vital for water testing, medico-legal investigations, and identifying victims of major disasters (Wagar, 2016;

Takács, 2020; Huang, 2022). These approaches not only aid in understanding the transmission of infectious illnesses but also contribute to forensic investigations after mass casualty incidents. (Wagar, 2016; Huang, 2022). The construction of high-standard laboratories for microbiological studies, such as skin banks for burn care, illustrates the relevance of microbiology in disaster response and management. (Huang, 2022).

Disaster microbiology is a multidisciplinary field that includes parts of microbiology, public health, and disaster management. This discipline seeks to increase our ability to avoid, detect, and respond effectively to microorganism-caused disasters by using microbiological tools, research, and preparation strategies.

Natural Disasters

Natural disasters have long been recognized as indicators of public health emergencies, notably those involving infectious illnesses (Waring and Brown, 2005). Hurricanes, tsunamis, floods, and tornadoes can all expose humans to microorganisms that they would not normally come into contact with. Where incidents result in harm, there may be an increased risk of infected wounds and blood infections. For example, storms in Joplin, Missouri, resulted in many cases of mucormycosis due to infection of individuals with the fungus *Apophysomyces trapeziformis* following storm damage (Austin et al., 2014; Neblett et al., 2011; Weddle et al., 2012). Anthropogenic climate change leads to rapid environmental changes. Climate change increases the likelihood of tropical cyclones, severe temperatures, storms, droughts, and forest fires. Climate change may accelerate the development of bacteria in response to stresses such as heat waves, drought, chemical or toxic waste, and wildfires, as well as increasing disaster-mediated exposure to pathogenic microorganisms. Microbial stress response factors, which allow microorganisms to withstand environmental changes, can also alter how they interact with the mammalian immune system. Furthermore, if heatwaves become more frequent and longer lasting, microorganisms will be selected for their capacity to survive and endure at higher temperatures (Casadevall et al., 2019).

Extreme Heat

Heat waves and severe heat occurrences are often quite stressful for species. Other natural calamities related to extreme heat include droughts, forest fires, dust storms, and tropical cyclones. Urban regions are more prone to heat because of the "heat island" effect caused by a lack of green space, ventilation, and construction materials (Saverino et al., 2021; Metzger et al., 2010). Heat can increase thermotolerance in bacteria, allowing them to survive at human body temperatures if they do not already have this capacity (Garcia-Solache and Casadevall, 2010). The increasing thermotolerance of microorganisms poses a significant challenge to the emergence of fungal pathogens. This is because many fungi grow best under 37°C or human body temperature. In fact, this thermal separation zone may be one reason why fungal diseases are rare, especially in people with weakened immune systems (Robert and Casadevall, 2009). While the immune system, a fungal barrier, weakened during the human immunodeficiency virus (HIV) epidemic and the introduction of immunosuppressive medical treatments in the twentieth century, the twenty-first century might witness an increase in fungal illnesses as fungus species with harmful potential to adapt to greater temperatures and overcome mammalian thermal barriers. (Casadevall, 2020). Global warming has been hypothesized as a factor for Candida auris to develop temperature tolerance and transform from a purely environmental niche to a niche that can cause human disease. (Casadevall et al., 2019).

Forest Fires

Wildfires can also spread microorganisms, particularly those present in the soil, and living near flames has been associated with an increased risk of upper and lower respiratory tract infections (Kobziar and Thompson, 2020; Yao, Eyamie, and Henderson, 2016; Van Leuken et al., 2016). More importantly, forest fires significantly affect the bacterial and fungal composition of the soil. As a consequence, forest fires gradually select stress-resistant spores and heat-tolerant microorganisms (Whitman et al., 2019). Since the structure of the bacterial spore and the adaptations to heat stress facilitate microbial infection at human body temperatures, such adaptations can be problematic for humans. Fire smoke can aerosolize and convey living fungi and bacteria, and its microbiological concentration is substantially higher than that of the surrounding air (Kobziar et al., 2018). Fire smoke has the potential to transport plant diseases, as well as beneficial soil and plant microorganisms, across great distances. The authors refer to the study of how bacteria interact with fire and smoke as "pyroerobiology" (Kobziar et al., 2018).

Floods

Floods can be caused by a variety of natural disasters, including tsunamis, rainstorms, and tropical cyclones. All these natural phenomena cause the typically dry land/soil to become saturated with water from rivers, streams, the ocean, marshes, or rainwater runoff. This can cause damage by invading homes, businesses, and roadways. Floods harmed an estimated 35 million people globally in 2020 (Ritche et al., 2022). Flooding increases the risk of microbial infections by disturbing the structure of the soil and, as a result, bringing soil-dwelling microorganisms into contact with flood victims. Tropical cyclones have been linked to respiratory infections, gastrointestinal diseases, and other infectious diseases. (Campanella, 1999; Zheng et al., 2017). Flooding and water damage to dwellings can induce respiratory problems by increasing the growth of toxin-producing molds, as seen in many homes in New Orleans after Hurricane Katrina (Crook and Burton, 2010; Cummings et al., 2008). Following Hurricane María, filamentous fungi, such as *Aspergillus* spp., thrived in water-damaged dwellings in Puerto Rico (Vélez-Torres et al., 2022). Molds developing in damp or water-damaged dwellings have been linked to an increased risk of asthma in both children and adults (Sahakian et al., 2008). Such molds may also be found in other buildings and are linked to "sick building syndrome," in which people become unwell for several causes, including persistent exposure to fungal toxins (Li and Yang, 2004; Saijo et al., 2011). During the 2004 Indian Ocean tsunami and the 2011 Japanese tsunami, *Fusarium* spp., *Mucor* spp., *Aspergillus fumigatus*, and *Scedosporium apiospermum* were discovered to be present (Benedict and Park, 2014; Kawakami et al., 2012). The 2011 earthquake and tsunami in Japan were linked to an increase in cases of legionellosis caused by inhaling soil-contaminated water droplets containing *Legionella* spp. bacteria and *Aspergillus fumigatus* (Takahashi et al., 2012). Furthermore, floods can cause wastewater treatment plants to overflow, resulting in human waste spills and additional waste-contaminated flood water (Ten Veldhuis et al., 2010). Although there is considerable dread after natural disasters, the rotting remains of victims seldom transmit illness or contaminate water systems (de Goyet, 2004; Morgan, 2004; Watson, Gayer, and Connolly, 2007). Floodwaters can transfer bacteria to an ecosystem where they can carve out a niche and cause diseases decades after the disaster has occurred. The tsunami created by the Great Alaska Earthquake in 1964 is thought to have transferred the *Cryptococcus gattii* fungus from its aquatic niche in the Pacific Ocean to land in the Pacific Northwest and Vancouver Island, resulting in outbreaks beginning in 1999 (Engelthaler and Casadevall, 2019). Flooding in tropical locations can allow disease vectors, such as malaria-carrying *Anopheles* spp. mosquitoes or *Aedes* spp. mosquitoes that cause yellow fever, dengue, and Zika, to multiply and spread additional vector-borne illnesses (Kondo et al., 2002; Nosrat et al., 2021; Kaya and Şahinöz, 2021). Mosquito-borne illnesses currently kill more than 700,000 people each year, despite considerable mitigation efforts; the added cost of natural disasters and the potential expanded breeding area might increase these deaths and the number of people at risk worldwide (de Souza and Weaver, 2024; Obradovic et al., 2022; Socha et al., 2022; Mojahed, Mohammadkhani, and Mohamadkhani, 2022).

Earthquakes

Earthquakes are natural disasters that have far-reaching consequences on many elements of the ecosystem, including microbiology. Earthquakes can cause considerable changes in the hydrological system, mixing shallow water with deep groundwater (Hosono et al., 2019). This mixing can cause chemical, isotopic, and microbiological changes, influencing the microbial diversity and community dynamics in the impacted locations (Lipus et al., 2022). Earthquake-induced environmental changes pique the attention from academics from several disciplines, including biology, microbiology, and ecology, as evidenced by studies on disaster resilience and water supply management (Babar, 2007; Deng, 2021). Furthermore, the link between earthquakes and microbiology has been investigated in the context of public health, as evidenced by research looking at the consequences of earthquakes in COVID-19 instances (Durmaz, 2023). The aftermath of seismic occurrences can affect a variety of characteristics, including age, gender, and time, which are evaluated with microbiological data to better understand the larger effects of earthquakes on microbial activity. Furthermore, the effect of geological activity on microbial life has been investigated in dynamic subsurface rift ecosystems, stressing the interdependence of geological, geochemical, and microbiological processes (Lipus et al., 2022). Understanding the relationship between earthquakes and microbiology is critical to fully analyzing the environmental effects of seismic occurrences. Researchers can gain insight into the larger ecological repercussions of earthquakes by studying changes in microbial communities, as well as creating disaster management and environmental conservation measures. This multidisciplinary approach, which includes geology, hydrogeology, biology, and microbiology, is critical to understanding the complex processes that occur in the aftermath of earthquakes.

Human-Caused Disasters

Human-caused disasters, including wars, terrorism, and industrial mishaps, can have serious effects on public health and the environment. These calamities not only cause immediate physical and economic devastation but also have long-term impacts on communities (Sheikhbardsiri et al., 2017). In the field of microbiology, human-made disasters can have a substantial influence on microbial communities. For example, violent events such as mass shootings or bioterrorism occurrences could increase the risk of infectious illnesses due to open wounds, forced relocation, and the convergence of various populations (Smith and Casadevall, 2022). Other human-caused disasters, such as wastewater treatment mishandling, can have serious consequences for microorganisms and their interactions with people. Outbreaks of the gastrointestinal parasite *Cryptosporidium* have been linked to wastewater treatment plant failures, contaminating the water supply (Mac Kenzie et al., 1994). Other man-made disasters, particularly those caused by chronic and persistent human-made hazards, have strange and unnatural settings that result in a distinct microbiology in which microorganisms have the capacity to grow in harsh environments. For example, melanotic and radiophilic fungi colonized the damaged reactor at Chernobyl (Tugay et al., 2011; Zhdanova et al., 2000; Dadachova and Casadevall, 2008). Some of these sites have been researched by microbiologists, and the examination of their microbial flora has revealed novel insights. The link of a melanotic fungus with the damaged Chernobyl reactor prompted research that revealed that melanized fungi grew quicker when exposed to high radiation, a process that was proposed to involve radiation uptake by melanin and conversion into physiologically usable energy (Dadachova et al., 2007). Such disasters might modify the microbial landscape in ways that fall under the purview of "disaster microbiology," which has typically focused on natural disasters (Smith and Casadevall, 2022). In disaster management and public health, it is critical to acknowledge the interdependence between human activities and disaster risk. Addressing the human components of vulnerability and hazard is critical to proactively mitigate the effects of disaster (Raju et al., 2022). Recognizing the role of human agency in disaster generation and response, stakeholders can collaborate to improve preparedness, response, and recovery efforts in the face of both natural and human-made disasters.

The Future Role and Importance of Disaster Microbiology

Disaster microbiology is the study of how natural and human-caused disasters affect microbial ecosystems and infectious disease dynamics. Understanding these effects is critical to forecasting and mitigating infectious disease epidemics (Smith and Casadevall, 2022). Human-caused disasters, such as terrorism, war, and mass shootings, can expose people to viruses through wounds, bioweapons, or forced migration. Similarly, natural disasters can disturb microbial communities, posing health hazards (Benedict and Park, 2014). In disaster response, timely and precise microbial identification is critical to efficiently control infectious disease epidemics. Compact microbiology laboratories in disaster-affected communities can considerably enhance infectious disease treatment outcomes (Kouadio et al., 2014). Disaster microbiology research can also help forecast and prevent certain illnesses related to disasters, such as wound-associated mucormycosis from battle injuries or natural disasters (Walsh et al., 2019). Furthermore, disaster microbiology can assist populations in building resistance to infectious disease epidemics following disasters. Understanding how disasters affect microbial ecosystems allows public health officials to design ways to reduce the risk of disease transmission among susceptible groups (Cvetković and Ivkovic, 2022). Furthermore, the function of microbiology in crisis scenarios includes infection control measures in healthcare institutions experiencing internal disasters, highlighting the need for planning and response strategies (Lee et al., 2007).

In conclusion, disaster microbiology is critical in disaster management and public health response. By investigating the relationships between disasters and microbial ecosystems, researchers and practitioners can better plan and respond to infectious disease concerns in the aftermath of disasters.

Contributions of New Technologies to Disaster Microbiology

In the field of disaster management, the convergence of different disciplines is encompassed by emergency management, which includes strategies for pre-disaster and the efficient utilization of resources during and post disasters. This intricate process engages numerous local, national, and international stakeholders, with a growing significance placed on civil initiatives (Babaoğlu, 2017). Within this multi-faceted framework, technology and advancements in technology play a pivotal role as tools to mitigate the impact of disasters.

It has been emphasized that significant benefits can be derived from the integrated disaster management approach, which is globally acknowledged and encompasses the risk assessment and reduction, preparedness, response, and recovery phases, through the lens of technological advancements (Macit, 2019). Technological progress and the advent of big data facilitate risk assessment in the mitigation stage, forecasting, and early warning of disasters; disaster surveillance and identification in the preparedness phase, as well as the effective assessment of needs and coordination in the response phase (Yu et al., 2018).

In the field of disaster microbiology, novel technologies have the potential to offer various contributions to disaster microbiology.

Rapid Diagnostic Methods

Conventional microbiological diagnostic procedures can be time-consuming, and timeliness is crucial in instances of calamity. Recent advancements have introduced expedited diagnostic techniques, enhancing the ability to promptly and precisely identify microbes (Wong et al., 2015). These methodologies encompass polymerase chain reaction (PCR)-based approaches, isothermal amplification methods, rapid test kits based on antigens and antibodies, and molecular diagnostic instruments. These methodologies are adept at producing swift outcomes suitable for deployment in the field, furnishing vital data to healthcare practitioners in emergency scenarios, and are proving pivotal in the oversight of food and water reservoirs (Korkmaz and Merey, 2023).

Digital Health Records and Monitoring Systems

Through the use of electronic health records (EHR) and computational methodologies, these systems facilitate cost-efficient and extensive evaluations of disease risks post-disaster (Boland et al., 2021; Acosta and Irizarry, 2020). Moreover, real-time surveillance via wireless bodily sensors integrated with Internet of Things (IoT) capabilities can help identify crisis occurrences during disasters, thereby enhancing strategic planning and response protocols (Lokmic-Tomkins, 2023). In essence, digital health records and monitoring systems prove to be effective in overseeing health infrastructures, gauging the impacts on mortality, and analyzing disease ramifications after disasters and outbreaks (Kulkarni and Bakal, 2018).

Remote Sensing Technologies

Remote sensing technologies play a pivotal role in the detection of microbial spread in the context of disasters. The use of technologies such as the Autonomous Microbiological Stimulation Sensor (AMAS) enables real-time assessment of pathogens such as Total Coliform and *Escherichia coli* in water sources, thereby assisting in crisis management and regular environmental surveillance (Garg, 2022; Huynh et al., 2016). Furthermore, the application of optical sensors for remote pathogen identification and monitoring enhances the capacity to recognize and track pathogens within the environment (Choi and Kim, 2023).

Mobile Health Technologies

Microbiological mobile health technologies are essential in disaster response for swift diagnosis, prevention, and management of infectious diseases. These technologies include the deployment of mobile diagnostic facilities equipped with diverse platforms such as HemoUE301, i-STAT, and BIOFIRE FILMARRAY to facilitate efficient testing (Viswanathan et al., 2021; Marr et al., 2021). Crisis scenarios underscore the emergence of the field known as "disaster microbiology," as they lead to alterations in microbial populations, thereby instigating novel ecological interactions with potential implications for human health (Smith and Casadevall, 2022).

Several Emerging Technologies

Exhibit significant potential for enhancing the prompt identification and surveillance of viral and bacterial pathogens in disaster-prone regions. These advancements encompass the creation of high-throughput tools for simultaneous detection of various virus strains (Kastro et al., 2023), real-time monitoring of virus-laden droplets through nano-DIHM (Pal et al., 2023), and the utilization of GIS technology for the mitigation of pathogen-related crises (Son et al., 2023). Furthermore, there is a focus on the development of nanomaterial-based sensor platforms tailored for high-throughput detection purposes, particularly in virus detection (Akvolunteer and Denizli, 2022). An innovative dual-channel fiber-optic immunosensor has been innovated for the concurrent detection of SARS-CoV-2 and influenza A virus, delivering swift outcomes with exceptional accuracy and sensitivity (Yang et al., 2022). The primary objective of these technologies is to deliver economical, precise, and timely solutions to the supervision and control of viral and bacterial pathogens during emergencies.

Discussion and Conclusion

Disasters have the potential to change microbial populations, resulting in novel ecological interactions that could have adverse implications for various species, including humans. The presence of stressors related to disasters can influence the selection of microbes, unveiling novel biological mechanisms and potential outcomes, whether pathogenic or beneficial in nature. The concept of "disaster microbiology" is being presented as a nascent area of research due to the repercussions of disasters on microbial communities and their ecological repercussions (Smith and Casadevall, 2022). Given the increasing impact of anthropogenic climate change and human undertakings on the environment, investigations within the realm of disaster microbiology are expected to gain prominence within the broader domain of microbiology. Microbiologists assume a pivotal role in post-disaster scenarios through the provision of laboratory testing, situational evaluation, strategic planning, preventive measures, and management. They are indispensable for conducting risk evaluations, identifying epidemic and endemic maladies, assessing laboratory infrastructures, and imparting training to personnel in field environments. Microbiologists are instrumental in the establishment of adaptable laboratories, the implementation of infection control measures in evacuation centers, and the reinstatement of healthcare facilities post-disasters. Technological advancements in the context of disasters, particularly in novel technologies, are exceedingly crucial in overseeing the post-disaster phase and notably in mitigating both primary and secondary consequences, as well as impeding the transmission of effects, specifically concerning infectious maladies. Notably, the active use of swift diagnostic kits in mobile healthcare settings and the prompt identification of contamination in food and water reservoirs accentuate the significance of this emerging discipline.

In this particular instance, emergency preparedness strategies must evaluate microbial hazards both pre- and post-calamities, and devise novel technological approaches to alleviate such hazards. Specifically, there should be a focus on offering training and support for the implementation of these innovative technologies. Particularly, it is crucial to regularly reassess and enhance technologies utilized in ensuring the provision of uncontaminated water and maintaining food hygiene standards to reduce the likelihood of microbial pollution.

Disclaimer

None.

References

Acosta, R. J., & Irizarry, R. A. (2020). Monitoring health systems by estimating excess mortality. *medRxiv*, 2020-06.

Akgönüllü, S., & Denizli, A. (2022). Nano-sensors and nano-devices for biological disaster monitoring (virus/disease epidemics/animal plagues detections). In *Nanotechnology-Based Smart Remote Sensing Networks for Disaster Prevention* (pp. 43-57). Elsevier.

Austin, C. L., Finley, P. J., Mikkelson, D. R., & Tibbs, B. (2014). Mucormycosis: a rare fungal infection in tornado victims. *Journal of Burn Care & Research, 35*(3), e164-e171.

Babar, A. (2007). Environmental changes and natural disasters. https://doi.org/10.59317/9789389992182.

Babaoğlu, C. (2017): Afet Yönetimi Politikalarına Sivil Toplum Örgütlerinin KatılımıSorunsalı, İçinde Afet Yönetimi, Edt. Özgür Önder ve Murat Yaman, Bursa: Ekin Yayınevi, s.163-170.

Benedict, K., & Park, B. J. (2014). Invasive fungal infections after natural disasters. *Emerging Infectious Diseases, 20*(3), 349.

Bhattacharyya, S., Nandy, B., Chandra, R., & Das, P. (2020). Laboratory and field observations in emergency relief work in a city in Eastern India. *Eastern Journal of Medical Sciences, 5*(1), 1-3.

Boland, M. R., Davidson, L. M., Canelón, S. P., Meeker, J., Penning, T., Holmes, J. H., & Moore, J. H. (2021). Harnessing electronic health records to study emerging environmental disasters: a proof of concept with perfluoroalkyl substances (PFAS). *NPJ Digital Medicine, 4*(1), 122.

Campanella, N. (1999). Infectious diseases and natural disasters: the effects of Hurricane Mitch over Villanueva municipal area, Nicaragua. *Public Health Reviews, 27*(4), 311-319.

Castro, J. M., Sommerhage, F., Khanna, R., Childs, A., DeRoo, D., & Rajaraman, S. (2023). High-throughput microbead assay system with a portable, cost-effective Wi-Fi imaging module, and disposable multi-layered microfluidic cartridges for virus and microparticle detection, and tracking. *Biomedical Microdevices, 25(3), 21.*

Casadevall, A., Kontoyiannis, D. P., & Robert, V. (2019). On the emergence of *Candida auris*: climate change, azoles, swamps, and birds. *MBio, 10*(4), 10:10.1128/mbio.01397-19. doi: 10.1128/mbio.01397-19.

Casadevall, A. (2020). Climate change brings the specter of new infectious diseases. *The Journal of Clinical Investigation, 130*(2), 553-555.

Choi, J., & Kim, D. J. (2023). Editorial for the Special Issue "Disaster Monitoring Using Remote Sensing." *Remote Sensing, 15*(3), Article 751. https://doi.org/10.3390/rs15030751.

Crook, B., & Burton, N. C. (2010). Indoor moulds, sick building syndrome and building related illness. *Fungal Biology Reviews, 24*(3-4), 106-113.

Cummings, K. J., Cox-Ganser, J., Riggs, M. A., Edwards, N., Hobbs, G. R., & Kreiss, K. (2008). Health effects of exposure to water-damaged New Orleans homes six months after Hurricanes Katrina and Rita. *American Journal of Public Health, 98*(5), 869-875.

Cvetković, V. & Ivkovic, T. (2022). Social resilience to flood disasters: demographic, socio-economic and psychological factors of impact. *Academic Perspective Procedia, 5*(2), 299-317. https://doi.org/10.33793/acperpro.05.02.8356.

Dadachova, E., & Casadevall, A. (2008). Ionizing radiation: how fungi cope, adapt, and exploit with the help of melanin. *Current Opinion in Microbiology, 11*(6), 525-531.

de Goyet, C. D. V. (2004). Epidemics caused by dead bodies: a disaster myth that does not want to die. *Pan American Journal of Public Health, 15*(5), 297-9.

de Souza, W. M., & Weaver, S. C. (2024). Effects of climate change and human activities on vector-borne diseases. *Nature Reviews Microbiology*, 1-16.

Deng, Y. (2021). Building disaster resilience of water supply with household water treatment. *Water Environment Research, 93*(8), 1154-1156. https://doi.org/10.1002/wer.1614.

Dhar, P. (2022). Infectious Diseases and Their Management During Disasters. In *Management of Animals in Disasters* (pp. 229-244). Singapore: Springer Nature Singapore.

Durmaz, S. (2023). The 2020 İzmir earthquake and its effect on COVID-19 cases. *Microbiology Australia, 44*(4), 207-211. https://doi.org/10.1071/ma23056.

Engelthaler, D. M., & Casadevall, A. (2019). On the emergence of Cryptococcus gattii in the Pacific Northwest: Ballast tanks, tsunamis, and black swans. *MBio, 10*(5), 10-1128.

Garbern, S. C. (2024). Infectious Disease in a Disaster Zone. In *Ciottone's Disaster Medicine* (pp. 388-392). Elsevier.

Garcia-Solache, M. A., & Casadevall, A. (2010). Global warming will bring new fungal diseases for mammals. *MBio, 1*(1), 10-1128.

Garg, A. (2022). CoReS-Respiratory Strength Predicting Framework Using Noninvasive Technology for Remote Monitoring During Heath Disasters. In *Global Healthcare Disasters* (pp. 109-121). Apple Academic Press.

Hosono, T., Yamada, C., Shibata, T., Tawara, Y., Wang, C., Manga, M., & Shimada, J. (2019). Coseismic groundwater drawdown along crustal ruptures during the 2016 mw 7.0 Kumamoto earthquake. *Water Resources Research, 55*(7), 5891-5903. https://doi.org/10.1029/2019wr024871.

Huang, H. F. (2022). The Establishment of a High Standard Skin Bank in Taiwan as a Burn Care Pioneer. *Clin Surg, 7*(13), 1-5.

Huynh, V., Hausot, A., & Angelescu, D. E. (2016). An autonomous field sensor for Total Coliform and *E. coli* monitoring at remote sites. *OCEANS 2016 MTS/IEEE Monterey*, 1-5.

Iacono, G. & Nichols, G. (2017). Modeling the impact of environment on infectious diseases. https://doi.org/10.1093/acrefore/9780199389414.013.339.

Kawakami, Y., Tagami, T., Kusakabe, T., Kido, N., Kawaguchi, T., Omura, M., & Tosa, R. (2012). Disseminated aspergillosis associated with tsunami lung. *Respiratory Care, 57*(10), 1674-1678.

Kaya, N. & Şahinöz, T. (2021). Bulaşıcı Hastalık Bildirimlerinin Afet Yönetimi Açısından İncelenmesi. *Gümüşhane Üniversitesi Sağlık Bilimleri Dergisi, 10*(1), 21-31.

Kobziar, L. N., Pingree, M. R., Larson, H., Dreaden, T. J., Green, S., & Smith, J. A. (2018). Pyroaerobiology: the aerosolization and transport of viable microbial life by wildland fire. *Ecosphere, 9*(11), e02507.

Kobziar, L. N., & Thompson III, G. R. (2020). Wildfire smoke, a potential infectious agent. *Science, 370*(6523), 1408-1410.

Kondo, H., Seo, N., Yasuda, T., Hasizume, M., Koido, Y., Ninomiya, N., & Yamamoto, Y. (2002). Post-flood—infectious diseases in Mozambique. *Prehospital and Disaster Medicine, 17*(3), 126-133.

Korkmaz, R., & Merey, G. (2023). Gıda güvenliğinin önemi ve bakterilerin tespiti için yöntemler. *Journal of Health Sciences and Management, 3*(2), 47-53.

Kouadio, I., Aljunid, S., & Oshitani, H. (2012). Enhancing the response outcome of infectious disease treatment following natural disasters. *Expert Review of Anti-Infective Therapy, 10*(5), 531-532. https://doi.org/10.1586/eri.12.40.

Kulkarni, N. J., & Bakal, J. W. (2018, August). E-health: IoT-based system and correlation of vital stats in identification of mass disaster event. In 2018 Fourth International Conference on Computing Communication Control and Automation (ICCUBEA) (pp. 1-6). *IEEE.*

Lee, H., Lee, N., Chang, C., Chou, C., Wu, Y., Wang, L., & Ko, W. (2007). Outbreak ofacinetobacter baumanniibacteremia related to contaminated morphine used for patient-controlled analgesia. *Infection Control and Hospital Epidemiology, 28*(10), 1213-1217. https://doi.org/10.1086/520741.

Li, D. W., & Yang, C. S. (2004). Fungal contamination as a major contributor to sick building syndrome. *Advances in Applied Microbiology, 55*, 31-112.

Lipus, D., Jia, Z., Sondermann, M., Bussert, R., Bartholomäus, A., Wagner, D., & Kallmeyer, J. (2022). Microbial diversity and community dynamics in an active, high co2subsurface rift ecosystem. https://doi.org/10.1101/2022.10.24.511890.

Macit, İ. (2019): Bütünleşik Afet Yönetiminde Sendai Çerçeve Eylem Planının Beklenen Etkisi. *Artvin Çoruh Üniversitesi Doğal Afetler Uygulama Ve Araştırma Merkezi Doğal Afetler ve Çevre Dergisi, 5* (1), 175-186.

Mac Kenzie, W. R., Hoxie, N. J., Proctor, M. E., Gradus, M. S., Blair, K. A., Peterson, D. E., & Davis, J. P. (1994). A massive outbreak in Milwaukee of Cryptosporidium infection transmitted through the public water supply. *New England Journal of Medicine, 331*(3), 161-167.

Marr I, Francis JR, Stephens DP, Marshall K, Read DJ, Baird RW, & Coatsworth N. (2021). Development of a Mobile Laboratory for Sudden Onset Disasters. *Disaster Med Public Health Prep. 15*(2):170-180. doi: 10.1017/dmp.2019.128. Epub 2020 Apr 21. PMID: 32312350.

Metzger, K. B., Ito, K., & Matte, T. D. (2010). Summer heat and mortality in New York City: how hot is too hot? *Environmental Health Perspectives, 118*(1), 80-86.

Mojahed, N., Mohammadkhani, M. A., & Mohamadkhani, A. (2022). Climate crises and developing vector-borne diseases: a narrative review. *Iranian Journal of Public Health, 51*(12), 2664.

Morgan, O. (2004). Infectious disease risks from dead bodies following natural disasters. *Revista Panamericana De Salud Pública, 15*, 307-312.

Neblett Fanfair, R., Benedict, K., Bos, J., Bennett, S. D., Lo, Y. C., Adebanjo, T., & Park, B. J. (2012). Necrotizing cutaneous mucormycosis after a tornado in Joplin, Missouri, in 2011. *New England Journal of Medicine, 367*(23), 2214-2225.

Nosrat, C., Altamirano, J., Anyamba, A., Caldwell, J. M., Damoah, R., Mutuku, F., & LaBeaud, A. D. (2021). Impact of recent climate extremes on mosquito-borne disease transmission in Kenya. *Plos Neglected Tropical Diseases, 15*(3), e0009182.

Noviana, U., Hasinuddin, M., Suhron, M., & Endah, S. (2022). Exploring perception and role of nurses during Covid-19 pandemic: experiences of frontline workers in east java, Indonesia. *Malaysian Journal of Medicine and Health Sciences, 18*(6), 72-76. https://doi.org/10.47836/mjmhs.18.6.11.

Obradovic, Z., Smjecanin, E., Pindzo, E., Omerovic, H., & Cibo, N. (2022). A literature review on vector borne diseases. *International Journal of Medical Reviews and Case Reports, 6*(7), 27-27.

Pal, D., Amyot, M., Liang, C., & Ariya, P. A. (2023). Real-time 4D tracking of airborne virus-laden droplets and aerosols. *Communications Engineering, 2*(1), 41.

Raju, E., Boyd, E., & Otto, F. (2022). Stop blaming the climate for disasters. *Communications Earth & Environment, 3*(1). https://doi.org/10.1038/s43247-021-00332-2.

Ritchie, H., Rosado, P., & Roser, M. (2022). *Natural disasters. Our World in Data.*

Robert, V. A., & Casadevall, A. (2009). Vertebrate endothermy restricts most fungi as potential pathogens. The *Journal of Infectious Diseases, 200*(10), 1623-1626.

Sahakian, N. M., Park, J. H., & Cox-Ganser, J. M. (2008). Dampness and mold in the indoor environment: implications for asthma. *Immunology and Allergy Clinics of North America, 28*(3), 485-505.

Saijo, Y., Kanazawa, A., Araki, A., Morimoto, K., Nakayama, K., Takigawa, T., & Kishi, R. (2011). Relationships between mite allergen levels, mold concentrations, and sick building syndrome symptoms in newly built dwellings in Japan. *Indoor Air, 21*(3), 253-263.

Saverino, K. C., Routman, E., Lookingbill, T. R., Eanes, A. M., Hoffman, J. S., & Bao, R. (2021). Thermal inequity in Richmond, VA: the effect of an unjust evolution of the urban landscape on urban heat islands. *Sustainability, 13*(3), 1511.

Sheikhbardsiri, H., Yarmohammadian, M., Rezaei, F., & Maracy, M. (2017). Rehabilitation of vulnerable groups in emergencies and disasters: a systematic review. *World Journal of Emergency Medicine, 8*(4), 253. https://doi.org/10.5847/wjem.j.1920-8642.2017.04.002.

Smith, D. F., & Casadevall, A. (2022). Disaster microbiology—a new field of study. *MBio, 13*(4), e01680-22.

Socha, W., Kwasnik, M., Larska, M., Rola, J., & Rozek, W. (2022). Vector-borne viral diseases as a current threat for human and animal health—One Health perspective. *Journal of Clinical Medicine, 11*(11), 3026.

Son, S. J., Lee, M. Y., Kim, J. H., & Cho, S. H. (2023). Gis-based intelligent decision-making support system for the disaster response of infectious diseases. *The International Archives of the Photogrammetry, Remote Sensing and Spatial Information Sciences, 48*, 217-222.

Takahashi, T., Goto, M., Yoshida, H., Sumino, H., & Matsui, H. (2012). Infectious diseases after the 2011 Great East Japan earthquake. *Journal of Experimental & Clinical Medicine, 4*(1), 20-23.

Takács, K. (2020). Analysis of microbiological methods applicable to water testing in our country. *Hadmérnök, 15*(1), 221-231. https://doi.org/10.32567/hm.2020.1.15.

Ten Veldhuis, J. A. E., Clemens, F. H. L. R., Sterk, G., & Berends, B. R. (2010). Microbial risks associated with exposure to pathogens in contaminated urban flood water. *Water Research, 44*(9), 2910-2918.

Tugay, T. I., Zheltonozhskaya, M. V., Sadovnikov, L. V., Tugay, A. V., & Farfán, E. B. (2011). Effects of ionizing radiation on the antioxidant system of microscopic fungi with radioadaptive properties found in the Chernobyl exclusion zone. *Health Physics, 101*(4), 375-382.

Van Leuken, J. P. G., Swart, A. N., Havelaar, A. H., Van Pul, A., Van der Hoek, W., & Heederik, D. (2016). Atmospheric dispersion modelling of bioaerosols that are pathogenic to humans and livestock–A review to inform risk assessment studies. *Microbial Risk Analysis, 1*, 19-39.

Vélez-Torres, L. N., Bolaños-Rosero, B., Godoy-Vitorino, F., Rivera-Mariani, F. E., Maestre, J. P., Kinney, K., & Cavallin, H. (2022). Hurricane María drives increased indoor proliferation of filamentous fungi in San Juan, Puerto Rico: a two-year culture-based approach. *PeerJ, 10*, e12730.

Viswanathan, R., Chakrabarty, A., & Basu, S. (2021). Active support after natural disasters: a review of a microbiologist's role. *Transactions of The Royal Society of Tropical Medicine and Hygiene, 115*(1), 110-116.

Walsh, T., Hospenthal, D., Petraitis, V., & Kontoyiannis, D. (2019). Necrotizing mucormycosis of wounds following combat injuries, natural disasters, burns, and other trauma. *Journal of Fungi, 5*(3), 57. https://doi.org/10.3390/jof5030057.

Watson, J. T., Gayer, M., & Connolly, M. A. (2007). Epidemics after natural disasters. *Emerging Infectious Diseases, 13*(1), 1.

Wagar, E. (2016). Bioterrorism and the role of the clinical microbiology laboratory. *Clinical Microbiology Reviews, 29*(1), 175-189. https://doi.org/10.1128/cmr.00033-15.

Waring, S. C., & Brown, B. J. (2005). The threat of communicable diseases following natural disasters: a public health response. *Disaster Management & Response, 3*(2), 41-47.

Weddle, G., Gandy, K., Bratcher, D., Pahud, B., & Jackson, M. A. (2012). Apophysomyces trapeziformis infection associated with a tornado-related injury. *The Pediatric Infectious Disease Journal, 31*(6), 640-642.

Whitman, T., Whitman, E., Woolet, J., Flannigan, M. D., Thompson, D. K., & Parisien, M. A. (2019). Soil bacterial and fungal response to wildfires in the Canadian boreal forest across a burn severity gradient. *Soil Biology and Biochemistry, 138*, 107571.

Wong, G., Wong, I., Chan, K., Hsieh, Y., & Wong, S. (2015). A rapid and low-cost PCR thermal cycler for low-resource settings. *PLoS one, 10*(7), e0131701.

Yang, Y., Zhao, R., Wang, Y., Song, D., Jiang, B., Guo, X., & Hao, R. (2022). Rapid and universal detection of SARS-CoV-2 and influenza A virus using a reusable dual-channel optic fiber immunosensor. *Journal of Medical Virology, 94*(11), 5325-5335.

Yao, J., Eyamie, J., & Henderson, S. B. (2016). Evaluation of a spatially resolved forest fire smoke model for population-based epidemiologic exposure assessment. *Journal of Exposure Science & Environmental Epidemiology, 26*(3), 233-240.

Yu, M., Yang, C., & Li, Y. (2018): Big Data in Natural Disaster Management: A Review, *Geosciences, 8,* 165, 1-26.

Zhdanova, N. N., Zakharchenko, V. A., Vember, V. V., & Nakonechnaya, L. T. (2000). Fungi from Chernobyl: mycobiota of the inner regions of the containment structures of the damaged nuclear reactor. *Mycological Research, 104*(12), 1421-1426.

Zheng, J., Han, W., Jiang, B., Ma, W., & Zhang, Y. (2017). Infectious diseases and tropical cyclones in Southeast China. *International Journal of Environmental Research and Public Health, 14*(5), 494.

Chapter 22

National and International Organizations in Disaster Management

Ayşe Elkoca[1,*], PhD
and Ebubekir Kaplan[2], PhD
[1]Faculty of Health Sciences, Gaziantep Islam Science and Technology University, Gaziantep, Türkiye
[2]Faculty of Health Sciences, Van Yüzüncü Yıl University, Van, Türkiye

Abstract

While disasters can occur anywhere in the world at any time, they damage people, living spaces, nature, infrastructure, and the economy. In disasters, the elements that show us the capacity for disasters are the existence and capacity of national and international institutions, disaster training, economic, technological, academic, and human resources. National and international disaster coordination has a very important place in global disaster response. At the national level, coordination in our country is conducted by the Disaster and Emergency Management Presidency (AFAD). The United Nations and its sub-organizations provide most of the international disaster coordination. This coordination is most effective on issues such as adherence to guidelines, local health structures, ethical constraints, planning, and preparation. In this article, national and international institutions and organizations in disasters are included.

Keywords: disaster, disaster management, international organizations, national organizations

---

* Corresponding Author's Email: ayse.elkoca@gibtu.edu.tr.

In: Disaster Management in the Digital Age and Emerging Technologies
Editors: Ayşe Asiltürk and Ayşegül Sarıoğlu Kemer
ISBN: 979-8-89530-213-2

Introduction

Disasters leave deep, long-lasting, and difficult to recover from health, social, and economic scars on the society in which they occur (Kıran, 2023). According to the United Nations Office for Disaster Risk Reduction (UNDRR), disasters caused 74,000 deaths in 2023. Of these, 58,000 were caused by the earthquakes in Türkiye and Syria, which have been dubbed the disaster of the century. These earthquakes caused around $50 billion in damage (UNDRR, 2023).

Today, disasters are a major threat to public health worldwide because of their widespread impact. This situation has accelerated the development of the fields of disaster management, risk management, and disaster medicine. Many countries have begun to establish frameworks to reduce or eliminate the effects of disasters through national organizations with which they have established within themselves and international organizations with which they are partners (Kıran, 2023).

National Institutions and Organizations in Disasters

In 2009 dated and numbered "Law on the Organization and Duties of the Disaster and Emergency Management Presidency," the understanding and organization of disaster management were re-established, and the Disaster and Emergency Management Presidency (AFAD) was established under the Prime Ministry in order to carry out services related to disasters and emergencies and civil defense (ResmiGazete, 2009). In 2018, some amendments were made, and the current form was established with a decree with force of law (ResmiGazete, 2018). The aim of this presidency is to ensure coordination between institutions and organizations and to formulate policies before disasters (Doğan, 2019).

The mission of AFAD for 2019–2023 is "to carry out the necessary work for the effective management of processes related to disasters and emergencies, to ensure coordination among relevant institutions and organizations, and to produce policies in this field." Its vision is determined as "Creating a society resilient to disasters." (AFAD, 2023b).

The Ministry of Health has established National Medical Rescue Teams (UMKE), which are well-trained and equipped for possible disasters, especially earthquakes, in Türkiye and abroad. The teams are ready to provide medical rescue services for the debris as soon as possible, to ensure rapid and safe patient and injured transport, and to provide emergency treatment units and services after transport (SağlıkBakanlığı, 2024).

Disaster and Emergency Management Presidency (Turkish: Afet ve Acil Durum Yönetimi Başkanlığı, AFAD)

For the effective realization of services related to disasters and emergencies and civil defense at the national level;

- Taking the necessary precautions, preparation, and risk reduction before the occurrence of events,
- Ensuring coordination between the institutions and organizations carrying out the intervention to be carried out during the incident and the improvement works to be carried out after the incident,
- It covers the conduct and coordination of humanitarian aid operations at home and abroad, as well as the development and implementation of policy recommendations on these issues.

Central Organization of the Disaster and Emergency Management Presidency (ResmiGazete, 2018);

- High Council for Disaster and Emergency
- Disaster and Emergency Coordination Board
- Earthquake Advisory Board
- The central organization of the Presidency consists of the following service units:

General Directorate of Earthquake and Risk Mitigation

The duties of the General Directorate of Earthquake and Risk Mitigation are as follows (AFAD, 2023a).

- To make or have made disaster and emergency response, risk management, and mitigation plans to be implemented at the national level.
- Determines the principles of in-kind, cash, and humanitarian aid.
- Collect and evaluate information on disasters and emergencies occurring at home and abroad.

- Determines management strategies for disasters and emergencies.
- It determines disaster hazards and risks according to their types and scales and makes or has made maps of them.
- It determines the principles of earthquake preparedness and earthquake risk management and performs its activities.
- Ensures the execution of zoning, plan, and project procedures for places likely to be damaged in earthquakes and damaged places.
- Ensures the identification and effective use of all kinds of resources belonging to public, private, and non-governmental organizations and foreign persons and organizations that can be used in earthquake preparedness, response, and recovery phases.
- Ensures that the public is informed about earthquakes.

General Directorate of Disaster Response

The duties of the General Directorate of Disaster Response are as follows (AFAD, 2023a):

- Determines the standards of the fire brigade, search, and rescue services.
- Ensures coordination and cooperation among public institutions and organizations, fire brigades, and non-governmental organizations in response activities.
- Carries out activities for the planning, coordination, and execution of search and rescue services, carries out, coordinates, and monitors international emergency and humanitarian aid and response activities.
- Carries out or has carried out domestic and international disaster drills.
- Determines the civil resources that will be needed in mobilization and war preparation.
- Determines the measures to be taken and activities to be carried out against the dangers posed by chemical, biological, radiological, and nuclear substances, and ensures coordination among relevant ministries, public and private institutions, and organizations.

General Directorate of Accommodation and Construction Works

The duties of the General Directorate of Housing and Construction Works are as follows (AFAD, 2023a):

- Takes measures to ensure the normalization of life after disasters and emergencies.
- Provides temporary shelter in disaster and emergency areas.
- Establishing, maintaining, repairing, operating, and managing disaster logistics depots.
- Carries out services related to the urgent needs of disaster victims such as treatment, food, and shelter.
- It carries out and monitors the entitlement, debt, and collection procedures of disaster victims.
- It determines the construction methods of the permanent structures built for the disaster victims deemed to be beneficiaries, prepares their projects, includes them in the investment program, provides their appropriations, and ensures their construction.

General Directorate of Management Services

The duties of the General Directorate of Management Services are as follows (AFAD, 2023a):

- It carries out the manpower planning of the Presidency and determines the human resources policy.
- It carries out the planning, implementation, and evaluation of basic, preparatory, and specialization in-service training in order to increase the level of knowledge of the central and provincial staff of the Presidency
- It carries out documentation, publication, and archive services related to the training activities of the central and provincial staff of the Presidency.

Presidency of Guidance and Inspection

The duties of the Presidency of Guidance and Inspection are as follows (AFAD, 2023a):

- It carries out examinations, audits, and investigations on all kinds of activities and operations of the central units of the Presidency and the provincial disaster and emergency directorates and their personnel.
- It provides guidance, conducts studies, and prepares the necessary proposals in order to better realize the objectives of the Presidency

and to ensure that it operates according to the legislation, plan, and program.

- It uses the powers of examination, audit, and investigation granted to the Presidency by the legislation.

Department of Foreign Relations

The duties of the Department of Foreign Relations are as follows (AFAD, 2023a):

- Organizing relations, including international agreements, with organizations operating in the international arena on issues falling within the mandate of the Presidency, and ensuring the necessary liaison and coordination.
- In the relations with the European Union and in the legislative and harmonization works toward the European Union and in the works for economic and technical cooperation, it ensures the necessary communication and unity for the execution of the services of the Presidency in a fast, regular, effective, and efficient manner.

Legal Counsellor's Office

The duties of the Legal Counseling Office are as follows (AFAD, 2023a):

- It carries out the duties assigned to the legal units according to the provisions of Decree Law No. 659 dated 26/9/2011.

Press and Public Relations Consultancy

The duties of the Press and Public Relations Consultancy are as follows (AFAD, 2023a):

- It plans the activities of the Presidency related to press and public relations and ensures that these activities are carried out according to the determined procedures and principles.
- It takes necessary measures to finalize the applications for obtaining information in an effective, swift, and accurate manner according to the Law on Right to Information No. 4982 dated 9/10/2003.
- It ensures that communication, media, and public relations are carried out properly in disasters and emergencies.

Disaster and Emergency Management Presidency Provincial Organization (ResmiGazete, 2018):

- Provincial disaster and emergency directorates
- Disaster and Emergency Search and Rescue Directorates

Provincial Disaster and Emergency Directorates

The duties of the provincial disaster and emergency directorates are as follows (AFAD, 2023a):

- Determining the disaster and emergency hazards and risks of the province and making disaster and emergency preparations.
- Managing the provincial disaster and emergency management center and ensuring uninterrupted and secure communication.
- It determines the losses and damages incurred in disasters and emergencies.
- Carries out training activities related to disasters and emergencies.
- It accredits and certifies non-governmental organizations and volunteers in relation to disaster and emergency management.
- In disasters and emergencies, it establishes and manages warehouses for the necessary search and rescue materials and food, tools, equipment, and materials to be used to meet the shelter, nutrition, and health needs of the people.
- It carries out the duties related to mobilization and war preparations and civil defense services in provinces, as stipulated in the relevant legislation.
- It carries out services related to the detection, diagnosis, and purification of chemical, biological, radiological, and nuclear substances and other similar technological substances, and ensures cooperation and coordination among relevant institutions and organizations.
- Implementing the annual work programs determined by the Presidency, preparing annual activity reports, and submitting them to the approval of the Presidency.

Disaster and Emergency Search and Rescue Directorates

In the provinces determined by the Presidency, disaster and emergency search and rescue unity directorates have been established under the provincial disaster and emergency directorate. These directorates work under the provincial disaster and emergency directorate (AFAD, 2023a).

National Medical Rescue Team (Turkish: Ulusal Medikal Kurtarma Ekibi, UMKE)

UMKE is a medical rescue team consisting of health personnel who have received the necessary basic and advanced training and are equipped according to the needs to provide medical rescue and emergency health services to the accident victims/disaster victims who need emergency health services in disasters and emergencies that may occur at home and abroad, starting from the scene of the incident until they are transferred to an ambulance or a health facility (SağlıkBakanlığı, 2024).

In addition to medical rescue services, UMKE takes an active role in every stage of health services to be provided within the scope of disaster response and recovery activities by establishing surgical intervention units, emergency response units, mobile hospitals and personnel accommodation units, mobile command centers, and mobile incident scene management centers in the field according to the size of the disasters (SağlıkBakanlığı, 2024).

In the 50-bed mobile hospitals established by UMKE, specialist physicians and general practitioners in all branches, specially trained emergency, newborn, operating theater, intensive care, and ward nurses, laboratory, radiology, anesthesia, and orthopedics technicians/technicians, and other health personnel work (SağlıkBakanlığı, 2024).

Civil Society Organizations

Civil society organizations (CSOs) are extremely sensitive to natural disasters. In addition to the many organizations established in our country to assist in disasters, non-governmental organizations, whose main task is in other fields, also support relief efforts during and after disasters.

The main role of non-governmental organizations in disaster situations is to act under the guidance of central and local administrations and to assist these administrations. Almost all CSOs provide support work on a voluntary basis.

Search and Rescue Association (Turkish: Arama Kurtarma Derneği, AKUT)

In 1994, after 14 days of fruitless search for two university students who went missing in the Bolkar Mountains, a group of mountaineers who knew the mountains well but had limited knowledge of search and rescue came together and founded AKUT in 1995. In the same year, this group carried out the first rescue activity in Uludağ under the name of AKUT. The association completed its official establishment on 14 March 1996 under the name "AKUT Search and Rescue Association." AKUT, Türkiye's first search and rescue association, was founded with the aim of carrying out accurate and effective search and rescue activities in mountains and other difficult natural conditions. On 19 January 1999, AKUT was granted the status of "Public Benefit Association" by the decision of the Council of Ministers, in recognition of its efforts in the 1998 Adana-Ceyhan earthquake (AKUT, 2024b).

In cases of disappearance and accidents occurring in mountain and natural conditions, natural disasters such as earthquakes, floods, and major accidents, it is to reach people in trouble as soon as possible with amateur work and a professional approach, to create the appropriate conditions for help, and to carry out correct search and rescue work. After providing basic first aid support to the victims, to ensure their transfer to safe environmental conditions, to minimize the loss of life in such incidents, and to inform the society on search and rescue issues (AKUT, 2024a).

The Turkish Red Crescent (Turkish: Türk Kızılay)

The Turkish Red Crescent was established with the vision of ensuring that all people live humanely by providing resilience-building activities and humanitarian aid in both disasters and emergencies and in ordinary times (TürkKızılay, 2024).

The Red Crescent is a voluntary social service organization with a legal personality, subject to the provisions of private law, non-profit, free of charge, and working for public benefit (TürkKızılay, 2024). In addition to providing disaster-affected citizens with water, food, clothing, shelter, etc., it provides blood and blood products to disaster victims when necessary through the blood donations it collects (Doğan, 2019). The duties of the Red Crescent in disasters and ordinary periods are to provide assistance to the needy and vulnerable, to develop solidarity in society, to ensure a safe blood supply, and to reduce vulnerability (TürkiyeKızılayDerneğiTüzüğü, 2022).

Community Disaster Response Teams (Turkish: Toplum Afet Müdahale Ekipleri, TAME)

Community Disaster Response Teams are organized groups consisting of citizens whose sensitivity to disasters has been increased by informing them about the precautions to be taken in combating disaster risks, who have received the necessary training, who have been supported with the necessary equipment, and whose facilities and capabilities have been strengthened from the immediate aftermath of disasters until the arrival of professional teams (Kaya, 2013).

Fire Brigade Neigborhood Volunteers

In the event of a possible disaster, until the arrival of professional fire brigades and Civil Defense teams, it is the local people who will make the first intervention. In that respect, in order to prevent possible post-disaster fires of local people and to contribute to rescue activities, it is necessary to organize such formations and to have the necessary fire safety, extinguishing, and cooling training given by competent personnel in professional fire brigade units and to supervise the training received (Kaya, 2013).

Civil Defense Volunteers

Civil Defense Services are carried out in accordance with the provisions of Civil Defense Law No. 7126, the by-laws and regulations issued pursuant to this law, and Law No. 3152 on the Organization and Duties of the Ministry of

Interior (www.aliaga.gov.tr) "Directive on the Principles of Participation of Volunteers in Civil Defense Services" was put into force on May 5, 2000, for the participation of volunteer individuals and organizations in search, rescue, and aid activities under the coordination of civil defense elements in possible disaster events and in case of war. It is stated that the personnel of the police force to be assigned in disasters should receive the necessary training together with the Civil Defense Organization, and within the framework of the Annex-8th article of the Civil Defense Organization, voluntary persons and organizations may participate in civil defense services under the coordination of the Civil Defense Organization (Kaya, 2013). Those who want to work as volunteers in civil defense services will have the opportunity to receive the necessary training, to be informed against possible disasters, and to be included in the plans if they apply to the institutions performing duties in this regard.

Community Disaster Volunteer

It will help the officials and expert teams in a possible disaster, prevent malicious people and behaviors at the disaster scene, correctly identify those directly and indirectly affected by disasters, keep the solidarity power and morale of the society high, assist the people and teams coming from outside, and contribute to the return to normal activities. Volunteers consisting of people who will design possible disaster risks in the pre-disaster neighborhood and create plans, who will engage in post-disaster reconstruction and improvement activities, and who will do so entirely on their own will are called community disaster volunteers (Meryem, 2022).

Neighborhood Disaster Volunteers

By providing training, equipment, and organization, strengthening the possibility and capability of intervention on a neighborhood basis, especially in the critical hours after the disaster, strengthening the cooperation and coordination with professional teams, raising awareness of the people in the neighborhood about the measures to be taken against disaster risks, and increasing sensitivity, within the scope of the project, in each neighborhood; they are the structures that assist professional teams with approximately 50 trained volunteers who can make the first intervention immediately after the disaster and provide all necessary support (Meryem, 2022).

Disaster Mitigation Teams

Disaster Mitigation Teams, which are designed within the scope of the Istanbul Seismic Risk Mitigation and Emergency Preparedness Project (ISMEP) and supported to be formed in Istanbul province in particular, are local teams formed by people equipped with knowledge, skills, and equipment to respond to disasters in the places where they live and work. These teams are expected not only to respond to disasters (www.gapsel.org) but also to be in a structure that performs activities to reduce disaster damage in the neighborhood where they live and leads society in this regard (Kaya, 2013).

International Institutions and Organizations in Disasters

In a disaster, when a nation's ability to respond falls short, international assistance becomes necessary. The United Nations (UN) is commonly involved as the central coordinating body, with specialized UN agencies addressing the specific needs in a disaster. In addition to the UN, many organizations and associations are involved in international assistance.

UN System

The United Nations (UN), founded on 24 October 1945, is an international organization established to maintain world peace and security and to establish economic, social, and cultural cooperation among nations. The number of members of the organization, whose headquarters is located in New York and Türkiye is among its founding members, is 193. The aims of the organization are:

- To prevent wars and threats to peace
- Establishing friendly relations between countries
- To ensure international economic and social cooperation (TC.DışişleriBakanlığı, 2024).

Organizations Affiliated to the UN System

The humanitarian cluster system created by the UN is a coordination mechanism used by the UN and other humanitarian organizations to respond to crises. It is designed to organize humanitarian actors according to key sectors such as water, health, and food security. The aim is to increase the effectiveness and efficiency of the response, avoid duplication, and ensure that the needs of affected people are met in a timely and appropriate manner (OCHA, 2024).

United Nations Development Programme (UNDP)

UNDP works to eradicate poverty and reduce inequalities and exclusion. It also aims to assist countries in developing policies, leadership skills, partnership capabilities, and building institutional capacity to achieve sustainable development. It was prepared within the scope of the "Post-Earthquake Early Recovery Support Programme for Model Factories and Organized Industrial Zones," which was carried out and funded by the organization after the earthquake that hit 11 provinces in Türkiye on February 6, 2023, and was called the disaster of the century. It also contributes to the provision of accessible and gender-sensitive social services, the design of employment and entrepreneurship support programs, and the capacity building of municipalities in waste management following earthquakes (UNDPTürkiye, 2024).

United Nations Office for the Coordination of Humanitarian Affairs (UNOCHA)

It is an organization within the UN that was created to determine how to respond to situations requiring immediate action, such as civil war and natural disasters. In an emergency, at the request of the government of the affected country, it coordinates response efforts to ensure that aid reaches those most in need. Coordination not only reduces the duplication of response efforts but also ensures that the assistance provided is predictable and prioritized according to needs (OCHA, 2024). It also provides a platform for the coordination of international response activities in the immediate aftermath of a sudden-onset emergency or a rapid change in a complex emergency through Field Operations Coordination Centers (OSOCCs).

OSOCC has three main objectives (OCHA, 2024):

a) Create a link between international responders and the government of the affected country.
b) Provide a system to coordinate and facilitate international relief efforts in a disaster area, especially after an earthquake, where the coordination of many international urban search and rescue teams is critical to ensuring optimal recovery efforts.
c) Provide a platform for cooperation, coordination, and information management between international humanitarian organizations.

United Nations Children's Fund (UNICEF)

UNICEF is a community that strives to ensure that the international community fulfills its promises to children in line with the Convention on the Rights of the Child ratified by the United Nations in 1989 (UNICEF, 2024).

It carries out long-term, international development programs with seven main priorities: child health, education, child protection, mother-to-child HIV/AIDS, access to clean water, nutrition, and enabling children to participate in decisions affecting their lives. It carries out humanitarian aid work by providing support and protection to children in crises, extraordinary situations, and disasters. UNICEF's work is supported by voluntary contributions from individuals and organizations (UNICEF, 2024).

World Food Programme (WFP)

The World Food Programme (WFP) is an aid agency established to provide food aid for humanitarian purposes to populations facing sudden starvation due to natural disasters, wars, famine, or civil conflicts. Like FAO and IFAD, the World Food Programme is based in Rome, Italy. The WFP is the largest international organization in the world with over 20,000 employees (WFP, 2024).

World Health Organization (WHO)

WHO tries to reduce the effects of emergencies and disasters through inter-sectoral cooperation. In particular, it carries out efforts to develop and increase regional, national, and community mechanisms in cooperation with member countries and to strengthen the emergency management capacity of the health sector. Governments, other authorities, and regional organizations cooperate to ensure that limited resources are sent in a coordinated and cost-effective manner. The World Health Organization also carries out efforts to reduce immediate and long-term disabilities, deaths, and diseases associated with emergencies and disasters (Doğan, 2019).

Pan American Health Organization (PAHO)

PAHO is the specialized international health organization for the Americas. It serves as the Regional Office for the Americas of the World Health Organization (WHO), the United Nations' specialized health agency. It works with countries in the region to improve and protect people's health. PAHO cooperates technically with its member countries to combat communicable and non-communicable diseases and their causes, strengthen health systems, and respond to emergencies and disasters. PAHO is committed to ensuring that all people have access to the health services they need, when they need them, of good quality and without fear of falling into poverty. Through its work, PAHO promotes and supports the right of everyone to good Health (PAHO, 2024).

International Search and Rescue Advisory Group (INSARAG)

It was established in 1991 under the umbrella of the United Nations by international USAR (Urban Search and Rescue) teams, who are experts in their field and who operated together during the earthquakes in Mexico in 1958 and Armenia in 1988. It is an intergovernmental humanitarian network of disaster managers, government officials, non-governmental organizations, and USAR (Urban Search and Rescue) practitioners operating under the umbrella of the UN and contributes to the implementation of the International Strategy for Disaster Reduction (ISDR) within its mandate. INSARAG's role is to prepare, mobilize, and coordinate effective and principled international

USAR assistance to support a disaster-affected country in emergencies involving collapsed structures and to support capacity building at international, regional, and national levels (INSIRAG, 2024).

International Committee of the Red Cross (ICRC)

The Red Cross is the largest international organization. Its membership exceeds 250 million, and its annual budget is over 25 billion dollars. The high humanist ideals it serves are embraced almost all over the world. Today, there are Red Cross and Red Crescent organizations in 189 countries worldwide. The mission of the International Committee of the Red Cross is to protect human life and dignity and to assist victims of war and civil conflict throughout the world. The mission of the International Federation of Red Cross and Red Crescent Societies is to mobilize the power of humanism to improve the lives of vulnerable and helpless people all over the world.

North Atlantic Treaty Organization (NATO)

It is an international military alliance founded on the basis of the North Atlantic Treaty signed by 12 countries on April 4, 1949, with 19 more countries joining at different times. NATO was established according to Article 51 of the UN Charter, which affirms the inherent right of sovereign states to individual and collective defense. The aim of the Alliance is to promote "peaceful and friendly relations in the North Atlantic area." NATO is an intergovernmental organization whose position is above that of the member states. NATO can provide emergency aid after disasters and emergencies. Doctors working within NATO can provide services to disaster victims, and engineers can take part in reconstruction work. The other important role of NATO is to respond to possible chemical, biological, and nuclear terrorist attacks (Doğan, 2019).

NATO recognizes that in any disaster response, relief activities will be led jointly by the UN, in particular UNOCHA (United Nations Office for the Coordination of Humanitarian Affairs), and the competent authorities of the affected country. The Euro-Atlantic Disaster Relief Coordination Centre (EADRCC), which includes 26 Allies and 20 partner countries, is the Alliance's main crisis response mechanism. NATO also maintains a UNOCHA liaison officer for information when necessary (Medcalf, 2008).

Discussion and Conclusion

National and international solidarity and a holistic response to disasters are very important to humanity, which has been exposed to many disasters throughout history and has gained great experience in combating disasters. Although the types of disasters are different according to the countries, the effects they cause are largely similar. Whether in natural or man-made disasters, the joint approach and holistic response of institutions and organizations before, during, and after the disaster are of vital importance. While one pillar of this holistic intervention is rescue teams, another pillar is social aid and services. National and international organizations have a great role and importance in the execution of these aid activities. The physical, economic, social, and psychological effects of disasters are seen as problems that societies must overcome first. To minimize the damages of disasters, each society has developed different disaster management systems according to its geographical conditions, experiences, and socio-cultural structures. The understanding of intervention for disasters and disaster management systems is moving towards a common ground with globalization. In addition to this, interventions for disasters have started to be seen as an international problem rather than a national problem.

Disclaimer

None.

References

AFAD. (2023a). Duties and Authorisations. T.C.

AFAD. (2023b). Mission and Vision. T.C.

AFAD. (2024). Search and Rescue Equipment.

AKUT. (2024a). Mission and Values. Retrieved from https://www.akut.org.tr/misyon-degerlerimiz.

AKUT. (2024b). History. Retrieved from https://www.akut.org.tr/tarihce.

Bylaws of the Turkish Red Crescent Society. (2022). Duties of the Red Crescent. Madde 7- (1).

Doğan, H. (2019). Institutions and organisations in disasters. Afetlerde Acil Tıp Hizmetleri, 83-88.

INSIRAG. (2024). Background. Retrieved from https://www.insarag.org/about/background/.

Kaya, M. (2013). Volunteerism and disaster management in Turkish public administration. Institute of Social Sciences.

Kıran, Ş. (2023). National and International Legislation in the Context of Preparedness for Disasters and Emergencies in Health. In A. A. Durmuş, Mustafa (Ed.), *Health Management in Extraordinary Situations* (pp. 319).

Medcalf, J. (2008). Going global or going nowhere?: NATO's role in contemporary international security: Peter Lang.

Meryem, G. (2022). The importance of non-governmental organisations in disaster management. *Journal of Social, Humanities and Administrative Sciences* (JOSHAS), 7(44), 1613-1619.

Ministry of Health. (2024). National Medical Rescue Team Unit (UMKE).

OCHA. (2024). We Coordınate. Retrieved from https://www.unocha.org/we-coordinate.

Official Newspaper. (2009). Afet Ve Acil Durum Yönetimi Başkanlığının Teşkilat Ve Görevleri Hakkında Kanun. T.C.

Official Newspaper. (2018). Law on Certain Regulations Regarding Disaster and Emergency Management Presidency. T.C.

PAHO. (2024). Who We Are. Retrieved from https://www.paho.org/en/who-we-are.

TC. Ministry of Foreign Affairs. (2024). United Nations Organisation and Turkey.

Turkish Red Crescent. (2024). History.

UNDPTürkiye. (2024). Impact of the UN System.

UNDRR. (2023). Uncounted costs. UNDRR News.

UNICEF. (2024). For Every Child, What We Do.

Chapter 23

Disasters and Climate Change

Yakup Sarpdağı[1], PhD
and Metin Yıldız[2,*], PhD
[1]Faculty of Health Sciences, Van Yüzüncü Yıl University, Van, Türkiye
[2]Faculty of Health Sciences, Sakarya University, Sakarya, Türkiye

Abstract

Climate change is not only an environmental problem but also a global security threat with economic damage and social instability caused by increasing natural disasters. In particular, the increase in the use of fossil fuels and the consequent rise in greenhouse gas concentrations in the atmosphere cause temperature increases worldwide. This leads to more frequent and more severe weather events, jeopardizing water and food security, spreading water-borne diseases, and increasing health problems caused by air pollution. These impacts complicate the living conditions of communities living in particularly vulnerable areas and trigger waves of migration, further straining social and economic structures. Increasing investments in renewable energy sources, promoting energy efficiency, and abandoning fossil fuels are of critical importance in tackling this multidimensional crisis. Bringing the international community together to address and adapt to the impacts of global warming is a vital step in overcoming shared challenges. At the same time, comprehensive policies and strategies need to be developed to mitigate the social and economic impacts of climate change. These policies should support affected communities and provide the necessary infrastructure and resources to cope with consequences such as migration and displacement. The steps to be taken in this framework will lead to significant progress in tackling

* Corresponding Author's Email: metinyildiz@sakarya.edu.tr.

In: Disaster Management in the Digital Age and Emerging Technologies
Editors: Ayşe Asiltürk and Ayşegül Sarıoğlu Kemer
ISBN: 979-8-89530-213-2

the challenges posed by climate change and lay the foundation for a more equitable and sustainable world.

Keywords: climate change, disasters, policies

Introduction

The World Health Organization (WHO) recognizes climate change, ozone depletion, loss of biodiversity, soil degradation, and issues with food production systems as significant global environmental threats that impact life (WHO, 2020). Since the mid-1800s, human activities have significantly increased levels of greenhouse gases such as carbon dioxide, methane, and nitrous oxide in the atmosphere of the Earth, resulting in a rise in average temperatures. This warming has led to several adverse effects, including the degradation of soil, reduced agricultural land productivity, desertification, loss of biodiversity, ecosystem damage, dwindling freshwater supplies, ocean acidification, and harm to the stratospheric ozone layer (Rossati, 2017). According to 2019 data from the National Aeronautics and Space Administration (NASA), global temperatures have risen by 1.0 degrees since 1880, based on 136 years of records. It is reported that 17 of the 18 hottest years on record have occurred since 2001. Additionally, Arctic ice has been decreasing by 13.2% per decade, sea levels have been increasing by 3.2 millimeters annually, and the concentration of CO2 in the atmosphere has reached its highest level in 650,000 years (NASA, 2021). As global warming harms the health of the population, mainly through air pollution (mainly through increased CO2 emissions), governments around the world are working to reduce carbon emissions (Hussain and Rehman, 2021; Yoro and Daramola, 2020). Climate change has economic, social, environmental, and health impacts at the national and international levels (IPCC, 2021). A safe environment is a fundamental requirement for health.

The Impact of Climate Change on Natural Disasters

Climate change is recognized as an important determinant of public health (Benevolenza and DeRigne, 2019). Climate change, which has become a global threat, is one of the main issues on the agenda of international politics in recent years. The rise in global temperatures, coupled with the intensification of extreme weather events such as droughts, floods, and storms, underscores the urgent need to focus on climate change. These environmental threats, which concern the whole world and, in this sense, "know no borders," are now considered a security issue and are associated with different security issues (Keser et al., 2024). Climate change can be described as a global issue of human origin, stemming from the disruption of nature's self-renewal capabilities (Metin, 2023). According to the report of the IPCC (2021), since 1850, the last 40 years have been warmer than the previous 40 years (IPCC, 2021).

Indeed, historical observations on climate change reveal that meteorological conditions are constantly evolving, bringing with them a range of issues, including socio-economic, environmental, and social challenges (Öztürk, 2002). According to Aydın and Genç, in the last 50 years, climate change-related disasters have caused great economic damage and loss of life worldwide. In Africa, 731,724 people lost their lives in 1,692 disasters and USD 38 billion in damages occurred. In North America, Central America, and the Caribbean, 74,833 people died in 1,974 disasters, and USD 1,655 billion was lost. Hurricanes Harvey, Maria, and Irma in 2017 caused the most damage. This particularly affected poor countries that are economically vulnerable and dependent on nature. Many parts of South America, Latin America, South and Southeast Asia, and Africa were affected (Aydın and Genç, 2023). Flood disasters, forest fires, and hot weather fluctuations in Türkiye are the effects of climate change. Fires that broke out in 54 different provinces in Türkiye in 2021 affected approximately 150 thousand hectares of land (ÇŞB, 2021). Natural disasters wreak havoc in societies, causing significant loss of life, material damage, and shifts in power changes. Since the 1970s, although there have been advancements in early warning systems and recovery efforts post-disaster, the frequency of natural disasters around the world has risen markedly, along with the economic costs associated with them (Yamamura, 2015; Coronese et al., 2019). According to the World Meteorological Organization (WMO), 79% of the disasters that occurred between 1970 and 2019 were caused by climate change (WMO, 2020). These changes are recognized for their adverse effects on human health globally (WHO, 2018b). Climate change represents a growing threat today with serious implications for public health. The increased frequency and intensity of natural disasters, including extreme weather events, floods, droughts, and forest fires, highlight the severity of the issue.

Global Warming and Health Effects

Global warming refers to the ongoing rise in the average temperature of Earth's atmosphere and oceans. The average surface temperature of the Earth has increased by approximately 0.8°C (1.4°F), with about two-thirds of this increase occurring since 1980 (Singh and Purohit, 2014). The consequences of climate change are now clearly visible. They include more severe weather events such as scorching heat waves, extremely hot days, hurricanes, storms, floods, regional flooding, and extreme droughts. These events result in a range of impacts, including food insecurity, migration of people from damaged agricultural areas, air pollution, coral reef bleaching and mortality, and loss of aquaculture (Rom and Pinkerton, 2021).

The recently released Sixth Assessment Report of the IPCC sets out the factors attributable to human-induced climate change, highlighting a range of mechanisms that negatively impact physical and mental health globally. These impacts include increased temperature events leading to higher rates of mortality and morbidity, a rise in food and waterborne diseases, the widespread occurrence of vector-borne diseases, the emergence of new zoonotic diseases, and the degradation of water quality. Additionally, there is an exacerbation of health problems associated with temperature fluctuations and extreme weather events (Parmesan et al., 2022). The most important factor for humans and other living beings is, undoubtedly, health. Human health has a structure that can be easily affected and changed by any factor in the environment. The direct impact of global warming and climate change on human health is manifested through heat waves, floods, storms, and other extreme climatological events. Indirectly, global warming and climate change affect human health through increased exposure to infectious diseases, which can be linked to water use and food consumption. The problems arising from global warming and climate change do not affect human communities in the same way (Çeçen and Güvenç, 2022; Ciscar et al., 2011; Council, 2007). All these factors impact human health, leading to non-communicable diseases such as injuries sustained during natural disasters, malnutrition during periods of famine, and increased mortality due to complications in chronic patients during heat waves (Rossati, 2017). One of the primary challenges in linking mortality rates with climate change lies in establishing standard definitions for what constitutes a heatwave and what defines a cold snap (Xu et al., 2016; Yang et al., 2019). Heat waves killed more than 70,000 people in Europe in 2003 and more than 11,000 in Russia in 2010 (Robine et al., 2007; Shaposhnikov et al., 2014). A study conducted in France between 2000 and

2010 showed that 3.9% of total deaths were caused by cold and 1.2% by heat (Pascal et al., 2018).

Global warming is a major environmental concern, as it harms the health of human society (Lau et al., 2022). In addition to the direct effects of temperature increases on human health, global warming negatively impacts primary producers and livestock, potentially leading to malnutrition. This, in turn, can cause a host of health-related issues. The situation is further exacerbated by environmental pollution (Calleja-Agius et al., 2021). Global warming, worsened by environmental pollution, is linked to an increase in adverse health impacts.

Climate Refugees

Climate change is considered the greatest threat to public health in the coming decades, as the environmental changes it brings about are expected to lead to significant population changes (Cegolon et al., 2017). Climate change contributes to resource insecurity, increased social unrest, the spread of poverty, decreased physical security, and increased migration. These challenges are interconnected, exacerbating vulnerabilities across societies globally (Vural, 2018). Global climate change causes migration in different parts of the world, which directly or indirectly affects national and human security (Gönenç and Kibaroğlu, 2017). Climate change-induced migrations can be both temporary (e.g., people fleeing severe storms or droughts) and permanent (when a region, a municipality, a city, or even a nation becomes uninhabitable, such as in small island countries) (Docherty and Giannini, 2009). As climate change and environmental degradation become prominent causes of migration and displacement, the concept of a 'climate refugee' is gaining traction. This term is used to describe individuals who migrate due to environmental changes in their habitats, such as extreme weather events, rising sea levels, and degradation of land and resources (Hiraide, 2023). Climate refugees are people who have been forced to leave their homes or countries due to the impacts of severe climate events, and who have had to rebuild their lives elsewhere despite the conditions to which they have been exposed (Berchin et al., 2017). Climate change is intensifying extreme weather events, which in turn trigger migration and displacement. As a result, the issue of climate refugees is drawing increasing attention worldwide, highlighting the need for comprehensive strategies to address their challenges and support their resilience (Bettini, 2013; Reuveny, 2007). In conclusion, climate change has a significant impact on migration and displacement caused by environmental changes. Climate events such as increasing temperatures, severe storms, and droughts force people to leave their homes, thus creating the phenomenon of climate refugeeism.

Renewable Energy Sources in the Fight Against Climate Change

Energy is crucial to meeting the basic needs of people and facilitating economic activities, as well as shaping the social and economic development levels of countries. It plays a central role in everything from household comfort and security to national industrial and technological advancement (Bayrak and Esen, 2014). Today, increasing population growth and technological advancements are driving up the demand for energy. This surge in demand highlights several critical issues: the potential insufficiency of finite fossil fuel resources in the future, the high dependency on foreign energy, supply security concerns due to political instability in regions with strategic resources, and the global negative impacts of climate change. These factors collectively underscore the growing importance of energy in the global context (Öymen, 2020). With low-carbon economy policies, the use of renewable energy sources on a global scale is increasing its share day by day. Incentives in countries associated with the International Energy Agency (IEA) have successfully doubled the share of renewable energy sources from 1970 to 2002, reaching an annual growth rate of 5.7%. Driven by ongoing policies and initiatives, this share is projected to increase to 60% by the year 2030, demonstrating a significant change toward renewable energy in response to global energy and environmental challenges (İEA, 2015).

For climate change management efforts to sustain success, it is crucial to increase the proportion of renewable energy in electricity generation while concurrently ramping up its utilization in transportation, heating, and cooling endeavors (Güner and Turan, 2017). Over the years, numerous strategies, agreements, laws, and policies concerning renewable and green energy sources, climate change mitigation, greenhouse gas emissions, and sustainable development have been ratified on the international stage. In this evolving global landscape, the ongoing advancement of green energy technologies and market diversification assume a pivotal role in the global energy sector (Masini and Menichetti, 2013). More than 90% of countries worldwide are actively striving to expand their renewable energy sources (Owusu and

Asumadu-Sarkodie, 2016). What needs to be done for a sustainable energy future has been categorized under five headings by the International Renewable Energy Agency (IRENA) (IRENA, 2016). These include;

- Reinforcing commitment to renewable energy policies.
- Mobilizing investments in renewable energy.
- Establishing an institutional and technical infrastructure to facilitate the deployment of renewable energy.
- Demonstrating the advantages of renewable energy for sustainable development.
- Strengthening regional involvement and fostering international cooperation in renewable energy development.

Renewable energy sources naturally replenish themselves without depletion on Earth. They include bioenergy, hydropower, geothermal energy, solar energy, wind energy, and ocean energy (including tidal and wave energy) (Owusu and Asumadu-Sarkodie, 2016).

In conclusion, the utilization of renewable energy sources is paramount in the battle against climate change, as it curtails greenhouse gas emissions while fulfilling global energy demands. The broader adoption of these resources diminishes reliance on fossil fuels, concurrently yielding economic, social, and environmental advantages.

Convention and Protocol on Global Climate Change

In recent years, the increase in the number of environmental problems has increased the number of attempts to find solutions to these problems at the bilateral, regional, and global levels. At the same time, the fact that most of the environmental problems such as ozone depletion, greenhouse effect, desertification, pollution of seas and oceans, deforestation, etc., are international in nature, affecting not only the whole world but also space, and the consequences of transboundary movements have triggered the necessity of establishing close international cooperation in eliminating the problems (Sürmelioğlu, 2012). The escalating environmental degradation stems primarily from several factors: the rapid rise in the human population, the expanding reach of consumption patterns, and the exponential depletion of energy and various resources. On the one hand, the increase in the human population has led to an overall increase in global consumption. Conversely, the elevation in per capita resource and energy consumption, propelled by heightened global prosperity, has intensified pressure on the world's resources (Ünal, 2014).

Especially since the 1970s, it is seen that all these initiatives, both in terms of action and thought generation, have started to be made at the global level. In this context, it is evident that the founding of Greenpeace (Green Peace) in 1972, the emergence of environmentalist parties in several countries during these years, environmental programs conducted by the United Nations, and the inaugural environmental conference held in Stockholm in 1972 followed by the second in Rio in 1992 all signify significant milestones in the global environmental movement (Arı, 2018). It is stated that there are three cornerstones of climate change negotiations. These include the United Nations Framework Convention on Climate Change, initiated for signature at the Earth Summit in Rio de Janeiro in 1992 and enacted in 1994; the Kyoto Protocol opened for signature in 1997 and implemented in 2005; and, lastly, the 2015 Paris Climate Summit (COP21), marking a pivotal global stride in addressing climate issues (Güner and Turan, 2017).

United Nations Framework Convention on Climate Change

Concerns about climate change began to surface as a matter of public and political concern during the 1960s and 1970s, ultimately leading to the negotiation of the United Nations Framework Convention on Climate Change (UNFCCC) during the United Nations Conference on Environment and Development (UNCED) in 1992. The Convention comprises 26 articles and serves as a declaration asserting that alterations in the global climate and their adverse repercussions are a shared concern of humanity. Given the global scope of climate change, it is acknowledged that all nations bear common but differentiated responsibilities, and efforts should be made to achieve the highest level of unity in line with their social and economic circumstances. It was determined that necessary measures should be taken to comprehend and address climate change while safeguarding the environment for current and future generations (ÇŞB, 2016). The aim of the Convention is to facilitate the natural adaptation of ecosystems to climate change, prevent disruptions in food production, and mitigate anthropogenic harm to the climate system (to a degree that is preventable), thereby enabling sustainable economic development to persist (ÇŞB, 2016). Following the entry into force of the

UNFCCC in 1994, the Conference of the Parties became an annual occurrence. Among these gatherings, two stand out prominently: the Third Conference of the Parties in 1997 and the Twenty-First Conference of the Parties in 2015. The Kyoto Protocol was established during COP3, held in Kyoto, Japan, in 1997, becoming an integral component of the UNFCCC framework. Subsequently, the Paris Agreement was signed during COP21, marking a significant milestone in global climate action (Yıldız and Yaşar, 2009).

Kyoto Protocol

In response to doubts regarding the adequacy of the United Nations Framework Convention on Climate Change, efforts were made to address these concerns through the adoption of the Kyoto Protocol during the Third Conference of the Parties (COP3) in Kyoto, Japan, in 1997. This agreement, primarily aimed at curbing global warming by reducing the consumption of fossil fuels, came into effect in 2005 (Yıldız and Yaşar, 2009). The Kyoto Protocol includes steps to regulate the carbon market to meet obligations (Ediger, 2008). The difference between the Kyoto Protocol and the UNFCCC is that it includes binding obligations for industrialized countries in greenhouse gas mitigation. The Protocol aims to uphold the commitment to reduce greenhouse gas emissions by 5% below 1990 levels during the period from 2008 to 2012, followed by an 18% reduction in emissions between 2013 and 2020 (Kaya, 2020). However, in the period 2008-2012, 36 countries made commitments to achieve GHG emissions at 1990 levels, which was partially achieved. Second, in the 2013-2020 period, the parties' obligations to continue GHG mitigation were not accepted by some countries (Kaya, 2020).

Paris Climate Summit

The 21st Conference of the Parties (COP21) was held in Paris in 2015. The main topic of COP21 was how to proceed after the Kyoto Protocol expires in 2020. The need for an agreement to cover the period after 2020 was suggested by the conference. In 2015, the Paris Agreement was unanimously adopted by 195 parties to the UN Framework Convention on Climate Change. This agreement gained greater acceptance among the parties compared to the Kyoto Protocol. The Paris Agreement outlines measures to cap the global temperature increase to 2 degrees Celsius, with efforts to limit it even further, aiming for below 1.5 degrees Celsius if feasible. Moreover, the agreement entails financial assistance from developed nations to support developing countries in their efforts to mitigate greenhouse gas emissions (Kaya, 2020).

In Türkiye, the Climate Change Convention recognizes that greenhouse gas emissions increase with increasing industrialization. This framework divides countries into three categories, Annex-1, Annex-2, and non-Annex according to their level of development. Annex-1 countries have obligations to limit emissions, protect greenhouse gas sinks, and implement national climate policies. Annex-2 countries are obliged to provide financial resources and technology transfer to meet the needs of developing countries. Although Türkiye was initially included in both lists, at the COP7 meeting in 2001, it was removed from the Annex-II list due to a special circumstance but remained on the Annex-I list (ÇŞİDB, 2002). According to Özışık (2020), Türkiye has regularly participated in international negotiations on climate change, starting with the 1992 Rio Conference, during which time it ratified the Kyoto Protocol in 2009 and became a party to the protocol. In 2015, Türkiye signed the Paris Agreement adopted at the Conference of the Parties in Paris, although it has not yet been ratified. These agreements take important steps globally with goals such as reducing greenhouse gas emissions, limiting temperature rise, and providing financial support. Türkiye has also participated in international climate negotiations, strengthened its environmental policies, and developed various action plans.

Discussion and Conclusion

While climate change is recognized as an international security issue worldwide, the frequency and severity of natural disasters have increased in recent years, magnifying economic losses. Global warming causes an increase in atmospheric and ocean temperatures due to the use of fossil fuels and greenhouse gas emissions. This increase leads to many negative impacts, such as severe weather events, water and food-borne diseases, and environmental pollution. Climate change has substantial impacts on human health, including increased risks of heat waves, environmental pollution, and the proliferation of diseases. Additionally, it significantly contributes to migration and displacement, posing threats to both national and human security. Urgent actions and policies are imperative on a global scale to mitigate and adapt to climate change. Increasing the utilization of renewable energy sources and reducing the reliance on fossil fuels represent crucial steps in this endeavor. Moreover, at the international level, there is a pressing need to formulate strategies and policies for the management of climate change, fostering technological advancements, and enhancing global cooperation. These efforts will provide significant progress in combating climate change and lay the foundation for a sustainable future.

Disclaimer

None.

References

Arı, T. (2018). *Uluslararası İlişkiler Teorileri* (T. Arı (ed.); 9th ed.). Aktüel Yayınları.

Aydın, G., & Genç, F. N. (2023). Sürdürülebilir Kalkınma Bağlamında Afet ve İklim Değişikliği Politikaları. *Cankiri Karatekin Universitesi Iktisadi ve Idari Bilimler Fakultesi Dergisi*, *13*(2), 563–590. https://doi.org/10.18074/ckuiibfd.1176640.

Bayrak, M., & Esen, Ö. (2014). Türkiye'nin enerji açığı sorunu ve çözümüne yönelik arayışlar. *Atatürk Üniversitesi İktisadi ve İdari Bilimler Dergisi*, *28*(3), 139-158.

Benevolenza, M. A., & DeRigne, L. (2019). The impact of climate change and natural disasters on vulnerable populations: A systematic review of literature. *Journal of Human Behavior in the Social Environment*, *29*(2), 266–281. https://doi.org/10.1080/10911359.2018.1527739.

Berchin, I. I., Valduga, I. B., Garcia, J., & de Andrade Guerra, J. B. S. O. (2017). Climate change and forced migrations: An effort towards recognizing climate refugees. *Geoforum*, *84*(2017), 147–150. https://doi.org/10.1016/j.geoforum.2017.06.022.

Bettini, G. (2013). Climate Barbarians at the Gate? A critique of apocalyptic narratives on 'climate refugees.' *Geoforum*, *45*, 63–72. https://doi.org/10.1016/j.geoforum.2012.09.009.

Calleja-Agius, J., England, K., & Calleja, N. (2021). The effect of global warming on mortality. *Early Human Development*, *155*, 105222. https://doi.org/10.1016/j.earlhumdev.2020.105222.

Çeçen, Z., & Güvenç, F. (2022). Cilt: 4, Sayı: 1, 14-25. *SDU Healthcare Management Journal*, *4*(1), 14–25.

Cegolon, L., Heymann, W. C., & Lange, J. H. (2017). Climate change, emerging infections and blood donations. *Journal of Travel Medicine*, *24*(3), taw098. https://doi.org/10.1093/jtm/taw098.

Ciscar, J.-C., Iglesias, A., Feyen, L., Szabó, L., Van Regemorter, D., Amelung, B., Nicholls, R., Watkiss, P., Christensen, O. B., Dankers, R., Garrote, L., Goodess, C. M., Hunt, A., Moreno, A., Richards, J., & Soria, A. (2011). Physical and economic consequences of climate change in Europe. *Proceedings of the National Academy of Sciences, 108*(7), 2678–2683. https://doi.org/10.1073/pnas.1011612108.

Coronese, M., Lamperti, F., Keller, K., Chiaromonte, F., & Roventini, A. (2019). Evidence for sharp increase in the economic damages of extreme natural disasters. *Proceedings of the National Academy of Sciences, 116*(43), 21450–21455. https://doi.org/10.1073/pnas.1907826116.

Council, N. R. (2007). *Evaluating Progress of the U.S. Climate Science Program: Methods and Preliminary Results*. The National Academies Press.

ÇŞB, Ç. ve Ş. B. (2016). *Birleşmiş Milletler İklim Değişikliği Çerçeve Sözleşmesi.*

ÇŞİDB, Ç. Ş. ve İ. D. B. (2002). *Birleşmiş Milletler İklim Değişikliği Çerçeve Sözleşmesi.*

Docherty, B., & Giannini, T. (2009). Confronting a rising tide: a proposal for a convention on climate change refugees. *Harvard Environ. Law Rev, 33*(2009), 349–403.

Ediger, V. (2008). Küresel İklim Değişikliğinin Uluslararası İlişkiler Boyutu ve Türkiye'nin Politikaları. *Mülkiye, XXXII*(259), 133–157.

Gönenç, D., & Kibaroğlu, A. (2017). İklim Güvenliği Kavramının Türkiye İklim Politikasındaki Yeri. *Alternatif Politika, İklim Değişikliği ve Enerji Özel Sayı*, 1–26.

Güner, E. D., & Turan, E. S. (2017). Yenilenebilir Enerji Kaynaklarının Küresel İklim Değişikliği Üzerine Etkisi. *Doğal Afetler ve Çevre Dergisi, 1*(3), 48–55. https://doi.org/10.21324/dacd.285523.

Hiraide, L. A. (2023). Climate refugees: A useful concept? Towards an alternative vocabulary of ecological displacement. *Politics, 43*(2), 267–282. https://doi.org/10.1177/02633957221077257.

Hussain, I., & Rehman, A. (2021). Exploring the dynamic interaction of CO2 emission on population growth, foreign investment, and renewable energy by employing ARDL bounds testing approach. *Environmental Science and Pollution Research, 28*(29), 39387–39397. https://doi.org/10.1007/s11356-021-13502-8.

İEA. (2015). Energy Climate and Change, World Energy Outlook Special Report, International Energy Agency, France. In *Economic Survey*.

IPCC. (2021). *Climate Change 2021 The Physical Science Basis Summary for Policymakers.*

IRENA. (2016). *Renewable Capacity Statistics 2016, International Renewable Energy Agency.*

Kaya, H. E. (2020). Kyoto'dan Paris'e Küresel İklim Politikaları. *Uluslararası Sosyal ve Stratejik Araştırmalar Dergisi, 4*(10), 165–191.

Keser, A., Köse, Z., & Aliyev, P. (2024). Doğal Afetlerin Tetikleyicisi Olarak İklim Değişikliği İle Gıda Güvenliği Arasındaki İlişkinin İncelenmesi: Sahra Altı Afrika Örneği. *Akademik Yaklaşımlar Dergisi, 15*(1-Deprem Özel Sayısı-), 215–234. https://doi.org/10.54688/ayd.1412901.

Lau, H. C., Zhang, K., Bokka, H. K., & Ramakrishna, S. (2022). A Review of the Status of Fossil and Renewable Energies in Southeast Asia and Its Implications on the Decarbonization of ASEAN. *Energies, 15*(6), 2152. https://doi.org/10.3390/en15062152.

Masini, A., & Menichetti, E. (2013). Investment decisions in the renewable energy sector: An analysis of non-financial drivers. *Technological Forecasting and Social Change, 80*(3), 510–524. https://doi.org/10.1016/j.techfore.2012.08.003.

Metin, S. (2023). Afrika'da iklim değişikliğinin etkileri: İklim adaleti ve göç. *Akdeniz Havzası ve Afrika Medeniyetleri Dergisi*, *5*(2), 99–118.

NASA. (2019). *Analyses Reveal 2019 Second Aeronautics and Space Administration.* https://climate.nasa.gov/.

NASA. (2021). *Overview: Weather, global warming and climate change.* https://climate.nasa.gov/resources/global-warming-vs-climate-change/.

Owusu, P. A., & Asumadu-Sarkodie, S. (2016). A review of renewable energy sources, sustainability issues and climate change mitigation. *Cogent Engineering*, *3*(1), 1167990. https://doi.org/10.1080/23311916.2016.1167990.

Öymen, G. (2020). Yenilenebilir Enerjinin Sürdürülebilirlik Üzerindeki Rolü. *İstanbul Ticaret Üniversitesi Sosyal Bilimler Dergisi*, *19*(39), 1069–1087. https://doi.org/10.46928/iticusbe.769022.

Özışık, F. U. (2020). Türkiye'de Bir Kamu Politikası Olarak İklim Değişikliği: Tarihsel Gelişim, Uluslararası Müzakereler, Yapısal ve İdari Sorunlar Çerçevesinde Bir Değerlendirme. *Marmara Üniversitesi Siyasal Bilimler Dergisi*, *8*(1), 66–96. https://doi.org/10.14782/marmarasbd.712908.

Öztürk, K. (2002). Küresel İklim Değişikliği ve Türkiye'ye Olası Etkileri. *Gazi Eğitim Fakültesi Dergisi*, *22*(1), 47–65.

Parmesan, C., Morecroft, M. D., & Trisurat, Y. (2022). *Climate Change 2022:Impacts, Adaptation and Vulnerability.*

Pascal, M., Wagner, V., Corso, M., Laaidi, K., Ung, A., & Beaudeau, P. (2018). Heat and cold related-mortality in 18 French cities. *Environment International*, *121*, 189–198. https://doi.org/10.1016/j.envint.2018.08.049.

Reuveny, R. (2007). Climate change-induced migration and violent conflict. *Political Geography*, *26*(6), 656–673. https://doi.org/10.1016/j.polgeo.2007.05.0.01.

Robine, J.-M., Cheung, S. L. K., Le Roy, S., Van Oyen, H., Griffiths, C., Michel, J.-P., & Herrmann, F. R. (2007). Death toll exceeded 70,000 in Europe during the summer of 2003. *Comptes Rendus. Biologies*, *331*(2), 171–178. https://doi.org/10.1016/j.crvi.2007.12.001.

Rom, W. N., & Pinkerton, K. E. (2021). Introduction: Consequences of Global Warming to Planetary and Human Health. In *Climate Change and Global Public Health* (pp. 1–33). https://doi.org/10.1007/978-3-030-54746-2_1.

Rossati, A. (2017). Global Warming and Its Health Impact. *The International Journal of Occupational and Environmental Medicine*, *8*(1), 7–20. https://doi.org/10.15171/ijoem.2017.963.

Shaposhnikov, D., Revich, B., Bellander, T., Bedada, G. B., Bottai, M., Kharkova, T., Kvasha, E., Lezina, E., Lind, T., Semutnikova, E., & Pershagen, G. (2014). Mortality Related to Air Pollution with the Moscow Heat Wave and Wildfire of 2010. *Epidemiology*, *25*(3), 359–364. https://doi.org/10.1097/EDE.0000000000000090.

Singh, A., & Purohit, B. M. (2014). Public Health Impacts of Global Warming and Climate Change. *Peace Review*, *26*(1), 112–120. https://doi.org/10.1080/10402659.2014.876326.

Sürmelioğlu, D. (2012). Uluslararası Çevre Hukukunda Güncel Bir Sorun: Sınıraşan Çevresel Zararlar. In M. Saraçlı (Ed.), *Uluslararası Hukukta Güncel Sorun Alanları* (1st ed., p. 416). Murat Saraçlı.

Ünal, S. (2014). Yeşil Teori. In M. Şahin & O. Şen (Eds.), *Uluslararası İlişkiler Teorileri Temel Kavramlar* (1st ed., p. 352). Kripto Yayınları.

Vural, Ç. (2018). Küresel İklim Değişikliği ve Güvenlik. *Güvenlik Bilimleri Dergisi, 7*(1), 57–85.

WHO. (2020). *WHO Global Strategy on Health, Environment and Climate Change and Wellbeing Sustainably Through Healthy.*

WHO. (2018b). *Climate Change and Health in Small Island Developing States. A WHO Special Initiative. Geneva.*

WMO. (2020). *State Of Clımate Servıces. Rısk Informatıon and Early Warnıng Systems.*

Xu, Z., FitzGerald, G., Guo, Y., Jalaludin, B., & Tong, S. (2016). Impact of heatwave on mortality under different heatwave definitions: A systematic review and meta-analysis. *Environment International, 89–90*, 193–203. https://doi.org/10.1016/j.envint.2016.02.007.

Yamamura, E. (2015). The Impact of Natural Disasters on Income Inequality: Analysis using Panel Data during the Period 1970 to 2004. *International Economic Journal, 29*(3), 359–374. https://doi.org/10.1080/10168737.2015.1020323.

Yang, J., Yin, P., Sun, J., Wang, B., Zhou, M., Li, M., Tong, S., Meng, B., Guo, Y., & Liu, Q. (2019). Heatwave and mortality in 31 major Chinese cities: Definition, vulnerability and implications. *Science of The Total Environment, 649*, 695–702. https://doi.org/10.1016/j.scitotenv.2018.08.332.

Yıldız, D., & Yaşar, D. (2009). *Küresel Isıtılan Dünya ve Su* (D. Yıldız & D. Yaşar (eds.); 1st ed.). Truva Yayınları.

Yoro, K. O., & Daramola, M. O. (2020). CO_2 emission sources, greenhouse gases, and the global warming effect. In *Advances in Carbon Capture* (pp. 3–28). Elsevier.

Chapter 24

Natural Disaster Notifications and Early Warning Systems

Imran Aslan[1,*], PhD
and Özge Karanfil[2], PhD
[1]Department of Health Management, Faculty of Health Science, Bingöl University, Bingöl, Türkiye
[2]College of Administrative Sciences and Economics and School of Medicine, Koç University, İstanbul, Türkiye

Abstract

An increasing number of disasters, such as floods, are known to be linked to climate change. Some of the major hazards include earthquakes, water scarcity, limited access to clean water, psychological trauma, famine, and malnutrition. Early warning systems (EWS), which are based on advanced technologies such as artificial intelligence (AI), satellite observations, and advanced modeling tools, have been widely used for the prediction of disasters for early notification and mitigation of their impact on people and the environment. These technologies can help to collect data on soil moisture, land movement, rainfall intensity, detection of high-risk areas, etc. for effective disease impact management, comprehensive disaster mitigation, identification of regions prone to slope failure, landslides, and road damage, provision of critical flow for real-time modeling for treatment facilities or hospitals in case of disasters, and use of IoT (Internet of Things) controlled devices such as gates, valves, and pumps to prevent and combat natural disasters. Our

* Corresponding Author's Email: imranaslan@gmail.com.

In: Disaster Management in the Digital Age and Emerging Technologies
Editors: Ayşe Asiltürk and Ayşegül Sarıoğlu Kemer
ISBN: 979-8-89530-213-2

aim in this chapter is to define the types of key technologies for disaster/emergency preparedness and mitigation.

Keywords: artificial intelligence, disasters, disaster preparedness, early warning systems, remote sensing

Introduction

Disasters are categorized as natural disasters and man-made disasters which are a challenge to all aspects of life (Araz et al., 2020). Earthquakes, floods, volcanic tsunamis, cyclones, typhoons, hurricanes, heavy rainfall, snowmelt, glacial collapse, and water-related climatic disasters, posing either immediate or long-term risks to human health, the environment and infrastructure, have caused many deaths, injuries, and economic losses around the world, affecting millions of people. The impacts of major hydro-meteorological hazards such as tropical cyclones, heavy rainfall, heat waves, cold waves, fog and haze, thunderstorms, and lightning and flash floods on agriculture, tourism, shipping, energy, air quality, transport, and health need to be assessed by governments to reduce negative impacts.

In 2020, it was estimated that weather-related disasters caused about 15,000 deaths, affected 98 million people, and cost US$ 171 billion (UN Office for Disaster Risk Reduction, 2021). To reduce the impact of disasters on the environment and people, well-developed early warning systems (EWS) can be developed. To decrease risks in advance and take timely action, EWS integrates hazard monitoring, forecasting, risk assessment, communication, and disaster preparedness. The mandate of participating agencies and data sharing mechanisms, as well as accountability for false or over-warnings, need to be clarified in EWSs. The role of the local population in the design and operation of the EWS is of crucial importance for the success of the emergency preparedness plans (Dash, 2023). The integration of disruptive technologies such as the Internet of Things (IoT), drones, imaging, big data, machine learning, social media, robotics and blockchain, artificial intelligence (AI), big data, and smartphone applications can be applied to disaster management of floods, earthquakes, wildfires, and other natural disasters as a part of EWS (Munawar et al., 2022). Analytical tools are used in operational risk management to support decision-making and to develop different strategies (Araz et al., 2020). This chapter discusses disasters requiring EWS and their applications.

Early Warning Systems for Disasters

EWS includes the risk of knowing, followed by monitoring, disseminating, and communicating, and the ability to respond. Assessing, communicating, and preparing for disasters, as well as monitoring, forecasting, and predicting hazards, can reduce the impact of disasters before they occur. Information from research projects, institutions, organizations, and government agencies, such as warnings, reports, open-source databases, satellite imagery, tools, and model services, is available to EWSs (Girotto et al., 2023). The first step is to collect, assess, and share hazard and vulnerability data. This can be followed by the development of an effective hazard monitoring and warning service based on sound scientific and technological criteria. Later, advanced systems for communicating and disseminating natural hazard information/warnings can be prepared. Finally, community empowerment can enable an effective disaster response (Sahana et al., 2023).

Flood risk reduction efforts can greatly benefit from the involvement of non-governmental organizations (NGOs), which increase the willingness to take rapid preventive action before and after floods. National Seismic and Tsunami Early Warning Centre (NSTEWC), Tropical Cyclone Warning Centre (TCWC), Flood Forecast and Warning Centre (FFWC), and National Drought Monitoring Centre (NDMC) systems have been developed in many countries for disaster preparedness and response. Power outages, lack of TV/radio access, and poor mobile phone signals during natural disasters must be managed. Suggested solutions include using sirens and loudspeakers and organizing evacuation routes, especially for women, young children, and the elderly (Shah et al., 2023).

Simulations, dynamic 3D flood risk level maps, geospatial data, satellite observations and advanced modeling tools, unmanned marine vehicle and unmanned aerial vehicles (UAV), aircraft, helicopters, blimp data collection, artificial intelligence (AI) technologies such as Internet of Things (IoT) and machine learning, big data, digital terrain model, digital elevation model, 5G networks, hydrological, augmented reality, virtual reality, meteorological observations, satellite-based near real-time monitoring, radio frequency and electroconductive sensors, stationary surveillance cameras, ultrasonic and infrared sensors, ultrasonic-equipped radar, etc. can be used for EWS to manage disasters (Aslan & Goundar, 2022; Munawar et al., 2022).

Intelligent early warning platforms, called digital safety early warning systems, have recently gained popularity. Examples include fire warning applications and human-centered EWS. As a synergy between coastal ecology and disaster mitigation, marine ecosystem protection and restoration efforts can help prevent disasters. Before a disaster, AI tools can be used by power companies to predict the location of possible damage and the duration of power outages. During disasters, industrial robotics can be used with tool cameras, GPS, and inertial measurement units for search and rescue. The IoT can also be used for disaster management, such as routine monitoring of natural events, sending alerts, and alerting authorities. As part of remote data transmission to cloud servers, IoT systems can assist with data collection, analysis, early warning notifications, remote monitoring of events, victim location, and real-time data generation. AI used in weather forecasting can be used to predict and provide reliable information for timely decision-making in mitigation, preparedness, response, and recovery. The "Help Me" notification, first aid application, tsunami alert, rescue and disaster management reporting, and Emergency+ based on GPS are some of the smartphone applications that are used during disasters (Munawar et al., 2022).

Urbanization, land use change, environmental degradation, climate change, etc. are other factors that affect disaster risk. Predicted rainfall can be used to forecast impact-based flooding. Expected losses should be estimated on the basis of risks and their impacts (Dash, 2023). A fast information system can lead to fast and accurate decision-making during disasters machines or systems based on neutral algorithm AI technologies called new forms of automated processes - Zero Touch Technologies (ZTT) (Meilani & Hardjosoekarto, 2020). To predict the impact and consequences of the famine on people, the Famine Early Warning System Network (FEWS NET) has been developed. The Impact-based Forecast and Warning (IbFW) framework will be developed to predict the impact of disasters on transport, agricultural productivity, energy supply, and demand, health epidemics, and water resource management (Dash, 2023).

Earthquake Disaster

Prediction of impending disasters and early warning to the public can be efficient applications of AI for the safe evacuation of the affected region. Earthquake Early Warning Systems (EEWS) can be used to quickly alert affected areas so that actions can be taken, such as opening building exits, moving escalators toward exits, cutting power, slowing or stopping high-speed trains, etc. A seismic station will be in place for the detection of the initial energy at the surface for the estimation of the magnitude of the earthquake. A reporting system called "Toretter" was developed to estimate the location of earthquakes and transmit information faster than the metrology department and using smartphone accelerometer measurements of tilting movements can be way to detect earthquakes (Munawar et al., 2022). Machine learning and AI can be used for the identification, assessment, monitoring, early warning, and prediction of earthquake hazard chains. High-altitude remote chain hazard mechanisms, active tectonic features, and the application of big data, optical remote sensing, InSAR technology, and UAV antennas have been used to develop earthquake hazard models. Network design, station placement, and real-time processing are important factors affecting the estimation performance (Chen, 2023; Munawar et al., 2022; Tunç et al., 2023).

Earth's ionosphere, as part of our atmosphere, extends from an altitude of about 50km to over 1000km. Signals are sent from a satellite through the ionosphere to a receiver on the Earth's surface by the Global Navigation Satellite System (GNSS). For ionosphere-related research and applications, the Global Ionospheric Map (GIM) is produced by the International GNSS Services (IGS). Prior to a major earthquake, abnormal changes in geoelectricity, geomagnetism, crustal deformation, subsurface fluid, geochemistry, and other means are expected. Space exploration technology has been successful in the detection of ionospheric changes prior to earthquakes. Various changes have been measured through satellites by sharing services with dedicated lines and websites (www.leos.ac.cn) in China. CSES-01 is China's first satellite for geophysical measurements and stereoscopic monitoring of earthquakes, studying changes in the ionosphere before and after earthquakes (Huang et al., 2023). The EEWS issued an earthquake warning after the earthquake struck, and a tsunami warning was issued in Japan for the entire Tōhoku region when the Mw 9.0 earthquake struck at 14:46:23JST on 11 March 2011. Media: radio warnings, official word of mouth, and sirens were also used to warn people to move to higher ground as part of disaster preparedness for the tsunami (Naylor et al., 2013). BMKG's EEWS was another application in Indonesia (Meilani and Hardjosoekarto, 2020).

Flood Early Warning Systems

Disasters caused by sudden and extreme rainfall and flooding in urban areas have resulted in huge losses of human life and property. Therefore, the identification of the disaster risks of urban rainfall and flooding and the assessment of disaster severity can save lives and prevent losses (Li et al., 2024). Flood emergency response can be prepared through a flood hazard assessment. Geographical Information Systems (GIS) and Remote Sensing (RS) are mostly used for risk reduction. VR, AR, and 3D hydrodynamic flood simulation based on a virtual presentation of UAV data, dynamic visual simulation, and decision support systems are applied for flood risk management (Girotto et al., 2023; Waseem and Rana, 2023). Multi-Criteria Decision Analysis (MCDA) such as Fuzzy Logic and Genetic Algorithms for forecasting extreme weather events, increasing public awareness, optimizing evacuation routes, and implementing various risk management strategies have been used for flood hazard assessment. Slope, elevation, land use, Normalized Difference Vegetation Index (NDVI), Topographic Wetness Index (TWI), drainage density, and rainfall are used to produce flood hazard maps (Girotto et al., 2023; Waseem and Rana, 2023). The collection of data on soil moisture, land movement, and rainfall intensity can be used to estimate the risk of landslides and damage to roads (Girotto et al., 2023; Li et al., 2024).

Remote sensing techniques include water flows, water levels, rainfall and moisture, radio frequency and electric, pressure, ultrasound, and infrared sensors, fixed and visual surveillance, radar and light detection and range (LiDAR). Portable and local sensors with web-based platforms can predict events such as floods, which can be used to establish EWS. There are plans for post-flood monitoring to detect the spread of floodwater, which can be breeding grounds for insects or parasites, to assess the impact of floods, and to assess soil and water contamination. X, Facebook, Weibo, WeChat, and YouTube play an important role in informing citizens about the progress of disasters and floods by quickly gathering information from a variety of sources about flooded areas, traffic disruptions, infrastructure damage, people in distress and early signs of disease outbreaks, API-based data sources, Google Vision, mobile phone applications (email and SMS), IoT-based RS, LoRaWAN and LiDAR remote sensing, satellite and ground real-time data, and the Internet of Drone are useful technologies in flood disasters (Girotto et al., 2023; Li et al., 2024). Flood alerts can be sent via social media (Facebook and WhatsApp), Cell Broadcast, and SMS. Repeated SMS transmissions can be a source of congestion and disruption to phone calls. Other methods, such as cell broadcast, can be used to avoid congestion-related communication problems. There may also be damage to the communications system. Other systems such as loud sounds, five short bells, and warning lights can also be employed (Shah et al., 2021).

Tropical cyclones of high and very high intensity have become more frequent in recent years due to climate change. Flood EWS can be used to develop emergency preparedness and community-based adaptation strategies to prevent loss of life and reduce the economic impact of disasters. Cyclones have damaged forests. These forests are critical for human settlements and agricultural activities (Sahana et al., 2023). Many countries have developed projects to protect and restore coastal ecosystems, such as coastal wetlands and marine fisheries. Sea walls can be a means of protection for coastal dwellers and their property from sea-level rise. Natural coastal ecosystems such as mangroves, saltmarshes, oyster reefs, and seagrass beds can be effectively monitored to understand dynamic changes in stability and ecological functions (Zheng et al., 2023). Tsunami waves can cause flooding, so tsunami early warning systems have been installed in many countries, such as Indonesia (Meilani and Hardjosoekarto, 2020). For example, an audio mobile phone (AMP) message for an earthquake or tsunami warning was sent to people in Japan, allowing them to evacuate their homes before the first tsunami wave arrived. Effective evacuation protocols and shelters, such as temporary housing, should be included in preparedness plans (Naylor et al., 2013).

Fire Warning Applications

Fire disasters are a major cause of loss of human life, environmental and property damage. The detection of fire hazards is a very critical part of fire warning systems. For such risky situations, thermoelectric (TE) and thermoelectric cell (TEC) based fire warning materials are mainly used. The heating stage with temperatures lower than the threshold temperature, the development stage with temperatures from the threshold temperature to the ignition point temperature, the violent burning stage, and the recession stage are the stages of fire development. The best early warning stage is the development stage. Gas/smoke sensors, resistance shift triggering at high temperatures, chromic change to detect CO and NO_2 gases, phase/shape transition working well at low temperatures, and heat to electricity conversion at a temperature difference are the mechanisms of the fire early warning system. The fire alarm is activated by signals from gas or smoke concentration sensors (Ding et al., 2023). The technological growth of wireless sensor networks (WSNs) has also expanded their applications in global wildfires. A taxonomy of some of the significant wildfire detection techniques using WSNs from the literature is reviewed by Dogra et al. (2020).

Outbreaks

The provision of information and risk assessments have been carried out for an early response to the pandemic (Aslan and Goundar, 2022). To reduce the risk of outbreaks, a sufficient lead time for contingency plans for potential disease outbreaks is required. Vector-borne diseases such as malaria and tetanus can be transmitted through the spread of infectious diseases (Girotto et al., 2023). During the COVID-19 pandemic, spatial tracing was used to detect the first case and the initial source of infection. To make effective plans as part of the early warning index system and the optimization of infectious diseases, simulation and prediction of the future evolution trend of the COVID-19 epidemic was carried out (Xiong et al., 2021).

European Environment and Epidemiological System (EES), using GIS, satellite, and model data from different sources to analyze environmental and climate-induced pathogens, and The European Environment and Epidemiology System (E3), which uses GIS, satellites, remote sensing, and modeling tools are some EWS for detecting disease outbreaks. Early Warning, Alert and Response System (EWARS) was tested in outbreaks in Brazil, Malaysia, and Mexico, using mobile phones/laptops/data hubs to collect, report, and manage infectious disease data to detect disease outbreaks during emergencies (Girotto et al., 2023). The IoT, characterized by Industry 4.0, cloud computing, fog computing, mobile computing, wearable sensors, and big data, has improved disease detection and monitoring and provided flexibility for patients to be treated at home rather than in a hospital with high accuracy (Aslan, 2021). Self-triage, self-scheduling, and avoidance of unnecessary face-to-face care have been achieved with the help of coronavirus symptom checkers (Aslan and Goundar, 2022). Advanced technologies used during the pandemic included cloud-based ubiquitous computing, big data analytics, machine learning, autonomous systems, smart robots, 3D printing, and virtual and augmented reality. Attempts were made to develop an intelligent diagnosis of medical imaging and temperature measurement technology for the accurate diagnosis and protection of healthcare workers based on computer vision and infrared technology, which can be supported by AI. Data from COVID-19 patients were analyzed for epidemiological characteristics of the pandemic, host susceptibility and immune responses, risk factors for severe disease, and routes of transmission in support of policymakers (Aslan and Goundar, 2022; Lee and Trimi, 2021).

Open-Pit Mining Operations

Predicting landslide risk can save many lives, so open pit mines need long-term monitoring of physical properties to estimate the instability condition of the mine rock mass by means of EWS platforms. A remote system for early warning of landslide disasters can be developed through real-time monitoring by integrating Global Navigation Satellite Systems (GNSS) and GIS. AI, IoT, machine learning methods, and cloud computing can be applied in open pit mines for disaster prevention (Haiping et al., 2018; Song et al., 2023). UAVs, TLS (terrestrial laser scanning) such as the 3D laser scanning technology, satellites and radars as non-contact remote sensing data, displacement monitoring, and microseismic monitoring can monitor open pit mines (Song et al., 2023). Some safety early warning applications that combine and visualize geological exploration are visualization platforms based on safety early warning applications. A high-speed corporate network, mining software, a processing platform, a three-dimensional geological model creation (3DGM) and monitoring data mining, monitoring data for transportation and processing, GIS system concerned with data transmission in the platform known as WebGIS, data storage management and human-computer interaction to the browser are required for the development of a EWS for open-pit mine.

Geological Hazards

Geological hazards that require multidisciplinary approaches include landslides, rockfalls, avalanches, and mudflows. Water storage and heavy rainfall can lead to these events. The use of GIS and RS can be a tool for their mitigation. Observational data collection, mechanism analysis, monitoring and early warning, risk assessment models, and technology applications are focused to reduce hazards and estimate their occurrence. Field surveys, exploratory tunnels, field boreholes, and Global Navigation Satellite Systems can help to obtain data on these high-risk locations. Remote sensing and GIS technology, detailed field surveys, 3D laser scanning and high-density electrical methods, and multi-temporal remote sensing image analysis are technological methods for assessing the situation. Soil water content, cohesion, and friction angles can be measured to estimate the risk of catastrophic events by means of Monte Carlo simulation, wavelet analysis, genetic algorithms, and Elman algorithms (Xu, 2023), Shuguang et al., 2022; Torrence and Compo (1998).

Coastal Hazards

The coastline (within 40 m) and more densely populated areas are threatened by six major natural hazards: Cyclones, storm surges, tsunamis, floods, earthquakes, and coastal erosion. In these regions, fisheries and aquaculture are important economic activities. Some of the most vulnerable coastal areas, especially during the monsoon season, are in India, Bangladesh, Vietnam, and Indonesia. RS and GIS techniques will be used to assess coastal vulnerability. Sentinel satellite data such as high-resolution ArcGIS base layer and Google Earth imagery, Global Seismic Hazard Map for earthquakes and LANDSAT satellite data can assist researchers. Synthetic Aperture Radar (SAR) data and rainfall data from the State Water Resources Data Centre and Meteorological Department reports are used for flood-prone areas to produce flood hazard layers. The pre-flood and post-flood Sentinel 1 SAR image datasets are used to apply the change detection and thresholding methodology. Flooding scenarios are modeled at different depths using Global Mapper software, with the layer determining the vulnerability type (Abarna et al., 2023; Panda et al., 2023).

Glacier-Related Disasters

Glacier collapse leading to secondary hazards such as debris flows, flash floods, landslides, lake and river damming are potential threats to population and transport, mainly related to global warming and extreme precipitation melting glaciers. Glacier retreats, permafrost degradation, lake expansion, increases in runoff and sediment flux, and intensification of water circulation are among the major risk factors. The Tibetan Plateau and the Indian Himalayas, Iceland, Central Asia, and the European Alps are possible locations for glacier collapse (An et al., 2022). The main tools used to monitor glaciers include active synthetic aperture radar (SAR) imagery, helicopter and UAV, RS interpretation, and seismic, hydrological, and meteorological observations. Long-term automated monitoring and EWS can mitigate the effects of glacier collapse. This can prevent potential loss of life and reduce economic and material losses. Radar water level sensors and weather stations can be used to record water levels and meteorological conditions (An et al., 2022).

Deploying Key Technologies to Manage Disasters

The main types of disasters and the associated technologies used are categorized in Table 1 below. This also includes the main sources of input data used by these technologies and the main areas of application.

Table 1. Major technologies in disaster management

Disaster Type	Technologies Employed	Data Sources	Key Domain of Use	Reference(s)
Floods	GIS and Global Positioning Systems (GPS), AI	Sensor data, social media, and GIS data	Detection and prediction of floods, risk models, monitoring	Waseem and Rana, 2023; Li et al., 2024; Shah et al., 2021; Girotto et al., 2023
Earthquakes	GPS, MySQL database, AI	Data from satellite signals, Global Ionospheric Map (GIM)	Forecasting, navigating, and timing of earthquakes.	Chen, 2023; Huang et al., 2023; Naylor et al., 2013; Meilani and Hardjosoekarto, 2020
Fires	GPS, GIS, RS, AI, WSN, Google Earth Engine	Forest surveys, data from WSN, TE, and TEC, signals from Gas/smoke sensors	Fire detection and monitoring, occurrence prediction, and modeling	Alkhatib et al., 2023; Chandra et al., 2005; Abid, 2021; Dogra et al., 2021, Smys and Raj, 2021.
Outbreaks	EWS, Spatial tracing, EWARS, GloFAS, mobile phones/laptops	Data from weather stations and satellites, GIS	Special tracing, prediction trends, planning	Girotto et al., 2023; Xiong et al., 2021.
Geological hazards	RS image analysis, GIS technology, Detailed field surveys, 3D laser scanning, and high-density electrical methods	Soil water content, cohesion, and friction angle	Estimation risks of events	Xu, 2023; Shuguang et al., 2022

Table 1. (Continued)

Disaster Type	Technologies Employed	Data Sources	Key Domain of Use	Reference(s)
Coastal hazards	RS, GIS, ArcGIS base layer and Google Earth, Global Seismic Hazard Map, Satellite	SAR data, rainfall data	Reports, change detection, flood rise scenarios, and estimations	Abarna et al., 2023; Panda et al., 2023
Glacier-related disasters	Helicopters UAV, RS, SAR, radar water-level sensors	Glacier residues, geothermal factors, and seismic activity	Estimation and detections	An et al., 2022; Lu et al., 2022

Discussion and Conclusion

Key technologies such as WSN, GIS, GPS, AI, machine learning, UAVs for various types of disasters, including floods, earthquakes, forest fires, and outbreaks, are effective technologies for managing the emergency and mitigating losses. Non-governmental organizations (NGOs), public awareness and risk perception management, and disaster preparedness are also critical factors.

A major challenge in assessing and coordinating the availability of funds and donations is the uncertainty of supply and demand for relief supplies. Relief logistics require effective techniques to optimize post-disaster preparedness using digital technologies, including distribution of relief items, location of relief facilities and temporary shelters, integrated distribution of relief items and location of shelters, and transportation of affected populations (Kamyabniya et al., 20-24). As seen in the Maras earthquake and Hatay 2023, Türkiye, disorganized relief logistics can be a source of desperation and chaos.

References

Araz, O. M., Choi, T. M., Olson, D. L., & Salman, F. S. (2020). Role of analytics for operational risk management in the era of big data. *Decision Sciences*, *51*(6), 1320-1346. https://doi.org/10.1111/deci.12451.

Aslan, I. (2021). Technologies and Applications of Internet of Things (IoT) in Healthcare. In S. Goundar, & P. Rayani (Eds.), Applications of Big Data in Large- and Small-Scale Systems (pp. 307-330). *IGI Global.* https://doi.org/10.4018/978-1-7998-6673-2.ch018.

Aslan, I., & Goundar, K. (2022). Emerging Technologies for Healthcare during the COVID-19 Pandemic. In *Applications of Artificial Intelligence, Big Data and Internet of Things in Sustainable Development* (pp. 203–239). CRC Press. https://doi.org/10.1201/9781003245469-13.

Abarna, R., George, S. L., Balasubramani, K., Yuvaraj, S., Shekhar, S., Gnanappazham, L., & Prasad, K. A. (2023). Estimating built-up risk from multi-natural hazards: A case study of Northern coastal plains of Tamil Nadu. *Natural Hazards Research, 3*(1), 49-65. https://doi.org/10.1016/j.nhres.2023.01.001.

Abid, F. (2021). A Survey of Machine Learning Algorithms Based Forest Fires Prediction and Detection Systems. *Fire Technology, 57*(2), 559–590. https://doi.org/10.1007/s10694-020-01056-z.

Alkhatib, R., Sahwan, W., Alkhatieb, A., & Schütt, B. (2023). A Brief Review of Machine Learning Algorithms in Forest Fires Science. *Applied Sciences, 13*(14), Article 14. https://doi.org/10.3390/app13148275.

An, B., Wang, W., Yang, W., Wu, G., Guo, Y., Zhu, H., ... & Yao, T. (2022). Process, mechanisms, and early warning of glacier collapse-induced river blocking disasters in the Yarlung Tsangpo Grand Canyon, southeastern Tibetan Plateau. *Science of the Total Environment, 816*, 151652.

Böhm, G., & Pfister, H. (2005). Consequences, morality, and time in environmental risk evaluation. *Journal of Risk Research, 8*(6), 461–479. https://doi.org/10.1080/1366987050064143.

Chandra, S. (2005). Application of Remote Sensing and GIS Technology in Forest Fire Risk Modeling and Management of Forest Fires: A Case Study in the Garhwal Himalayan Region. In P. van Oosterom, S. Zlatanova, & E. M. Fendel (Eds.), *Geo-information for Disaster Management* (pp. 1239–1254). Springer. https://doi.org/10.1007/3-540-27468-5_86.

Chen J. (2023) Great potential of study on earthquake disaster chains in the new era: Summary of the 2023 academic annual meeting of the Earthquake Disaster Chain Committee. *Natural Hazards Research, 3*(4), 664-665, https://doi.org/10.1016/j.nhres.2023.12.003.

Dash, B. (2023). Gap between impact-based and impact forecast and warning: Implications for people-centric early warning system (EWS) in India. *Natural Hazards Research.* In Press, *Journal Pre-proof,* https://doi.org/10.1016/j.nhres.2023.09.005.

Ding, Z., Du, C., Long, W., Cao, C. F., Liang, L., Tang, L. C., & Chen, G. (2023). Thermoelectrics and thermocells for fire warning applications. *Science Bulletin, 68*(24), Pages 3261-3277. https://doi.org/10.1016/j.scib.2023.08.057.

Dogra, R., Rani, S., & Sharma, B. (2021). A Review to Forest Fires and Its Detection Techniques Using Wireless Sensor Network. In G. S. Hura, A. K. Singh, & L. Siong Hoe (Eds.), *Advances in Communication and Computational Technology* (pp. 1339–1350). Springer Nature. https://doi.org/10.1007/978-981-15-5341-7_101.

Girotto, C. D., Piadeh, F., Bkhtiari, V., Behzadian, K., Chen, A. S., Campos, L. C., & Zolgharni, M. (2023). A critical review of digital technology innovations for early warning of water-related disease outbreaks associated with climatic hazards. *International Journal of Disaster Risk Reduction, 104151* .https://doi.org/ 10.1016/j.ijdrr.2023.104151.

Haiping, X. I. A. O., Lanlan, C. H. E. N., Guangli, G. U. O., Xiaosheng, L. I. U., & Qimin, H. E. (2018). Scheme Design of GNSS+ GIS Remote Landslide Real-time Monitoring System. *Bulletin of Surveying and Mapping, 5,* 116._doi: 10.13474/j.cnki.11-2246.2018.0156.

Huang, J., Wang, Q., Yan, R., Lin, J., Zhao, S., Chu, W., ... & Huang, H. (2023). Pre-seismic multi-parameters variations before Yangbi and Madoi earthquakes on May 21, 2021. *Natural Hazards Research*, *3*(1), 27-34. https://doi.org/ 10.1016/j.nhres.2022.12.002.

Kamyabniya, A., Sauré, A., Salman, F. S., Bénichou, N., & Patrick, J. (2024). Optimization models for disaster response operations: a literature review. *OR Spectrum*, 1-47. https://doi.org/10.1007/s00291-024-00750-6.

Lee, S. M., & Trimi, S. (2021). Convergence innovation in the digital age and in the COVID-19 pandemic crisis. *Journal of Business Research*, *123*, 14–22. https://doi.org/10.1016/j.jbusres.2020.09.041.

Li, H., Han, Y., Wang, X., & Li, Z. (2024). Risk perception and resilience assessment of flood disasters based on social media big data. *International Journal of Disaster Risk Reduction*, *101*, 104249. https://doi.org/10.1016/j.ijdrr.2024.104249.

Naylor A., Walker J. F., Suppasri A. (2013). Lessons learned from the tsunami disaster caused by the 2011 great East Japan earthquake and improvements in JMA's tsunami warning system. *International Journal of Disaster Risk Reduction, 31,* 302-310, https://doi.org/10.1016/j.ijdrr.2018.05.022.

Meilani, N. L., & Hardjosoekarto, S. (2020). Digital weberianism bureaucracy: Alertness and disaster risk reduction (DRR) related to the Sunda Strait volcanic tsunami. *International Journal of Disaster Risk Reduction*, *51*, 101898. https://doi.org/ 10.1016/j.ijdrr.2020.101898.

Munawar, H. S., Mojtahedi, M., Hammad, A. W., Kouzani, A., & Mahmud, M. P. (2022). Disruptive technologies as a solution for disaster risk management: A review. *Science of the total environment*, *806*, 151351. https://doi.org/10.1016/j.scitotenv.2021. 151351.

Panda, G. K., Mishra, M., & Chatterjee, U. (2023). Risk assessment of coastal erosion for Odisha coast along Bay of Bengal, India using coastal hazard wheel model. *Geo Journal*, *88*(1), 109–134. https://doi.org/10.1007/s10708-021-10552-w.

Sahana, M., Patel, P. P., Rehman, S., Rahaman, M. H., Masroor, M., Imdad, K., & Sajjad, H. (2023). Assessing the effectiveness of existing early warning systems and emergency preparedness towards reducing cyclone-induced losses in the Sundarban Biosphere Region, India. *International Journal of Disaster Risk Reduction*, *90*, 103645. https://doi.org/10.1016/j.ijdrr.2023.103645.

Smys, S., & Raj, J. S. (2020). Assessment of Fire Risk and Forest Fires in Rural Areas Using Long Range Technology. *Journal of Electronics and Informatics*, *2* (1), 38–48. https://doi.org/10.36548/jei.2020.1.005.

Shah, A. A., Wu, W., Gong, Z. *et al.* Multidimensional six-stage model for flood emergency response in schools: a case study of Pakistan. *Nat Hazards 105,* 1977–2005 (2021). https://doi.org/10.1007/s11069-020-04386-x.

Shah, A. A., Ullah, A., Khan, N. A., Khan, A., Tariq, M. A. U. R., & Xu, C. (2023). Community social barriers to non-technical aspects of flood early warning systems and NGO-led interventions: The case of Pakistan. *Frontiers in Earth Science, 11,* https://doi.org/10.3389/feart.2023.1068721.

Lu, S., Wu, L., Cheng, X., Xu, G., Shen, F., Li, C., Hu, X., Yang, B., Zhang, H., & Li, X. (2022). Spatial–temporal variations in natural disasters during the Ming and Qing dynasties (1368–1911 AD) in the ancient Huizhou region, eastern China. *Frontiers in Earth Science, 10.* https://doi.org/10.3389/feart.2022.960113.

Lu, Y., Zhang, Z., Kong, Y., & Hu, K. (2022). Integration of optical, SAR and DEM data for automated detection of debris-covered glaciers over the western Nyainqentanglha using a random forest classifier. *Cold Regions Science and Technology, 193*, 103421. https://doi.org/10.1016/j.coldregions.2021.103421.

Song, Z., Li, X., Huo, R., & Liu, L. (2023). Intelligent early-warming platform for open-pit mining: Current status and prospects. *Rock Mechanics Bulletin, 100098.* https://doi.org/10.1016/j.rockmb.2023.100098.

Tunç, S., Tunç, B., Budakoğlu, E., Çaka, D., Nof, R. N., & Barış, Ş. (2023). Implementation of the EPIC Earthquake Early Warning System in the Bursa Province (Türkiye) and Its Surroundings. *Applied Sciences, 13*(8), 4985. https://doi.org/10.3390/app13084985.

Torrence, C., & Compo, G. P. (1998). A Practical Guide to Wavelet Analysis. *Bulletin of the American Meteorological Society, 79*(1), 61–78. https://doi.org/10.1175/1520-0477(1998)079<0061:APGTWA>2.0.CO;2.

UN Office for Disaster Risk Reduction. (2021). Risk Reduction 2020 the non-Covid year in disasters: global trends and perspectiveshttp://hdl.handle.net/2078.1/245181 (Accessed Date: 04.03.2024).

Xu, C. (2023). An introduction to "Application of Novel High-Tech Methods to Geological Hazard Research." *Natural Hazards Research, 3,* 2, June 2023, Pages 353-35 https://doi.org/10.1016/j.nhres.2023.05.001.

Waseem, H. B., & Rana, I. A. (2023). Floods in Pakistan: A state-of-the-art review. *Natural Hazards Research, 3*(3), Pages 359-373. https://doi.org/10.1016/j.nhres.2023.06.005.

Zheng, J., Xu, W., Tao, A., Fan, J., Xing, J., & Wang, G. (2023). Synergy between coastal ecology and disaster mitigation in China: Policies, practices, and prospects. *Ocean & Coastal Management, 245,* 106866. https://doi.org/10.1016/j.ocecoaman.2023.106866.

Xiong, L., Hu, P., & Wang, H. (2021). Establishment of epidemic early warning index system and optimization of infectious disease model: Analysis on monitoring data of public health emergencies. *International Journal of Disaster Risk Reduction, 65,* 102547. https://doi.org/10.1016/j.ijdrr.2021.102547.

Chapter 25

Emotional Labor of Nurses During Disasters

Ayşegül Sarıoğlu Kemer[1,*], PhD
and Ayşe Asiltürk[2], PhD
[1]Department of Emergency Aid and Disaster Management, School of Applied Sciences, Trabzon University, Trabzon, Türkiye
[2]Department of Management Information Systems, School of Applied Sciences, Trabzon University, Trabzon, Türkiye

Abstract

Emotional labor is defined as professionals managing their emotions while performing their jobs and exhibiting the emotional outputs desired by their organizations (Mesmer-Magnus et al., 2012). Emotional labor manifests itself in three different ways: surface acting, deep acting, or the expression of genuine emotions. It is influenced by the social and personal identity of the employee, shaping individuals' service roles, task effectiveness, and self-expression (Ashforth and Humphrey, 1993). According to the literature, emotional labor strategies enhance the effects of leadership (Humphrey et al., 2015). Emotional labor is intensely experienced in the aftermath of situations affecting large populations, such as floods, landslides, and earthquakes. The concept of a disaster is defined as an event or situation that affects people and living beings, causes loss of life, and occurs acutely. Despite technological innovations pushing the boundaries of imagination in our rapidly evolving and changing world, humanity has no control over natural disasters such as earthquakes, volcanic eruptions, floods, droughts, or storms, nor can these disasters be prevented. A recent example highlighting human helplessness in the face of natural disasters occurred on February 6, 2023,

* Corresponding Author's Email: aysegulsarioglu@msn.com.

In: Disaster Management in the Digital Age and Emerging Technologies
Editors: Ayşe Asiltürk and Ayşegül Sarıoğlu Kemer
ISBN: 979-8-89530-213-2

with the earthquake centered on Kahramanmaraş, Türkiye, affecting a vast area and approximately 15 million people. Türkiye, due to its geographical structure and four-season climate characteristics, frequently faces hazards such as earthquakes, floods, flash floods, landslides, etc. Nurses can take on roles in disaster response, including evacuating victims, providing initial medical care, determining priorities, providing psychosocial support, and many other interventions. Disasters represent special situations where nurses assume responsibility for the intense implementation of emotional labor behaviors and may even feel obligated to apply them. There are both individual and organizational impacts of emotional labor. Increasing social support in nursing interventions will reduce the exhausting aspects of emotional labor for members of the profession. It is expected that new technologies developed to enhance emotional labor will not only improve nurses' mental health and job performance but also increase the effectiveness of nursing interventions.

Keywords: disasters, emotional labor, nurses, technology

Introduction

In recent years, due to the increasing frequency and diversification of disasters worldwide, the field of emergency aid and disaster management has gained growing significance. Alongside efforts to prevent disasters, interventions in disasters inevitably caused by various reasons have become high-priority issues for institutions, managers, experts, and researchers.

Emergency aid and disaster management teams inherently operate in highly uncertain, stressful situations requiring non-routine problem-solving based on information (Schaafstal et al., 2001). Particularly in teams involved in disaster zones, such as search and rescue teams, considering the high levels of emotional arousal experienced by workers, to what extent can these individuals maintain their professional masks in reflecting their emotions to the people with whom they interact? Moreover, how are the health and job performance of these workers affected while all this is happening? Can the combination of organizational behavior and psychology with technology improve the health status and disaster response performance of workers? Compared to established research fields, emergency aid and disaster management emerge as a niche area in terms of addressing organizational behavior issues with its interdisciplinary structure.

In this context, the question of what can be said about the emotional labor displayed by teams working in disaster areas, their interactions with the individuals and their families whom they are trying to rescue, and their communication with their team members in the face of the situations and scenes they encounter has been identified as the research question of this section.

The Concept and Scope of Emotional Labor

In recent years, researchers and theorists have begun to delve deeper into the realm of emotions in the workplace (Grandey, 2000). One of these concepts, emotional labor, is defined as employees managing their emotions while performing their duties to exhibit the emotional displays demanded by their organizations (Mesmer-Magnus et al., 2012). Emotional labor involves not only the effort employees exert to suppress their own emotions while performing their jobs but also self-care (Edward et al., 2017). Within the scope of occupational health psychology, the concept of emotional labor, introduced by the sociologist Arlie Hochschild (1983), refers to the "occupational requirements that distance workers from their emotions." In the early 2000s, emotional labor was conceptualized as emotional regulation strategies associated with performance and well-being, approached from a psychological perspective (Grandey and Melloy, 2017).

Emotional labor manifests itself in surface acting, deep acting, or the expression of genuine emotions. Influenced by employees' social and personal identities, emotional labor in service roles can affect task effectiveness and self-expression (Ashforth and Humphrey, 1993). Also known as deep acting, natural emotional labor is associated with positive relationships between employees' organizational commitment, job satisfaction, job performance, and the satisfaction of service recipients such as customers, patients, and disaster victims (Humphrey et al., 2015). For instance, studies conducted in healthcare settings conceptualize emotional labor as the ability to recognize others' emotions and manage one's own emotions within the caregiving role, emphasizing the need for recognition, support, and supervision in coping with emotional demands (Riley and Weiss, 2016). In cases of emotional labor dissonance, employees may experience attitude, health, and performance-related issues (Mesmer-Magnus et al., 2012). Indeed, emotional intelligence, emotional exhaustion, and emotional labor burnout are acknowledged as factors influencing emotional labor (Edward et al., 2017). Emotional labor has been strongly associated with negative outcomes in psychological well-being and mental health among workers. High emotional labor among female workers has been found to increase depressive symptoms, while high job autonomy among male workers can buffer against the detrimental effects of emotional labor (Han et al., 2018).

Different perspectives have been put forth regarding the definition and operationalization of emotional labor, highlighting the lack of a general framework to guide research (Grandey, 2000). In effectively measuring emotional labor, psychometric tools such as the Discrete Emotions Emotional Labor Scale have been developed, focusing on emotional expression behaviors to distinguish between genuine, surface, and deep-acting displays (Glomb and Tews, 2004). Emotional labor, as a process of monitoring and reducing discrepancies between emotional expressions and display rules, is achieved through the use of emotional regulation strategies that impact job performance (Diefendorff and Gosserand, 2003). The literature suggests that emotional labor strategies can enhance leadership effectiveness (Humphrey et al., 2015).

The management of mass emergencies such as earthquakes involves the decision-making and implementation of high-risk decisions directing relief operations following a disaster and/or calamity. The emergency aid and disaster management teams involved in this process, as a whole, comprise multiple teams from different organizations with diverse organizational goals and cultures, aiming to minimize the adverse effects of the emergency (Schaafstal et al., 2001). In their study on firefighters working in high-risk and high-stress workgroups, Fullerton et al. (1992) examined the psychological responses of two groups of firefighters following rescue operations, revealing experiences of "fear of the unknown, helplessness and guilt, identification with victims, and physiological responses." Shuler and Sypher (2000), demonstrating the central role of emotion regulation in their study, found that emotional labor for 911 operators was an integral and rewarding aspect of their work, contributing to the formation of organizational cohesion.

Disaster Process and Emotional Labor

The concept of disaster is defined as an acute event or situation that disrupts the lives and life cycles of humans or other living beings, causing destruction and damage, leading to injuries and loss of life, and requiring assistance at national or international levels. Common characteristics of disasters include their sudden occurrence and affecting a large part of the population.

Furthermore, during disasters, the loss of life and property damage generally cannot be prevented through humanitarian intervention. Disasters are most commonly known as earthquakes, floods, landslides, erosion, avalanches, tornadoes, war, terrorist attacks, nuclear leakage, industrial accidents, and many other situations. Despite technological innovations pushing the boundaries of imagination in a rapidly evolving and changing world, humans have no influence over the occurrence of natural events such as earthquakes, volcanic eruptions, floods, droughts, or storms that cause natural disasters.

The most significant initiative that can be taken regarding natural disasters is to make plans to minimize the damages of disasters on society and effectively implement them. The example that best illustrates humanity's helplessness in the face of natural disasters is the earthquake that occurred in Kahramanmaraş, Türkiye on February 6, 2023, affecting a wide area and approximately fifteen million people. Disasters cause destructive and enduring damage that is often difficult to compensate for. In the absence of an effective preparation and planning process before disasters, adverse conditions affecting multiple cities simultaneously arise (Demir et al., 2024; Wijayasundara, 2021).

With rapid and unbalanced population growth, unchecked settlement, and disregard for ecological balance, the destruction caused by unprepared demolitions is increasing, and humanity is facing united disaster scenarios. Regardless of the cause, activities and disaster management cycles related to all natural disaster events can be grouped under five main headings: (1) Mitigation, (2) Preparedness, (3) Response and First Aid, (4) Recovery, and (5) Reconstruction stages (Sawalha, 2020; Varol and Gültekin, 2016).

Activities carried out by expert or volunteer teams after natural disasters must be conducted in a disciplined and orderly manner. This discipline is of great importance in reducing the scale of the disaster. The efforts of each individual in the field, physical exertion, long hours of exhausting work, and meticulous work come into play. Furthermore, managing emotions as an unseen aspect of effort represents emotional labor (Dworznik-Hoak, 2022; Kranke et al., 2022).

The term emotional labor was first addressed by Arlie Russell Hochschild in her book 'The Managed Heart: Commercialization of Human Feeling.' Hochschild defines emotional labor as the management of emotions by a worker to create the expected facial and bodily appearance for a specific purpose. According to Hochschild, workers are actors, service recipients are audiences, the service provided is performance, and the working environment is stage (İrigüler and Güler, 2016; Hochschild, 1983). Another definition by

Basım and Beğenirbaş (2012) states that emotional labor is "the impression created in the inner world of the person by the consciously undertaken working process."

The concept of emotional labor is considered in three dimensions: surface acting, deep acting, and genuine expression. In surface acting, there is a disparity between the emotional state displayed by the worker and their actual feelings. The worker conceals their true emotions. For example, bureaucrats and actors are among those who exhibit the best surface acting, whereas children typically exhibit the worst surface acting. According to research conducted by Hochschild with flight attendants, as cited by Çevik Tekin (2021), flight attendants try to calm passengers and manage a difficult situation inside the plane by displaying calm behavior, even though they may be as anxious as the passengers themselves, indicating that they hide their emotions and display different behavior (Çevik Tekin, 2021).

Deep acting, on the other hand, involves genuinely experiencing and feeling the behavior that the worker wishes to exhibit. In this dimension, there is an effort by the worker to truly feel the emotion they are portraying. What sets deep acting apart from surface acting is the alignment of behavior and emotions with the rules of conduct related to the job (Zeng et al., 2021; Çevik Tekin, 2021).

Genuine expression, on the other hand, is the ability of the worker to express their emotional state through their behavior without any compulsion or pressure. In this behavioral strategy, the emotions felt by individuals correspond to the emotions they reflect (Anaç and Özyer, 2021).

Emotional Labor and Nursing During Disaster

The current World Disaster Report, published in 2022, focuses on the COVID-19 pandemic. Titled "Trust, Equity, and Local Action: Lessons from the COVID-19 pandemic to avert the next global crisis," the report highlights that the most frequent disasters in 2020 and 2021 were climate-related. Particularly, the COVID-19 pandemic was prevalent during the 2020-2021 period. Among the countries with the highest death toll during this period were India and China, while the United States experienced the highest number of disaster cases. The report also mentions that through the Turkish Red Crescent, more than 30,000 food and hygiene packages were distributed during the COVID-19 period, with 10,000 Red Crescent volunteers serving in the field. It is estimated that the COVID-19 pandemic, which has claimed the lives of 6.5 million people, will generate a cost of approximately 13.8 trillion US dollars by the end of 2024.

Due to its geographical structure and the ability to accommodate all four climate characteristics, Türkiye frequently faces disaster risks such as earthquakes, floods, flash floods, and landslides. As a natural consequence of disasters, loss of life and property occurs, and a large part of society can be negatively affected both financially and spiritually. It is vital for all healthcare workers to be knowledgeable about disasters and their aftermath to minimize their impact on individuals. The high number of nurses among healthcare professionals makes them even more crucial in this field. Disasters are extremely challenging processes for both patients and their families. In this regard, the role and scope of responsibility of nurses are broad and uncertain. Qualified nurses are greatly needed during disaster processes, which result in both material and human losses (WDR, 2022).

Nurses can assume responsibilities such as transporting and evacuating disaster victims, providing initial interventions, determining priorities, offering psychosocial support, and many other interventions. The reason nurses undertake such critical tasks is to ensure the rapid normalization of the health care system or society in the aftermath of a disaster. Nurses are the largest group of healthcare professionals serving in all stages of disasters (preparation, response, and recovery). Nurses who prioritize disasters and disaster literacy feel motivated to intervene in disasters according to their competencies (Liou et al., 2020). Disaster processes bring uncertainty, frequent changes, and intense guidance responsibilities regarding care to nurses. Expanding responsibilities, increasing task numbers, and assigning new roles without guidance are common challenges nurses face during this process. Despite such difficult situations, it is known that nurses support the implementation of high-quality care (Jackson and Nowel, 2021).

Nurses are expected to expend more emotional labor than usual during disaster processes due to high levels of stress, increased workload, and lack of guidance (Liu et al., 2020; Korkmaz et al., 2021). Emotional labor can be defined as performing a job by adopting certain emotions and expressions and making an effort to evoke a feeling in individuals (Öz and Baykal, 2018). Nurses particularly engage in intense communication with patients and their families during the clinical process after a disaster. During this process, nurses are expected to approach their nursing interventions and observations with empathy and establish a sense of trust in the patient. Additionally, nurses are expected to have a high level of disaster knowledge, support psychological treatment after disasters, contribute to disaster management and leadership, and participate in disaster drills, which are roles specific to disaster periods (Hirohara et al., 2019; Korkmaz et al., 2021; Kimin et al., 2022).

As evident, disasters are particular situations where emotional labor behavior is intensively applied by nurses within the scope of their responsibilities and sometimes even felt as obligatory. In these processes, nurses are required to carry out their interventions with zero error, quickly and effectively, while also establishing trust, showing empathy, being understanding, smiling, and being patient with the patient. Formulating this behavioral equation and putting it into practice is essentially the manifestation of emotional labor in life. In this regard, nurses are healthcare professionals who are required to have a high level of emotional labor (Kim, 2020).

Nurses attempting to display emotions they do not genuinely feel or internalize will ultimately lead to negative emotions. Emotional labor, especially with its superficial role-playing aspect, leads to burnout, stress, anxiety, pain, musculoskeletal disorders, and heart problems (Kim et al., 2022; Ceylan, 2021). The individual effects of emotional labor are paralleled by its organizational impacts. Burnout, presenteeism, stress, decreased organizational commitment, intention to leave the job, and decreased performance are potential effects of emotional labor in organizations (Sevinç Altaş and Özişli, 2021; Song et al., 2021; Theodosius et al., 2021).

Emotional labor is a factor that affects performance within the organizational system, with employees suppressing or masking their own emotions and projecting different emotions onto their surroundings. From an organizational perspective, emotional labor affects variables such as "emotional display rules, autonomy, and social support." Emotional display rules can be defined as standardized employee behaviors established by institutions. Nurses are expected to adhere to emotional display rules, such as being impartial, maintaining control, showing empathy, and not showing grief regarding emotional labor (Kirk et al., 2021). The relationship between autonomy and emotional labor is negative. Particularly in professions requiring high levels of emotional labor, autonomy is thought to be threatened (Grandey et al., 2015). Increasing social support in nursing interventions will reduce the draining aspects of emotional labor for professionals. Additionally, there are significant effects of social support on the connection between emotional labor and factors such as job satisfaction and intention to leave the job (Kim et al., 2011; Penekli, 2022).

The Utilization of Technology in Developing Emotional Labor

As digitalization and the widespread adoption of digital technologies impact various dimensions of work life, there is curiosity about the outcomes that will arise from the combination of organizational behavior and psychology with technology. Botella et al. (2012) discusses the concept of positive technologies by merging positive psychology, which focuses on emotional quality, participation/achievement, and commitment, with information and communication technologies. Positive technologies encompass technologies designed to structure, enhance, and/or modify the quality of experience to build and enhance the power and resilience of individuals, organizations, and societies. Examples include independent and social virtual reality, m-health, smartphone applications, exergames, video games, social technologies, and so forth, which are highlighted for their functions in reducing individuals' psychological burden and aiding their development even in challenging and complex times (Riva et al., 2020). Positive technologies are classified according to their purposes as follows (Botella et al., 2012):

1. Hedonic positive technologies are devices that alter mood, aiming to provide individuals with positive and enjoyable experiences.
2. Eudaimonic positive technologies aim for individuals to experience engaging and self-fulfilling experiences.
3. Social/interpersonal positive technologies aim to enhance connections between individuals, teams, and organizations.

Lee et al. (2019) emphasizes that smart positive computing systems equipped with digital technologies such as mobile, wearable, and the Internet of Things could be utilized to address health and well-being issues such as stress and depression. Baños et al. (2017) indicate that in the near future, a synergistic interaction between positive technologies and the field of positive emotions will lead to exponential growth in understanding and supporting positive emotions. Gaggioli et al. (2014) suggest that a free mobile platform/application developed for the self-management of psychological stress, utilizing 3D biological feedback and stress tracking, holds promise in reducing perceived stress levels. Such studies reflect the need for research on technologies applicable to professions and workers experiencing high levels of stress, such as emergency aid and disaster management teams. Future research is expected to focus on the development of emotional labor along with the use of new technologies, which are anticipated to enhance both the mental health and job performance of nurses, as well as the effectiveness of nursing interventions.

Discussion and Conclusion

Literature examining the emotional labor profile of nurses during disaster processes reveals various findings. In a study conducted with Taiwanese nurses, it was found that nurses perceived their disaster preparedness at a low level, leading to a lack of motivation to participate in disaster management (Liou et al., 2020). In a qualitative study involving eight nurse managers, it was noted that they struggled with the uncertainty and guidance during the disaster period (Covid-19), had to take on new roles without any training, and supported patients/employees individually without being supported themselves (Jackson and Nowel, 2021). Nurse managers reported that their experiences during the disaster process were influenced by changes in role and perceived support.

In a study involving 303 nurses and analyzed using structural equation modeling, the relationship between three sub-dimensions of emotional labor, stress, and burnout was examined. It was found that surface-acting increased stress and burnout (Kim, 2020). Another study investigating the impact of emotional labor on burnout among public health nurses found a negative relationship between high emotional labor and perceived health status and organizational support, increasing burnout (Kim et al., 2022). In another study involving a group of 118 individuals, including novice nurses, a positive relationship was found between the surface-acting dimension of emotional labor and burnout and the intention to leave the current job (Theodosius et al., 2021). According to a different assessment, there was a parallel increase between nurses' emotional labor strategies and presenteeism. Additionally, burnout played a completely mediating role in the relationship between emotional labor and presenteeism (Song et al., 2021).

A study conducted with 504 nurses from Türkiye obtained the highest score from the "deep acting" sub-dimension, which signifies displaying emotions in appropriate behaviors by internalizing them. The lowest score was obtained from the "suppression behavior" sub-dimension, which signifies hiding true feelings by suppressing emotions (Öz and Baykal, 2018). In another qualitative study, emotion rules that nurses should adhere to during emergencies were identified. According to the research, nurses are expected to adhere to four basic rules: "being fearless, being calm, not feeling resentment, and not being overly sorrowful" (Kirk et al., 2021). In a study analyzing emotional labor among nurses based on emotional well-being and personality structure, it was found that deep emotional labor nurtured the psychological well-being of workers, while surface emotional labor harmed psychological well-being. As superficial emotional labor, occurring only at the behavioral level, increases, life satisfaction decreases. Conversely, deep emotional labor, which aligns emotions with behavior, reduces symptoms of depression (Korkmaz, 2021).

Nurses are authorized and responsible healthcare professionals tasked with planning, implementing, supervising, and evaluating patient care (Nursing Law, 2007). In performing these duties, they are required to engage in significant emotional labor. The emotional labor behavior exhibited by nurses can have both positive and negative consequences. As the discrepancy between the emotions felt and displayed by nurses increases, a decrease in service quality and patient satisfaction is expected. Emotional labor is a fundamental component that ensures both the patient's safety and the comfort of family members. In other words, emotional labor is an invisible bond established between nurses and patients (Fouquereau et al., 2019). It is crucial for the patient's concerns to be alleviated and their morale to be kept high during the treatment process. Therefore, regardless of the conditions they are in, nurses expend emotional labor to control their emotions and fulfill professional requirements to the best of their ability (Korkmaz et al., 2021).

Disclaimer

None.

References

Altaş, S. & Özişli, Ö. (2021). The relationshıp emotional labor between job satısfaction and individual performance. *Journal of International Social Research, 14* (76), 862-873.

Anaç, A. S., & Özyer, K. (2021). Emotional labor: A conceptual review. *Journal of Organizational Behavior Studies, 1*(1), 51-64.

Ashforth, B., & Humphrey, R. (1993). Emotional Labor in Service Roles: The Influence of Identity. *Academy of Management Review, 18*, 88-115. https://doi.org/10.5465/AMR.1993.3997508.

Baños, R., Carrillo, A., Etchemendy, E., & Botella, C. (2017). Positive Technologies for Understanding and Promoting Positive Emotions. *The Spanish Journal of Psychology, 20*. https://doi.org/10.1017/sjp.2017.42.

Basım, H. N., & Beğenirbaş, M. (2012). Çalışma yaşamında duygusal emek: Bir ölçek uyarlama çalışması. *Journal of Management and Economics, 19*(1), 77-90. https://dergipark.org.tr/en/download/article-file/146079.

Botella, C., Riva, G., Gaggioli, A., Wiederhold, B., Raya, M., & Baños, R. (2012). The Present and Future of Positive Technologies. *Cyberpsychology, Behavior and Social Networking, 15* 2, 78-84. https://doi.org/10.1089/cyber.2011.0140.

Ceylan, A. K. (2021). Doktor ve hemşirelerin duygusal emek özelliklerinin incelenmesi. *Uluslararası Anadolu Sosyal Bilimler Dergisi, 5*(1), 44-59. https://doi.org/10.47525/ulasbid.83137.

Çevik Tekin, İ. (2021). Emotional labor. M. Ak (Ed.), In Literary Analysis and Scales on Organizational Behavior (Örgütsel Davranış Üzerine Literatürel Analiz ve Ölçekler) (p: 35-54). Ankara: Nobel Akademik Publishing.

Demir, A., Celebi, E., Ozturk, H., Ozcan, Z., Ozocak, A., Bol, E., Sert, S., Sahin, F. Z., Arslan, E., Utkucu, M., & Mert, N. (2024). Destructive impact of successive high magnitude earthquakes occurred in Türkiye's Kahramanmaraş on February 6, 2023. *Bulletin of Earthquake Engineering,* 1-27. https://doi.org/10.1007/s10518-024-01865-5.

Diefendorff, J., & Gosserand, R. (2003). Understanding the emotional labor process: A control theory perspective. *Journal of Organizational Behavior, 24,* 945-959. https://doi.org/10.1002/JOB.230.

Dworznik-Hoak, G. (2022). Emotional labor during disaster coverage: exploring expectations for emotional display. *Journalism Practice, 16*(5), 864-882. https://doi.org/10.1080/17512786.2020.1816858.

Edward, K., Hercelinskyj, G., & Giandinoto, J. (2017). Emotional labour in mental health nursing: An integrative systematic review. *International Journal of Mental Health Nursing, 26,* 215–225. https://doi.org/10.1111/inm.12330.

Erkal, T., & Değerliyurt, M. (2009). Disaster management of Turkey. *Eastern Geographical Review, 14*(22), 147-164.

Erol Korkmaz, H. T. (2021). Sağlık sektöründe duygusal emeğin kişilik ve psikolojik esenlikle ilişkileri. *Mersin Üniversitesi Sosyal Bilimler Enstitüsü e-Dergisi, 5* (1), 59-83.

Fouquereau, E., Morin, A. J., Lapointe, É., Mokounkolo, R. & Gillet, N. (2019). Emotional labour profiles: associations with key predictors and outcomes. *Work & Stress, 33*(3): 268-294. https://doi.org/10.1080/02678373.2018.1502835.

Fullerton, C., McCarroll, J., Ursano, R., & Wright, K. (1992). Psychological responses of rescue workers: fire fighters and trauma. *The American Journal of Orthopsychiatry, 62* 3, 371-8. https://doi.org/10.1037/H0079363.

Gaggioli, A., Cipresso, P., Serino, S., Campanaro, D., Pallavicini, F., Wiederhold, B., & Riva, G. (2014). Positive technology: a free mobile platform for the self-management of psychological stress. *Studies in Health Technology and Informatics, 199,* 25-9. https://doi.org/10.3233/978-1-61499-401-5-25.

Glomb, T., & Tews, M. (2004). Emotional labor: A conceptualization and scale development. *Journal of Vocational Behavior, 64*, 1-23. https://doi.org/10.1016/S0001-8791(03)00038-1.

Grandey, A. (2000). Emotion regulation in the workplace: a new way to conceptualize emotional labor. *Journal of Occupational Health Psychology, 5,* (1), 95-110. https://doi.org/10.1037/1076-8998.5.1.95.

Grandey, A. A., Rupp, D., & Brice, W. N. (2015). Emotional labor threatens decent work: A proposal to eradicate emotional display rules. *Journal of Organizational Behavior, 36*(6), 770-785. https://doi.org/10.1002/job.2020.

Grandey, A., & Melloy, R. (2017). The State of the Heart: Emotional Labor as Emotion Regulation Reviewed and Revised. *Journal of Occupational Health Psychology,* 22, 407–422. https://doi.org/10.1037/ocp0000067.

Han, K., Shin, C., Yoon, H., Ko, Y., Kim, Y., & Han, C. (2018). Emotional labor and depressive mood in service and sales workers: Interactions with gender and job autonomy. *Psychiatry Research, 267*, 490-498. https://doi.org/10.1016/j.psychres.2018.06.044.

Hirohara, M., Ozaki, A., & Tsubokura, M. (2019). Determinants and supporting factors for rebuilding nursing workforce in a post-disaster setting. *BMC Health Services Research, 19,* 1-10. https://doi.org/10.1186/s12913-019-4765-y.

Hishan, S. S., Ramakrishnan, S., Qureshi, M. I., Khan, N., & Al-Kumaim, N. H. S. (2020). Pandemic thoughts, civil infrastructure and sustainable development: Five insights from COVID-19 across travel lenses. *Journal of Talent Development and Excellence, 12*(2), 1690-1696.

Hochschild, A. R. 1983. The Managed Heart: Commercialization of the Human Feeling. London: University of California Press, Ltd.

Högberg, L. (2013). Root causes and impacts of severe accidents at large nuclear power plants. *Ambio, 42*(3), 267-284. https://doi.org/10.1007/s13280-013-0382-x.

Humphrey, R., Ashforth, B., & Diefendorff, J. (2015). The bright side of emotional labor. *Journal of Organizational Behavior, 36*, 749-769. https://doi.org/10.1002/JOB.2019.

International Federation of Red Cross and Red Crescent Societies, (2022).World Disasters Report 2022: Trust, Equity and Local Action, 2022. Available from: https://www.ifrc.org/sites/default/files/2023-03/2022_IFRC-WDR_EN.0.pdf.pdf.

İrigüler, F., & Güler, M. (2016). Emotional labor of tourist guides: how does ıt affect their job satisfaction and burnout levels? *Journal of Yaşar University, 11*(42), 113-123. https://doi.org/10.19168/jyu.28146.

Jackson, J., & Nowell, L. (2021). 'The office of disaster management'nurse managers' experiences during COVID-19: A qualitative interview study using thematic analysis. *Journal of Nursing Management, 29*(8), 2392-2400.

Kim, H. J., & Kim, J. H. (2011). Emotional labor, social support, and depressive symptoms of clinical nurses in a province, Korea. *Korean Journal of Occupational Health Nursing, 20*(3), 308-318.

Kim, J. S. (2020). Emotional labor strategies, stress, and burnout among hospital nurses: a path analysis. *Journal of Nursing Scholarship, 52*(1), 105-112. http://dx.doi.org/10.1111/jnu.12532.

Kim, M.-N., Yoo, Y.-S., Cho, O.-H., Hwang, K.-H. Emotional labor and burnout of public health nurses during the Covid-19 pandemic: mediating effects of perceived health status and perceived organizational support. *International Journal of Environmental Research and Public Health 19* (1), 1-12. https://doi.org/10.3390/ijerph19010549.

Kimin, A., Nurachmah, E., Lestari, F., & Gayatri, D. (2022). Factors affecting nurses' ability to provide effective care in a disaster response: a review. *Journal of Public Health Research, 11*(2), http://dx.doi.org/10.4081/jphr.2021.2732.

Kirk, K., Cohen, L., Edgley, A., & Timmons, S. (2021). "I don't have any emotions": An ethnography of emotional labour and feeling rules in the emergency department. *Journal of Advanced Nursing, 77*(4), 1956-1967.

Korkmaz, Z., Arpacı, P. & Balcı, Ö. (2021). Emotional labor and burnout in nurses during the pandemic period. *International Social Sciences Studies Journal, 7* (89), 4367-4376. http://dx.doi.org/10.26449/sssj.3504.

Kranke, D., Mudoh, Y., Weiss, E. L., Hovsepian, S., Gin, J., Dobalian, A., & Der-Martirosian, C. (2022). 'Emotional preparedness': A nuanced approach to disaster readiness among social workers. *Social Work Education, 41*(5), 860-873. https://doi.org/10.1080/02615479.2021.1900099.

Lee, U., Han, K., Cho, H., Chung, K., Hong, H., Lee, S., Noh, Y., Park, S., & Carroll, J. (2019). Intelligent positive computing with mobile, wearable, and IoT devices: Literature review and research directions. *Ad Hoc Networks, 83*, 8-24. https://doi.org/10.1016/J.ADHOC.2018.08.021.

Liou, S. R., Liu, H. C., Tsai, H. M., Chu, T. P., & Cheng, C. Y. (2020). Relationships between disaster nursing competence, anticipatory disaster stress and motivation for disaster engagement. *International Journal of Disaster Risk Reduction, 47*, 101545.

Liu, Q., Luo, D., Haase, J. E., Guo, Q., Wang, X. Q., Liu, S., Xia, L., Liu, Z., Yang, J., & Yang, B. X. (2020). The experiences of health-care providers during the COVID-19 crisis in China: A qualitative study. *The Lancet Global Health, 8*(6), e790–e798. https://doi.org/10.1016/ S2214-109X (20)30204-7.

Mesmer-Magnus, J., DeChurch, L., & Wax, A. (2012). Moving emotional labor beyond surface and deep acting. *Organizational Psychology Review, 2*, 53 - 6. https://doi.org/10.1177/2041386611417746.

Nursing Law (Hemşirelik Kanunu) (2007). TC Sağlık Bakanlığı (1954). TC Resmi Gazete, 2, 05.

Önsüz, M., & Işıktekin Atalay, B. (2015). Disaster logistics. *Osmangazi Journal of Medicine, 37*(3), 1-6. https://dergipark.org.tr/en/download/article-file/190605.

Öz, S. D., & Baykal, Ü. (2018). Hemşirelerin duygusal emek davranışı ve etkileyen faktörler. *Florence Nightingale Journal of Nursing, 26*(1), 1-10. https://doi.org/10.26650/FNJN.387142.

Penekli, H. (2022). Covid-19 sürecinde hemşirelerde duygusal emek, sosyal destek ve tükenmişlik. *ERÜ Sağlık Bilimleri Fakültesi Dergisi, 9*(2): 20-27.

Riley, R., & Weiss, M. (2016). A qualitative thematic review: emotional labour in healthcare settings. *Journal of Advanced Nursing, 72*, 1, 6-17. https://doi.org/10.1111/jan.12738.

Riva, G., Mantovani, F., & Wiederhold, B. (2020). Positive Technology and COVID-19. *Cyberpsychology, Behavior and Social Networking*. https://doi.org/10.1089/cyber.2020.29194.gri.

Sawalha, I. H. (2020). A contemporary perspective on the disaster management cycle. *Foresight, 22*(4), 469-482. https://doi.org/10.1108/FS-11-2019-0097.

Schaafstal, A., Johnston, J., & Oser, R. (2001). Training teams for emergency management. *Comput. Hum. Behav., 17*, 615-626. https://doi.org/10.1016/S0747-5632(01)00026-7.

Shuler, S., & Sypher, B. (2000). Seeking Emotional Labor. *Management Communication Quarterly, 14*, 50 - 89. https://doi.org/10.1177/0893318900141003.

Song, J., Liu, F., Li, X., Qu, Z., Zhang, R., & Yao, J. (2021). The effect of emotional labor on presenteeism of Chinese nurses in tertiary-level hospitals: the mediating role of job burnout. *Frontiers in Public Health, 9*, 733458. https://doi.org/10.3389/fpubh.2021.733458.

Theodosius, C., Koulouglioti, C., Kersten, P., & Rosten, C. (2021). Collegial surface acting emotional labour, burnout and intention to leave in novice and pre-retirement nurses in the United Kingdom: A cross-sectional study. *Nursing Open, 8*(1), 463-472.

Varol, N., & Gültekin, T. (2016). Disaster anthropology. *Electronic Journal of Social Sciences, 15* (59), 1431-1436. https://doi.org/10.17755/esosder.263244.

Wijayasundara, N. D. (2021) Disaster preparedness in Sri Lankan university libraries: before Covid-19. *Journal of the Australian Library and Information Association, 70* (3), 246-262. https://doi.org/10.1080/24750158.2021.1955319.

Zeng, Z., Wang, X., Bi, H., Li, Y., Yue, S., Gu, S., & Xiang, G. (2021). Factors that influence perceived organizational support for emotional labor of Chinese medical personnel in Hubei. *Frontiers in Psychology, 12,* 684830. https://doi.org/10.3389/fpsyg.2021.684830.

Chapter 26

The Status of Disasters in Social Science Research in Türkiye: A Review of Social Science Doctoral Theses

Nejdet Özberk*, PhD
Nevşehir Hacı Bektaş Veli University, Nevşehir, Türkiye

Abstract

The dominance of physical and engineering perspectives and the technocratic focus on disaster risk reduction have always been a real problem. Until the late 1990s, disasters were generally ignored by social scientists in Türkiye and left to the natural sciences (geologists, seismologists, and climatologists). The appropriation of the problem by the social sciences (sociology, geography, and anthropology) or disaster management as a field of research is quite recent. Natural and theological orthodoxy dominates disaster thinking, research, practices, policies, and popular discourse. In this perspective, referred to in the literature as "disaster orthodoxy," disaster is seen as an isolated, rare, and exceptional event outside of predictive patterns, the normality of daily life, and human understanding. This conceptual framework also applies to disaster research in Türkiye, where disaster is not accepted as a social science field in higher education institutions but is limited to the narrow field of science-engineering and health sciences. From a natural science perspective, disaster is not even considered a social phenomenon in the Turkish community of natural scientists. Social science research on disasters in Türkiye suffers from the same dominant approach. This scoping review examines the current state of disaster social science in Türkiye based on 79 doctoral theses from 1996 to 2021 at the Thesis

* Corresponding Author's Email: nozberk@nevsehir.edu.tr

In: Disaster Management in the Digital Age and Emerging Technologies
Editors: Ayşe Asiltürk and Ayşegül Sarıoğlu Kemer
ISBN: 979-8-89530-213-2

Center of the Turkish Higher Education Board (YÖK). The study reveals that disasters are often viewed as isolated natural events outside of society, history, and politics, and treated as a matter and field of superficial technocratic intervention. The lack of a critical disaster social science perspective and the weakness of social understanding in the country are key indicators of the naturalization of disasters and technocratic practice. Research highlights the need for studies in anthropology, theology, philosophy, and more critical sociology and political science.

Keywords: disaster, doctoral thesis, social science, scoping review

Introduction

From the 1940s to the 1960s, natural hazards scientists focused on physical, technical, and engineering knowledge, neglected social causality, and viewed the physical and human worlds as separate and disasters as external natural events (Remes et al., 2021; Dynes and Drabek, 1994). Even Burton and Kates (1963), two of the leading proponents of the first social science disaster approach from geography, defined natural disasters as harmful elements of the physical environment caused by "alien forces." This approach has led to a focus on physical preferences rather than social science analysis and policy recommendations in disaster management. Critical disaster scientist Cannon (2008: 356) argues that natural science work on disasters is typically conducted by specialists in their narrow scientific field, with little awareness of the social processes involved. The natural sciences focus on specific types of natural hazards within their narrow field, such as earthquakes and hurricanes. However, this science generally lacks awareness of the social processes involved and rarely attempts to create a science of disasters that combines social and natural processes: "Little of this science has much awareness of the social processes involved, and it is rare that it makes any attempt to construct a science of disasters in which the social and natural processes can be understood in combination." Unlike the natural and engineering sciences, which focus on natural hazards as external factors, environmental risks in the social sciences are not natural phenomena, but the result of social, economic, and cultural processes (Kendra and Nigg, 2014).

Disasters are not natural hazards but social processes caused by human vulnerabilities (Chmutina and von Meding, 2019). From the 1940s to the present, two theoretical perspectives have been developed: behavioralist and structuralist (Montz and Tobin, 2011). Behavioralists focus on the responses of individuals and groups to natural hazards, while structuralists view disasters as part of social structures. The vulnerability approach emerged in the 1970s and initiated a shift in the approach to disasters (O'Keefe et al., 1976). This social approach has led to a "social turn" in social thought, research, policy, practice, and popular political discourse, moving from an understanding of natural hazards (Burton et al., 1978) to an understanding of social vulnerability (Hewitt, 1983; Blaikie et al., 1994). Disasters are now attributed to the social vulnerability of a particular community, rather than to natural hazards such as earthquakes, floods, and landslides (Tierney, 2007).

This research aims to identify and synthesize the scope of scientific literature on disasters in Türkiye in terms of social science doctoral dissertations that use "disaster" as a lens. The data and key findings presented identify major gaps in the literature, opportunities for theoretical, methodological, and practical advances, and suggest future research directions.

Social Science Matters: The Role of Social Science in Disaster Research

Disaster studies are an important part of social science disciplines, especially after the "social turn" in the 1980s and 1990s (Aronsson-Storrier and Dahlberg, 2022). Social sciences are now considered more important than physical and engineering sciences in understanding and researching disasters and developing preventive policies. Even a new discipline called "disaster studies" has emerged at the intersection of geography, sociology, anthropology, political science, and development studies (Alexander, 1997). Disaster studies provide an excellent context for exploring key social science concepts such as politics, risk, preparedness, governmentality, power, social capital, class, race, gender, identity, justice, sustainability, trust, and security. Disaster studies help us apply key social science theories to the field, deepen our understanding of basic human behavior, and describe disasters as human-centered phenomena that transcend disciplinary boundaries (Burger et al., 2019). They also show the bigger picture of disasters as social and political phenomena. The essence of social science is to discover how human and social agencies respond to new circumstances, challenges, and changes (Reinhard and Ross, 2019). The importance of disaster social science cannot be denied, as the formation of disaster and disaster preparedness awareness and culture largely depends on the socialization of the phenomenon. Social sciences play a key role in the socialization of disasters and the idea of "natural" disasters by bringing different theoretical and critical perspectives to the production of knowledge, understanding, research, and development of preventive policies. They problematize and socialize not only the disaster itself but also its study, placing it in broader religious, political, technological, and institutional contexts. Social sciences contribute to disaster research with methods different from the physical and engineering sciences (Peek et al., 2020; Nigg, 1995).

Materials and Methods

This research aims to better understand how the issue of disaster is problematized in relevant social science research. One of the qualitative research methods, the scoping review, is a new type of literature review that was used in the study instead of the typical systematic literature review (Arksey and O'Malley, 2005; Levac et al., 2010; Pham et al., 2014). Daudt et al. (2013: 8) indicate that "scoping studies aim to map the literature on a particular topic or research area and provide an opportunity to identify key concepts, gaps in research, and types and sources of evidence to inform practice, policy, and research." Scoping reviews, one of several forms of knowledge synthesis, was identified as the most appropriate research method for mapping social science literature on disasters for several reasons. First, scoping reviews are the most useful technique to review literature that has not been reviewed previously. Second, scoping reviews are useful for research topics that cross-disciplinary and methodological boundaries, as is undoubtedly the case in disaster literature. Third, scoping reviews are useful for identifying research/literature gaps and emerging research priorities. Scoping research is particularly important in practice, education, research, and policymaking: it guides practice through a broad assessment of evidence and debates; it allows students and researchers to see the diversity and complexity of knowledge on the topic; it provides invaluable information to policymakers; and it provides an excellent basis for setting research and policy priorities for research and projects when it identifies a knowledge gap in the field (Peterson, 2017: 3-5). It improves conceptual clarity and identifies gaps in the literature, providing valuable information for future research.

This study uses the framework for scoping reviews first developed by Arksey and O'Malley (2005: 22-23) and further refined by Levac et al. (2010),

Pham et al. (2014), and Westphaln et al. (2021). Arksey and O'Malley (2005) formulated scoping reviews as a six-step process:

(1) Definition of research questions,
(2) Determination of relevant studies,
(3) Selection of studies to be included or excluded from the review,
(4) Data graphic,
(5) Collating, synthesizing, summarizing, and reporting results,
(6) Consultation (optional stage).

In this study, four specific research questions guided the scoping process:

(1) How are disasters understood in doctoral dissertations?
(2) How are disasters conceptualized in dissertations?
(3) How are disasters studied in dissertations?
(4) What are the neglected issues and gaps in current dissertations on disasters?

In this context, the main questions to be answered are: What research has been done on disasters in different social science fields? What are the main characteristics of these studies? What are their epistemological, theoretical, and methodological approaches to the topic? What kind of knowledge have they produced? First, a search strategy was determined to identify relevant studies and to select studies to be included or excluded in the review: identification, screening, eligibility, and inclusion/exclusion (Pham et al., 2014: 375). The YÖK Thesis Database was used as the data source.

Initially, a search strategy was determined to identify relevant studies and select studies to be included or excluded in the review: identification, screening, eligibility, and inclusion/exclusion (Pham et al., 2014: 375). The YÖK Thesis Database was used as the data source. (This database, the National Thesis Center, is a central automation system where all old and new graduate theses prepared in universities in Türkiye are uploaded). The search criteria for the detailed search module of the database were developed, and a systematic search of the dissertation archive was conducted. The database search strategy was developed from the research questions and concept definitions. At the beginning of the scoping study, criteria were established to include and exclude the most appropriate dissertations for the research question. The search criteria we developed based on the basic concepts of disaster terminology are as follows: The terms "disaster," "natural disaster," and "earthquake" were identified in the title, abstract, or main text of the theses that mentioned the keywords. The methodological approaches of the theses, if any, especially the use of the term natural disaster in the disaster terminology of the thesis author, paying attention to the use in the definitions of concepts, theoretical framework and methodology, ontology, and epistemology were determined. In determining the relevant studies, the source and time interval spontaneously emerged within the framework of the research question. In terms of inclusion and exclusion criteria, theses with a clear empirical, theoretical, or practical social science focus on disasters were identified. Dissertations written in English were also included.

Findings

The YÖK Theses Database categorizes theses into three groups: Science, Social Sciences, and Medical. The database was searched using three keywords: "disaster," "natural disaster," and "earthquake" in titles and abstracts. A total of 128 records were identified for analysis in the social group. 79 dissertations were subjected to a scoping review. Many dissertations were excluded due to their scientific group or irrelevance, which did not answer the research question and did not meet the eligibility criteria. After the sample was narrowed, in-text scanning of the remaining dissertations as part of the thematic analysis allowed us to see how disasters were approached, how concepts were used, and to capture their context. The sample was limited to doctoral theses in the social sciences, under the assumption that both conceptual distinctions and analytical awareness would be highest and that they would better reflect the international literature. The remaining dissertations were subjected to a thematic analysis to understand how disasters were approached and how concepts were used. Data were evaluated in four main categories: general characteristics, understanding of the disaster, conceptualization of the -disaster, and approach to the disaster. Data were converted into data graphs through scoping review stages, classified, and mapped according to the main areas and years. The findings were summarized in terms of key issues and themes, and the key findings were synthesized and interpreted. The dissertations demonstrated the scope, nature, and distribution of existing disaster social science studies, identified dominant research areas, and highlighted significant gaps, weaknesses, and problems. The distribution of dissertations by year of publication is shown in Figure 1. It shows that disaster research has followed a fluctuating course since the Great Marmara earthquake of 1999, except for the theses on Disaster Management in 2020-2021. Disaster social science studies are not limited to a specific discipline; they include a wide range of academic disciplines, with the exception of anthropology, theology, and philosophy.

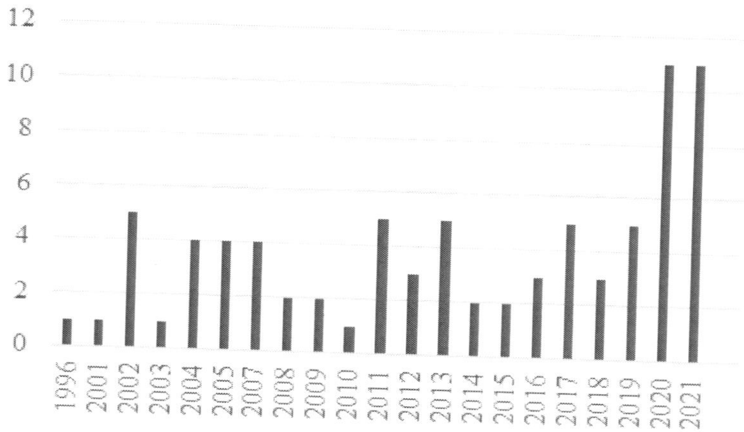

Figure 1. Distribution of theses by year.

A brief summary of Figure 2, which shows the distribution of dissertations according to the relevant social sciences, shows that the fields of business administration and public administration dominate. Twelve dissertations were written in the field of disaster management. The first theses in this field, which were institutionalized at the beginning of the 2010s, were written in the last two years. Disaster management was opened as an undergraduate program about ten years ago and as a graduate program six years ago. Although the concepts of perception, adaptation, individual resilience, and technical nature bear strong traces of the resilience paradigm, it can be said that studies in this field are of relative quality in terms of theory and methodology. Seven dissertations were written in the field of geography. This is the only field of Ph.D. dissertations submitted before 2000. Contrary to the contributions of geography to the world disaster research literature, it can be argued that it is one of the shallowest fields in Türkiye. The field relies heavily on the physical geography approach, which manifests the nature paradigm based on science and engineering, and isolates disasters from their social content (Yavan and Anli, 2018). In general, research in the fields of economics, labor economics, finance, and business administration does not problematize disasters as a social problem, but rather research that proposes technical solutions/models for the phenomenon that is seen as a technical problem by the dominant political economy actors and corporations. Most of the 14 dissertations in the field of public administration, which are not based on theory and methodology, have the characteristics of a legislative review, not of disaster research. The multiplicity of theses in the field of public administration (urbanization and environmental problems) indicates that disasters are seen as a management problem and not as a problem of social vulnerability. Five theses have been written in the field of psychology. Most of the dissertations in the field of psychology, which are mostly submitted in English (at METU), are on post-traumatic developmental/stress disorders, anxiety and depression, psychological resilience, coping strategies, and factors predicting earthquake prevention behavior. Sociological theses on topics such as disaster and capitalism, urbanization, women, social and spatial inequality (class), social change, and vulnerability had the highest level of literature awareness. In various fields, the naturalness of disasters is emphasized. However, as Arksey and Malley (2005) point out, synthesis is also important, as scoping research is not a concise summary of data from many sources. By synthesizing the literature, the study has produced the following results.

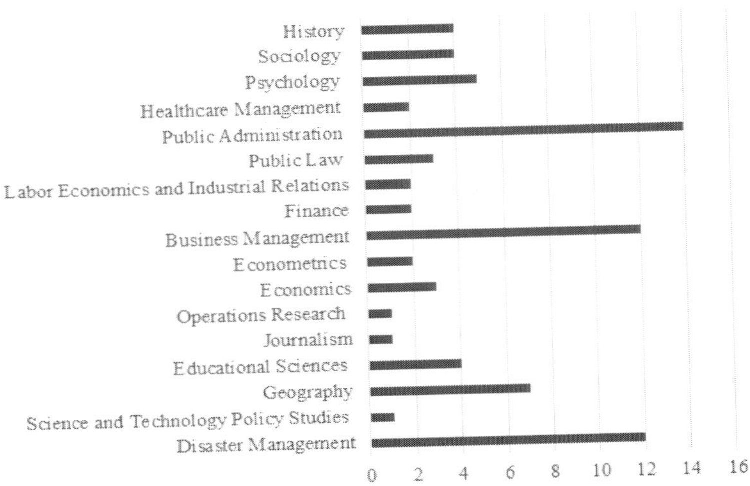

Figure 2. Distribution of theses according to related social sciences.

A Weak Disaster: Ontology and Epistemology

The main finding of the research is that disaster social science theses reflect the prevailing hierarchy and division between the social and natural sciences. Due to the dominance of science and engineering approaches, disaster social science theses often view disasters as natural phenomena, such as earthquakes, and are synonymous with them. Disasters are seen and experienced outside of society and are interpreted as natural inevitabilities. The focus is on environmental rather than social phenomena, ignoring the relationship between disaster risk, development, and the environment. They highlight major events, such as the Istanbul and Van earthquakes, and view disasters as high-magnitude, low-frequency natural events. They also overlook the relationship between disasters and the political economy of development, viewing disasters as a central, technical, administrative, and financial problem at the national level, while ignoring disaster risk management at the local level. Empirical research has not been conducted in collaboration with local people, resulting in isolated studies without contact with local social actors. The lack of an ontological or epistemological approach to disasters and the failure to critique social, political, and economic systems are a significant problem.

Lack of Theoretical and Methodological Precision

The second important finding of our research is the absence of a paradigmatic perspective that methodologically approaches disasters in terms of hazards and, in particular, vulnerability. The lack of appropriate knowledge and application of the basic principles of scientific knowledge production constitute methodological weaknesses. Furthermore, the lack of theoretical grounding, especially in terms of the theoretical perspective, is problematic. Inadequate research program and lack of a critical social science perspective. Despite claims of scientificity, there is a lack of data, purpose, and theory; there is also a lack of research that does not analyze its findings theoretically or in the light of theory; it does not provide theoretical explanations, models, concepts, or analyses.

A Dominant Orthodox Technocratic Perspective

The third finding reveals that disaster social science dissertations often use physical science research terminology rather than social science terminology in all areas. The study shows that there is an incredibly weak disaster social science perspective in the theses, and the dominant environmental determinist orthodoxy's emphasis on scientific hierarchy and terminology is dominated by the "physicalist" technocratic perspective (Yavan and Anli, 2018). A finding of Açıkalın's (2018: 99) articles is also valid for doctoral dissertations: Indifference to the importance of social knowledge: "Lack of understanding of the importance of social knowledge in solving problems due to overreliance on technical knowledge." It is an apolitical paradigmatic view that does not take into account the social and physical production of the disaster, is hazard-oriented, and blames nature. Even research that claims to be critical is, in fact, produced within the mainstream disaster paradigm and is uncritical. Disasters are identified with extreme natural events as if they were natural phenomena themselves. While few of these provide clear definitions of the concepts of natural hazard ("natural disaster"), social vulnerability, and disaster, many fail to recognize the epistemological and methodological distinctions of these concepts and generally maintain conceptual hegemony. Eighty years after the West, in 1999, the first disaster sociological study was carried out following the Marmara earthquake in Türkiye. Additionally, the language and methods of physical sciences predominate in geography dissertations, as opposed to geography, which is the birthplace of disaster social science in the West.

Neglecting Key Areas and Issues

In the West, disaster studies grew out of three core social sciences: sociology, geography, and anthropology. One of the most significant findings is the absence of anthropology and social anthropology, which highlights how little is known about disaster safety and preparedness. This is the fact that the black box of the relationship between disaster and culture has never been opened: the absence of a culture of disaster preparedness or a culture of safety behind the scenes (Hoffman and Oliver-Smith, 2002). There is also a lack of studies on the temporal and spatial dimensions of disasters and on the cultural memory of disasters. This lack of research indicates the inadequacy of disaster anthropology and the failure to establish the context of disaster and culture in Türkiye. The field lacks research that focuses on understanding lived experiences, including human behavior, local knowledge, language, religious practices, social representations of disasters, relationships between society and nature, and ontological and epistemological understandings of space and time in disaster risk reduction. On the other hand, despite the dominance and pervasiveness of religious discourse on disasters, the phenomenon of disasters is not the subject of theological and philosophical studies. Despite the strong discourse on disaster explanations, there is a lack of disaster research in theology, religious studies, and philosophy, especially in the context of religious and social beliefs. This lack of research hinders understanding of society's perceptions of disaster hazards, scientific risk assessment, and disaster preparedness, and there is also no study of disaster and folk beliefs. Sociology is the only field that includes class and gender as an analytical category and has a theoretical perspective, showing characteristics of political ecology research. Social knowledge of disasters is mainly concentrated on crisis periods, with limited diversity and focus on the Marmara and Van earthquakes/regions. There is also a lack of connection with urban scales, upper and lower scales, and research on non-crisis regions and places (Açıkalın, 2018).

Discovery and the Rise of Resilience

The study reveals that public administration and business management have significantly dominated disaster research, accounting for half of the theses. This indicates a process of economicization, privatization, marketization, and individualization of disasters, which is seen as a management problem and an issue focused on economic losses. However, none of the theses in public administration deal with the perspective of vulnerability and the political interpretation of the "natural" phenomenon. This proves that the disaster is seen as a management problem on the one hand, and as an issue focused on economic losses on the other. Similarly, in business management, disaster research tends to focus on orthodox technical solutions such as earthquake insurance, real estate valuation, occupational safety, disaster logistics, public finance, information technologies, supply chain, and inventory management. An understanding that implies that society's preparedness for disasters is possible by becoming part of the market consisting of specialized institutions, municipalities, banks, insurance, real estate, and construction companies. It points out that disaster preparedness is produced as an activity that can only be carried out as an actor with property/capital in the market and by coming together with other capitalist actors (Aslan, 2020). It seems that there is no dissertation in the field of political ecology that places the responsibility for disasters on the power centers that produce vulnerability, the government, especially the contracting and real estate companies, and the political economy forces behind them. This approach does not see or criticize the complementarity between disaster and capitalism, except in the discipline of sociology.

Discussion and Conclusion

According to Gaillard (2022) and Bankoff (2001), the natural sciences hold a hegemonic position over the social sciences in disaster research worldwide. The hegemony of the North over the South and of the technocratic natural paradigm over the social paradigm is also evident. Although rejected by disaster scientists, the understanding of a "natural disaster" that sees disaster as an event rather than a process has maintained its dominance in science, law, and policy (Bankoff and Hilhorst, 2022). Achievements in the physical and biological sciences, according to Bates et al. (1991: 288), have led some people to believe that more research in these fields will suffice to provide adequate knowledge and that new technologies will provide ready-made solutions to future disaster problems. However, the implementation of solutions takes place in specific political, economic, and socio-cultural contexts. Natural hazard mitigation programs implemented without understanding these social dimensions are doomed to failure. Defined by geographer Kenneth Hewitt (1983) as the "dominant view" and sociologist Kathleen Tierney (2019) as "disaster orthodoxy," the conventional view of disaster is the understanding that disasters are unexpected, unfortunate events that suddenly strike societies and cause the disruption of normal functions and social structures. Orthodoxy refers to any ontological or epistemological idea or body of knowledge that its proponents or adherents claim is the only reasonable and correct course of action; the only truth or path to truth for a particular subject; an exclusive interpretation of truth. It is more than a collection of concepts; it is also a particular relationship based on power. This dominant perspective, based on institutional mechanisms such as education, science, culture, political, economic, and social power relations, largely exonerates the powerful, as well as state institutions, organizations, and corporate actors. The orthodox view of disaster can be read in the language of power and corporations, in religious discourse, in the media and social media, in educational curricula at different levels, in disaster research (dissertations, journals, books), and in institutional documents that guide disaster policy.

Prior to 2000, the disaster agenda in Türkiye was dominated by theology in the popular discourse and technical debates at the scientific level. Arslan (2022) highlights two changes in disaster research in Türkiye since the 1999 Izmit earthquake. These changes have led to a reduction in earthquake preparedness for individual measures and financial resource development practices. Furthermore, the interventions of political and market actors have changed earth sciences research and applied techniques towards a project-oriented and risk management axis, transforming disaster research into a new process that determines the context of disaster research in Türkiye. It should be added that a significant part of current disaster practice is carried out under the guidance of the World Bank's technocratic climate change, disaster-resilient cities, schools, and earthquake and flood projects. These studies are in line with the dominant agenda of the current disaster orthodoxy. The liberalization-neoliberalization of disaster management is a decentralized, individually participatory discourse that shifts from state-based to market-based governance. In contrast to structural factors that contribute to vulnerability, the emphasis is on individual responsibility and choice. This neoliberal resilience paradigm emphasizes individual responsibility in disaster management, focusing on power relations, agency, values, and local knowledge. This approach denies the role of the state and instead sees resilience as an individual issue. It places more emphasis on NGOs and civil society than on the state and local governments. It sees vulnerability and social responsibility as voluntary, development as sustainability, and disasters as governance and stakeholder issues. By reducing vulnerability to a voluntary condition resulting from weak individual choices, this conservative approach upholds the status quo and encourages individual responsibility. Resilience is prioritized over government involvement in disaster management, and a neoliberal perspective focuses on maintaining existing social structures and relationships.

Finally, the theses we have examined show that the issue of risk and disaster in Türkiye is approached with technocratic and "physicalist" theories and concerns about humanitarian intervention in disaster social science research. This is significantly related to the immaturity of critical thinking in the social sciences. Overall, several hypotheses can be proposed to explain why the Turkish disaster research community has been slow to socialize risks and disasters. The first is that the dominant paradigm in political and civil society on the issue of hazard, risk, and disaster is not questioned.

The dominant theocratic discourse and the predominance of technocratic practice. Disasters are not accepted as a social science subject and as a social science research area in higher education. A tradition of institutional and bureaucratic state that has little interest in risk prevention and mitigation. A dominance that both stifles critical voices in academia and hinders critical studies, a rigid disciplinary approach that excludes multidisciplinary and interdisciplinary research. The third is the distance of academia from critical approaches and submission to the hegemony of positivist social science paradigms developed mainly in the US and Europe. Ignorance of the social science perspective of decolonial disaster studies focused on Latin America and Africa. Closure of critical ideas and thoughts, alternative barriers to influence. The dominant ways of thinking prioritize physicalistic engineering solutions to natural hazards and see disaster management as a hierarchical response process, giving excessive attention and importance to structural measures and technical and engineering interventions rather than prevention and risk reduction. In this context, social research and a thinking community on disasters has not been created. The absence of a tradition of collective work, communication, and intellectual debate in the academy. A social science approach trapped in disaster risk management, such as road dependency and the status quo that prioritizes intervention rather than risk, two patterns that are difficult to break. In practice, a technocratic and bureaucratic perspective that prioritizes crisis management over risk reduction dominates disaster management and research. It is a disaster research community that is disconnected from the international disaster social science research agenda. It is this linguistic and intellectual distance from the outside world that contributes to the lack of diffusion and integration of new ideas and concepts.

The research revealed that critical social science perspectives on disasters have not been developed in Türkiye, which could influence theory, methodology, and policy. The perception, thinking, policy, and practice of disasters are determined by the dominance of the physical and engineering sciences and theological discourse. Disasters and disaster problems are viewed from a theocratic and technocratic perspective, with basic sciences proposing technical solutions and political theological discourse emphasizing the need to understand and research the phenomenon of disasters. The paradigm shift from disaster as a natural hazard to disaster as social vulnerability is not reflected in the social science research in Türkiye. The orthodox traditional understanding, which sees disasters as natural and technical problems, denies the importance of social knowledge in solving problems by relying too much on technical knowledge. The agenda of disaster research and implementation is still shaped by the hegemonic policies of the World Bank, infrastructure, financial and insurance actors, and neoliberal policies. However, there are signs that this situation is beginning to change with new studies and publications. There is an urgent need to develop a more collaborative approach to disaster research and education that reconceptualizes disaster and risk and integrates risk into routine discussions of development and management. In terms of disaster policy, the opportunities that disaster social science can offer to focus on pre-disaster risk management remain to be explored. The prevalence of conceptual inconsistencies related to disasters suggests the need to focus on issues and problems that provide greater depth and clarity in understanding and applying disaster concepts. Additionally, there is still much to be researched about disasters, with priority areas for future studies that will transform our understanding of disasters.

Disclaimer

None.

References

Açıkalın, O. (2018). Türkiye'ye ilişkin sosyal boyutlu deprem yazımına eleştirel bir bakış [A critical look at the social earthquake literature on Turkey]. *Dirençlilik Dergisi, 2* (2). 85-104.

Alexander, D. (1997). The Study of Natural Disasters, 1977-1997: Some Reflection on a Changing Field of Knowledge. *Disasters, 21* (4), 284-304.

Arksey, H., & O'Malley, L. (2005). Scoping studies: towards a methodological framework. *International Journal of Social Research Methodology, 8*(1), 19-32.

Aronsson-Storrier, M., & Dahlberg, R. (Eds.). (2022). Defining disaster: disciplines and domains. Edward Elgar Publishing.

Arslan, O. (2022). Fay Hatları Konuşabilir Mi? Beklenen Büyük İstanbul Depreminin Topluma Havale Ettikleri [Can Fault Lines Talk? What the Expected Great Istanbul Earthquake Has Delegated to the Society]. IstanbuLab.

Bankoff, G. (2019). Remaking the world in our own image: vulnerability, resilience and adaptation as historical discourses. *Disasters, 43*(2), 221–239.

Bankoff, G. (2001). Rendering the World Unsafe: 'Vulnerability' as Western Discourse. *Disasters, 25* (1), 19-35.

Bankoff, G., Hilhorst, D. (2022). Why Vulnerability Still Matters: The Politics of Disaster Risk Creation. Routledge.

Bates, F. L., Dynes, R. R., & Quarantelli, E. L. (1991). The importance of the social sciences to the International Decade for Natural Disaster Reduction. *Disasters, 15*(3), 288-289.

Blaikie, P., Cannon, T., Davis, I., & Wisner, B. (2014). At risk: Natural Hazards, People's Vulnerability and Disasters. Routledge.

Burger, A., Oz, T., Kennedy, W. G., & Crooks, A. T. (2019). Computational social science of disasters: Opportunities and challenges. *Future Internet, 11*(5), 103.

Burton, Ian, Robert W. Kates, and Gilbert F. White. (1978). The Environment as Hazard. Oxford University Press.

Burton, I., & Kates, R. W. (1963). The perception of natural hazards in resource management. *Nat. Resources J.,* 3, 412.

Cannon, T. (2008). Vulnerability, "innocent" disasters and the imperative of cultural understanding. *Disaster Prevention and Management: An International Journal, 17*(3), 350-357.

Chmutina, K., & Von Meding, J. (2019). A dilemma of language: "Natural disasters" in academic literature. *International Journal of Disaster Risk Science, 10*, 283-292.

Daudt, H. M., Van Mossel, C., & Scott, S. J. (2013). Enhancing the scoping study methodology: a large, inter-professional team's experience with Arksey and O'Malley's framework. *BMC Medical Research Methodology, 13*, 1-9.

Dynes, R. R., & Drabek, T. E. (1994). The structure of disaster research: Its policy and disciplinary implications. *International Journal of Mass Emergencies & Disasters, 12*(1), 5-23.

Gaillard, J. C. (2022). The Invention of Disaster: Power and Knowledge in Discourses on Hazard and Vulnerability, Routledge.

Hewitt, K. (Ed.). (2019). Interpretations of calamity: From the viewpoint of human ecology. Routledge.

Hoffman, S. M. and Oliver-Smith A. (2002). Catastrophe and Culture: The Anthropology of Disaster, School of American Research Press.

Kendra, J., & Nigg, J. (2014). Engineering and the social sciences: historical evolution of interdisciplinary approaches to hazard and disaster. *Engineering Studies, 6*(3), 134-158.

Levac, D., Colquhoun, H., & O'brien, K. K. (2010). Scoping studies: advancing the methodology. *Implementation Science, 5*, 1-9.

Montz, B. E., & Tobin, G. A. (2011). Natural hazards: An evolving tradition in applied geography. *Applied Geography, 31*(1), 1-4.

Nigg, J. M. (1995). Social science approaches in disaster research: Selected research issues and findings on mitigating natural hazards in the urban environment. *Urban Disaster Mitigation: The Role of Engineering and Technology*, 303-310.

O'keefe, P., Westgate, K., & Wisner, B. (1976). Taking the naturalness out of natural disasters. *Nature, 260*(5552), 566-567.

Peek, L., Champeau, H., Austin, J., Mathews, M., & Wu, H. (2020). What methods do social scientists use to study disasters? An analysis of the social science extreme events research network. *American Behavioral Scientist, 64*(8), 1066-1094.

Peterson, J., Pearce, P. F., Ferguson, L. A., & Langford, C. A. (2017). Understanding scoping reviews: Definition, purpose, and process. *Journal of the American Association of Nurse Practitioners, 29*(1), 12-16.

Pham, M. T., Rajić, A., Greig, J. D., Sargeant, J. M., Papadopoulos, A., & McEwen, S. A. (2014). A scoping review of scoping reviews: advancing the approach and enhancing the consistency. *Research Synthesis Methods, 5*(4), 371-385.

Reinhardt, G. Y., & Ross, A. D. (2019). Expanding social science through disaster studies. *Social Science Quarterly, 100*(7), 2523-2529.

Remes, J. A., Horowitz, A. D., & Horowitz, A. (Eds.). (2021). Critical disaster studies. University of Pennsylvania Press.

Tierney, K. (2019). *Disasters: A sociological approach.* John Wiley & Sons.

Tierney, K. J. (2007). From the margins to the mainstream? Disaster research at the crossroads. *Annu. Rev. Sociol., 33*, 503-525.

Westphaln, K. K., Regoeczi, W., Masotya, M., Vazquez-Westphaln, B., Lounsbury, K., McDavid, L., Lee, H., Johnson, J. & Ronis, S. D. (2021). From Arksey and O'Malley and Beyond: Customizations to enhance a team-based, mixed approach to scoping review methodology. *MethodsX, 8*, 101375.

Yavan, N., & Anli, C. K. (2018). (Un)making human geography in Turkey under the dominance of environmental determinism. *Journal of Thinking, 13*, 77-98.

Index

A

B

C

D

E

M

N

O

P

R

S

T

U

V

W